Lecture Notes in Bioinformatics 9683

Subseries of Lecture Notes in Computer Science

More information about this series at http://www.springer.com/series/5381

Anu Bourgeois · Pavel Skums
Xiang Wan · Alex Zelikovsky (Eds.)

Bioinformatics Research and Applications

12th International Symposium, ISBRA 2016
Minsk, Belarus, June 5–8, 2016
Proceedings

 Springer

Editors
Anu Bourgeois
Department of Computer Science
Georgia State University
Atlanta, GA
USA

Xiang Wan
Department of Computer Science
Hong Kong Baptist University
Kowloon Tong
Hong Kong SAR

Pavel Skums
Centre for Disease Control and Prevention
Atlanta, GA
USA

Alex Zelikovsky
Georgia State University
Atlanta, GA
USA

ISSN 0302-9743   ISSN 1611-3349   (electronic)
Lecture Notes in Bioinformatics
ISBN 978-3-319-38781-9   ISBN 978-3-319-38782-6   (eBook)
DOI 10.1007/978-3-319-38782-6

Library of Congress Control Number: 2016938416

LNCS Sublibrary: SL8 – Bioinformatics

Printed on acid-free paper

This Springer imprint is published by Springer Nature
The registered company is Springer International Publishing AG Switzerland

Preface

The 12th edition of the International Symposium on Bioinformatics Research and Applications (ISBRA 2016) was held during June 5–8, 2016, in Minsk, Belarus. The symposium provides a forum for the exchange of ideas and results among researchers, developers, and practitioners working on all aspects of bioinformatics and computational biology and their applications.

There were 77 submissions received in response to the call for papers. The Program Committee decided to accept 42 of them for publication in the proceedings and for oral presentation at the symposium: 22 for Track 1 (an extended abstract) and 20 for Track 2 (an abridged abstract). The technical program also featured invited keynote talks by five distinguished speakers: Dr. Teresa M. Przytycka from the National Institutes of Health discussed the network perspective on genetic variations, from model organisms to diseases; Prof. Ion Mandoiu from the University of Connecticut spoke on challenges and opportunities in single-cell genomics; Prof. Alexander Schoenhuth from Centrum Wiskunde and Informatica spoke on dealing with uncertainties in big genome data; Prof. Ilya Vakser from the University of Kansas discussed genome-wide structural modeling of protein–protein interactions; and Prof. Max Alekseyev from George Washington University spoke on multi-genome scaffold co-assembly based on the analysis of gene orders and genomic repeats.

We would like to thank the Program Committee members and external reviewers for volunteering their time to review and discuss symposium papers. Furthermore, we would like to extend special thanks to the steering and general chairs of the symposium for their leadership, and to the finance, publicity, local organization, and publication chairs for their hard work in making ISBRA 2016 a successful event. Last but not least, we would like to thank all authors for presenting their work at the symposium.

June 2016

Anu Bourgeois
Pavel Skums
Xiang Wan
Alex Zelikovsky

Organization

Steering Chairs

Dan Gusfield	University of California, Davis, USA
Ion Măndoiu	University of Connecticut, USA
Yi Pan	Georgia State University, USA
Marie-France Sagot	Inria, France
Ying Xu	University of Georgia, USA
Alexander Zelikovsky	Georgia State University, USA

General Chairs

Sergei V. Ablameyko	Belarusian State University, Belarus
Bernard Moret	École Polytechnique Fédérale de Lausanne, Switzerland
Alexander V. Tuzikov	National Academy of Sciences of Belarus

Program Chairs

Anu Bourgeois	Georgia State University, USA
Pavel Skums	Centers for Disease Control and Prevention, USA
Xiang Wan	Hong Kong Baptist University, SAR China
Alex Zelikovsky	Georgia State University, USA

Finance Chairs

Yury Metelsky	Belarusian State University, Belarus
Alex Zelikovsky	Georgia State University, USA

Publications Chair

Igor E. Tom	National Academy of Sciences of Belarus

Local Arrangements Chair

Dmitry G. Medvedev	Belarusian State University, Belarus

Publicity Chair

Erin-Elizabeth A. Durham	Georgia State University, USA

Workshop Chair

Ion Mandoiu   University of Connecticut, USA

Program Committee

Kamal Al Nasr	Tennessee State University, USA
Sahar Al Seesi	University of Connecticut, USA
Max Alekseyev	George Washington University, USA
Mukul S. Bansal	University of Connecticut, USA
Robert Beiko	Dalhousie University, Canada
Paola Bonizzoni	Università di Milano-Bicocca, Italy
Anu Bourgeois	Georgia State University, USA
Zhipeng Cai	Georgia State University, USA
D.S. Campo Rendon	Centers for Disease Control and Prevention, USA
Doina Caragea	Kansas State University, USA
Tien-Hao Chang	National Cheng Kung University, Taiwan
Ovidiu Daescu	University of Texas at Dallas, USA
Amitava Datta	University of Western Australia, Australia
Jorge Duitama	International Center for Tropical Agriculture, Columbia
Oliver Eulenstein	Iowa State University, USA
Lin Gao	Xidian University, China
Lesley Greene	Old Dominion University, USA
Katia Guimaraes	Universidade Federal de Pernambuco, Brazil
Jiong Guo	Universität des Saarlandes, Germany
Jieyue He	Southeast University, China
Jing He	Old Dominion University, USA
Zengyou He	Hong Kong University of Science and Technology, SAR China
Steffen Heber	North Carolina State University, USA
Jinling Huang	East Carolina University, USA
Ming-Yang Kao	Northwestern University, USA
Yury Khudyakov	Centers for Disease Control and Prevention, USA
Wooyoung Kim	University of Washington Bothell, USA
Danny Krizanc	Wesleyan University, USA
Yaohang Li	Old Dominion University, USA
Min Li	Central South University, USA
Jing Li	Case Western Reserve University, USA
Shuai Cheng Li	City University of Hong Kong, SAR China
Ion Mandoiu	University of Connecticut, USA
Fenglou Mao	University of Georgia, USA
Osamu Maruyama	Kyushu University, Japan
Giri Narasimhan	Florida International University, USA
Bogdan Pasaniuc	University of California at Los Angeles, USA
Steven Pascal	Old Dominion University, USA
Andrei Paun	University of Bucharest, Romania

Nadia Pisanti	Università di Pisa, Italy
Russell Schwartz	Carnegie Mellon University, USA
Joao Setubal	University of Sao Paulo, Brazil
Xinghua Shi	University of North Carolina at Charlotte, USA
Yi Shi	Jiao Tong University, China
Pavel Skums	Centers for Disease Control and Prevention, USA
Ileana Streinu	Smith College, USA
Zhengchang Su	University of North Carolina at Charlotte, USA
Wing-Kin Sung	National University of Singapore, Singapore
Sing-Hoi Sze	Texas A&M University, USA
Weitian Tong	Georgia Southern University, USA
Gabriel Valiente	Technical University of Catalonia, Spain
Xiang Wan	Hong Kong Baptist University, SAR China
Jianxin Wang	Central South University, China
Li-San Wang	University of Pennsylvania, USA
Peng Wang	Shanghai Advanced Research Institute, China
Lusheng Wang	City University of Hong Kong, SAR China
Seth Weinberg	Old Dominion University, USA
Fangxiang Wu	University of Saskatchewan, Canada
Yufeng Wu	University of Connecticut, USA
Minzhu Xie	Hunan Normal University, China
Dechang Xu	Harbin Institute of Technology, China
Can Yang	Hong Kong Baptist University, SAR China
Ashraf Yaseen	Texas A&M University-Kingsville, USA
Guoxian Yu	Southwest University, China
Alex Zelikovsky	Georgia State University, USA
Yanqing Zhang	Georgia State University, USA
Fa Zhang	Institute of Computing Technology, China
Leming Zhou	University of Pittsburgh, USA
Fengfeng Zhou	Chinese Academy of Sciences, China
Quan Zou	Xiamen University, China

Additional Reviewers

Aganezov, Sergey	Cirjila, Ionela	Leung, Yuk Yee	Previtali, Marco
Alexeev, Nikita	Cobeli, Matei	Li, Xin	Rizzi, Raffaella
Avdeyev, Pavel	Della Vedova,	Luo, Junwei	Sun, Sunah
Castro-Nallar,	Gianluca	Pei, Jingwen	Sun, Luni
Eduardo	Ghinea, Razvan	Peng, Xiaoqing	Wang, Weixin
Chen, Huiyuan	Ionescu, Vlad	Pham, Son	Zhang, June
Chen, Yong	Kuksa, Pavel	Postavaru, Stefan	Zhao, Wei
Chu, Chong	Lara, James	Preda, Catalina	Zhou, Chan

Contents

Next Generation Sequencing Data Analysis

An Efficient Algorithm for Finding All Pairs k-Mismatch Maximal
Common Substrings 3
  Sharma V. Thankachan, Sriram P. Chockalingam, and Srinivas Aluru

Poisson-Markov Mixture Model and Parallel Algorithm for Binning
Massive and Heterogenous DNA Sequencing Reads 15
  Lu Wang, Dongxiao Zhu, Yan Li, and Ming Dong

FSG: Fast String Graph Construction for De Novo Assembly of Reads Data . . . 27
  *Paola Bonizzoni, Gianluca Della Vedova, Yuri Pirola, Marco Previtali,
  and Raffaella Rizzi*

OVarCall: Bayesian Mutation Calling Method Utilizing Overlapping
Paired-End Reads 40
  *Takuya Moriyama, Yuichi Shiraishi, Kenichi Chiba, Rui Yamaguchi,
  Seiya Imoto, and Satoru Miyano*

High-Performance Sensing of DNA Hybridization on Surface
of Self-organized MWCNT-Arrays Decorated by Organometallic
Complexes 52
  *V.P. Egorova, H.V. Grushevskaya, N.G. Krylova, I.V. Lipnevich,
  T.I. Orekhovskaja, B.G. Shulitski, and V.I. Krot*

Towards a More Accurate Error Model for BioNano Optical Maps 67
  *Menglu Li, Angel C.Y. Mak, Ernest T. Lam, Pui-Yan Kwok, Ming Xiao,
  Kevin Y. Yip, Ting-Fung Chan, and Siu-Ming Yiu*

HapIso: An Accurate Method for the Haplotype-Specific Isoforms
Reconstruction from Long Single-Molecule Reads. 80
  *Serghei Mangul, Harry (Taegyun) Yang, Farhad Hormozdiari,
  Elizabeth Tseng, Alex Zelikovsky, and Eleazar Eskin*

Protein-Protein Interactions and Networks

Genome-Wide Structural Modeling of Protein-Protein Interactions. 95
  *Ivan Anishchenko, Varsha Badal, Taras Dauzhenka, Madhurima Das,
  Alexander V. Tuzikov, Petras J. Kundrotas, and Ilya A. Vakser*

Identifying Essential Proteins by Purifying Protein Interaction Networks 106
  Min Li, Xiaopei Chen, Peng Ni, Jianxin Wang, and Yi Pan

Differential Functional Analysis and Change Motifs in Gene Networks
to Explore the Role of Anti-sense Transcription 117
   Marc Legeay, Béatrice Duval, and Jean-Pierre Renou

Predicting MicroRNA-Disease Associations by Random Walking on
Multiple Networks 127
   Wei Peng, Wei Lan, Zeng Yu, Jianxin Wang, and Yi Pan

Progression Reconstruction from Unsynchronized Biological Data using
Cluster Spanning Trees 136
   Ryan Eshleman and Rahul Singh

Protein and RNA Structure

Consistent Visualization of Multiple Rigid Domain Decompositions
of Proteins.. 151
   Emily Flynn and Ileana Streinu

A Multiagent *Ab Initio* Protein Structure Prediction Tool for Novices
and Experts 163
   *Thiago Lipinski-Paes, Michele dos Santos da Silva Tanus,
   José Fernando Ruggiero Bachega, and Osmar Norberto de Souza*

Filling a Protein Scaffold with a Reference........................ 175
   Letu Qingge, Xiaowen Liu, Farong Zhong, and Binhai Zhu

Phylogenetics

Mean Values of Gene Duplication and Loss Cost Functions............ 189
   Paweł Górecki, Jarosław Paszek, and Agnieszka Mykowiecka

The SCJ Small Parsimony Problem for Weighted Gene Adjacencies 200
   *Nina Luhmann, Annelyse Thévenin, Aïda Ouangraoua, Roland Wittler,
   and Cedric Chauve*

Path-Difference Median Trees................................ 211
   Alexey Markin and Oliver Eulenstein

NEMo: An Evolutionary Model with Modularity for PPI Networks........ 224
   *Min Ye, Gabriela C. Racz, Qijia Jiang, Xiuwei Zhang,
   and Bernard M.E. Moret*

Multi-genome Scaffold Co-assembly Based on the Analysis of Gene Orders
and Genomic Repeats 237
   Sergey Aganezov and Max A. Alekseyev

Sequence and Image Analysis

Selectoscope: A Modern Web-App for Positive Selection Analysis
of Genomic Data.   253
  Andrey V. Zaika, Iakov I. Davydov, and Mikhail S. Gelfand

Methods for Genome-Wide Analysis of MDR and XDR Tuberculosis from
Belarus   258
  *Roman Sergeev, Ivan Kavaliou, Andrei Gabrielian, Alex Rosenthal,
  and Alexander Tuzikov*

Haplotype Inference for Pedigrees with Few Recombinations   269
  B. Kirkpatrick

Improved Detection of 2D Gel Electrophoresis Spots by Using Gaussian
Mixture Model   284
  Michal Marczyk

Abridged Track 2 Abstracts

Predicting Combinative Drug Pairs via Integrating Heterogeneous Features
for Both Known and New Drugs   297
  Jia-Xin Li, Jian-Yu Shi, Ke Gao, Peng Lei, and Siu-Ming Yiu

SkipCPP-Pred: Promising Prediction Method for Cell-Penetrating Peptides
Using Adaptive k-Skip-n-Gram Features on a High-Quality Dataset   299
  Wei Leyi and Zou Quan

CPredictor2.0: Effectively Detecting both Small and Large Complexes from
Protein-Protein Interaction Networks   301
  Bin Xu, Jihong Guan, Yang Wang, and Shuigeng Zhou

Structural Insights into Antiapoptotic Activation of Bcl-2 and Bcl-xL
Mediated by FKBP38 and tBid.   304
  Valery Veresov and Alexander Davidovskii

VAliBS: A Visual Aligner for Bisulfite Sequences   307
  *Min Li, Xiaodong Yan, Lingchen Zhao, Jianxin Wang, Fang-Xiang Wu,
  and Yi Pan*

MegaGTA: A Sensitive and Accurate Metagenomic Gene-Targeted
Assembler Using Iterative de Bruijn Graphs   309
  *Dinghua Li, Yukun Huang, Henry Chi-Ming Leung, Ruibang Luo,
  Hing-Fung Ting, and Tak-Wah Lam*

EnhancerDBN: An Enhancer Prediction Method Based on Deep Belief
Network.. 312
  Hongda Bu, Yanglan Gan, Yang Wang, Jihong Guan,
  and Shuigeng Zhou

An Improved Burden-Test Pipeline for Cancer Sequencing Data.......... 314
  Yu Geng, Zhongmeng Zhao, Xuanping Zhang, Wenke Wang, Xiao Xiao,
  and Jiayin Wang

Modeling and Simulation of Specific Production of Trans10,
cis12-Conjugated Linoleic Acid in the Biosynthetic Pathway 316
  Aiju Hou, Xiangmiao Zeng, Zhipeng Cai, and Dechang Xu

Dynamic Protein Complex Identification in Uncertain Protein-Protein
Interaction Networks 319
  Yijia Zhang, Hongfei Lin, Zhihao Yang, and Jian Wang

Predicting lncRNA-Protein Interactions Based on Protein-Protein Similarity
Network Fusion (Extended Abstract)............................. 321
  Xiaoxiang Zheng, Kai Tian, Yang Wang, Jihong Guan,
  and Shuigeng Zhou

DCJ-RNA: Double Cut and Join for RNA Secondary Structures
Using a Component-Based Representation 323
  Ghada Badr and Haifa Al-Aqel

Improve Short Read Homology Search Using Paired-End Read Information ... 326
  Prapaporn Techa-Angkoon, Yanni Sun, and Jikai Lei

Framework for Integration of Genome and Exome Data for More Accurate
Identification of Somatic Variants 330
  Vinaya Vijayan, Siu-Ming Yiu, and Liqing Zhang

Semantic Biclustering: A New Way to Analyze and Interpret Gene
Expression Data 332
  Jiří Kléma, František Malinka, and Filip Železný

Analyzing microRNA Epistasis in Colon Cancer Using Empirical Bayesian
Elastic Nets 334
  Jia Wen, Andrew Quitadamo, Benika Hall, and Xinghua Shi

Tractable Kinetics of RNA–Ligand Interaction 337
  Felix Kühnl, Peter F. Stadler, and Sebastian Will

MitoDel: A Method to Detect and Quantify Mitochondrial DNA Deletions
from Next-Generation Sequence Data 339
  Colleen M. Bosworth, Sneha Grandhi, Meetha P. Gould,
  and Thomas LaFramboise

TRANScendence: Transposable Elements Database and *De-novo* Mining
Tool Allows Inferring TEs Activity Chronology   342
    Michał Startek, Jakub Nogły, Dariusz Grzebelus, and Anna Gambin

Phylogeny Reconstruction from Whole-Genome Data Using Variable
Length Binary Encoding   345
    Lingxi Zhou, Yu Lin, Bing Feng, Jieyi Zhao, and Jijun Tang

Author Index   347

Next Generation Sequencing Data Analysis

An Efficient Algorithm for Finding All Pairs k-Mismatch Maximal Common Substrings

Sharma V. Thankachan[1], Sriram P. Chockalingam[2], and Srinivas Aluru[1(✉)]

[1] School of CSE, Georgia Institute of Technology, Atlanta, USA
sharma.thankachan@gatech.edu, aluru@cc.gatech.edu
[2] Department of CSE, Indian Institute of Technology, Bombay, India
sriram.pc@iitb.ac.in

Abstract. Identifying long pairwise maximal common substrings among a large set of sequences is a frequently used construct in computational biology, with applications in DNA sequence clustering and assembly. Due to errors made by sequencers, algorithms that can accommodate a small number of differences are of particular interest, but obtaining provably efficient solutions for such problems has been elusive. In this paper, we present a provably efficient algorithm with an expected run time guarantee of $O(N \log^k N + \mathsf{occ})$, where occ is the output size, for the following problem: Given a collection $\mathcal{D} = \{S_1, S_2, \ldots, S_n\}$ of n sequences of total length N, a length threshold ϕ and a mismatch threshold $k \geq 0$, report all k-mismatch maximal common substrings of length at least ϕ over all pairs of sequences in $\mathcal{D}$. In addition, we present a result showing the hardness of this problem.

1 Introduction

Due to preponderance of DNA and RNA sequences that can be modeled as strings, string matching algorithms have myriad applications in computational biology. Modern sequencing instruments sequence a large collection of short reads that are randomly drawn from one or multiple genomes. Deciphering pairwise relationships between the reads is often the first step in many applications. For example, one may be interested in finding all pairs of reads that have a sufficiently long overlap, such as suffix/prefix overlap (for genomic or metagenomic assembly) or substring overlap (for read compression, finding RNA sequences containing common exons, etc.). Sequencing instruments make errors, which translate to insertion, deletion, or substitution errors in the reads they characterize, depending on the type of instrument. Much of modern-day high-throughput sequencing is carried out using Illumina sequencers, which have a small error rate (< 1–$2\,\%$) and predominantly ($> 99\,\%$) substitution errors. Thus, algorithms that tolerate a small number of mismatch errors can yield the same solution as the much more expensive alignment/edit distance computations. Motivated by such applications, we formulate the following **all pairs k-mismatch maximal common substrings problem**:

© Springer International Publishing Switzerland 2016
A. Bourgeois et al. (Eds.): ISBRA 2016, LNBI 9683, pp. 3–14, 2016.
DOI: 10.1007/978-3-319-38782-6_1

Problem 1. *Given a collection $\mathcal{D} = \{S_1, S_2, \ldots, S_n\}$ of n sequences of total length N, a length threshold ϕ, and a mismatch threshold $k \geq 0$, report all k-mismatch maximal common substrings of length $\geq \phi$ between any pair of sequences in $\mathcal{D}$.*

Throughout this paper, for a sequence $S_i \in \mathcal{D}$, $|S_i|$ is its length, $S_i[x]$ is its xth character and $S_i[x..y]$ is its substring starting at position x and ending at position y, where $1 \leq x \leq y \leq |S_i|$. For brevity, we may use $S_i[x..]$ to denote the suffix of S_i starting at x. The substrings $S_i[x..(x+l-1)]$ and $S_j[y..(y+l-1)]$ are a k-mismatch common substring if the Hamming distance between them is at most k. It is maximal if it cannot be extended on either side without introducing another mismatch.

Before investigating Problem 1, consider an existential version of this problem, where we are just interested in enumerating those pairs (S_i, S_j) that contain at least one k-mismatch maximal common substring of length $\geq \phi$. For $k = 0$ and any given (S_i, S_j) pair, this can be easily answered in $O(|S_i| + |S_j|)$ time using a generalized suffix tree based algorithm because no mismatches are involved. For $k \geq 1$, we can use the recent result by Aluru *et al.* [1], by which the k-mismatch longest common substring can be identified in $O((|S_i| + |S_j|) \log^k (|S_i| + |S_j|))$ time. Therefore, this straightforward approach can solve the existential problem over all pairs of sequences in $\mathcal{D}$ in $\sum_i \sum_j (|S_i| + |S_j|) \log^k (|S_i| + |S_j|) = O(n^2 L \log^k L)$ time, where $L = \max_i |S_i|$. An interesting question is, can we solve this problem asymptotically faster?

We present a simple result showing that the existence of a *purely combinatorial* algorithm, which is "significantly" better than the above approach is highly unlikely in the *general setting*. In other words, we prove a conditional lower bound, showing that the bound is tight within poly-logarithmic factors. This essentially shows the hardness of Problem 1, as it is at least as hard as its existential version. Our result is based on a simple reduction from the boolean matrix multiplication (BMM) problem. This hard instance is simulated via careful tuning of the parameters – specifically, when L approaches the number of sequences and $\phi = \Theta(\log n)$. However, this hardness result does not contradict the possibility of an $O(N + \mathsf{occ})$ run-time algorithm for Problem 1, because occ can be as large as $\Theta(n^2 L^2)$.

In order to solve such problems in practice, the following seed-and-extend type filtering approaches are often employed (see [13] for an example). The underline principle is: if two sequences have a k-mismatch common substring of length $\geq \phi$, then they must have an exact common substring of length at least $\tau = \lceil \frac{\phi}{k+1} \rceil$. Therefore, using some fast hashing technique, all pairs of sequences that have a τ-length common substring are identified. Then, by exhaustively checking all such candidate pairs, the final output is generated. Clearly, such an algorithm cannot provide any run time guarantees and often times the candidate pairs generated can be overwhelmingly larger than the final output size. In contrast to this, we develop an $O(N)$ space and $O(N \log^k N + \mathsf{occ})$ expected run time algorithm, where occ is the output size. Additionally, we present the results of some preliminary experiments, in order to demonstrate the effectiveness of our algorithm.

2   Notation and Preliminaries

Let Σ be the alphabet for all sequences in $\mathcal{D}$. Throughout the paper, both $|\Sigma|$ and k are assumed to be constants. Let $\mathsf{T} = S_1 \$_1 S_2 \$_2 \ldots S_n \$_n$ be the concatenation of all sequences in $\mathcal{D}$, separated by special characters $\$_1, \$_2, \ldots, \$_n$. Here each $\$_i$ is a unique special symbol and is lexicographically larger than all characters in Σ. Clearly, there exists a one to one mapping between the positions in T (except the $\$_i$ positions) and the positions in the sequences in $\mathcal{D}$. We use $\mathsf{lcp}(\cdot, \cdot)$ to denote the longest common prefix of two input strings and $\mathsf{lcp}_k(\cdot, \cdot)$ to denote their longest common prefix while permitting at most k mismatches. We now briefly review some standard data structures that will be used in our algorithms.

2.1   Suffix Trees, Suffix Arrays and LCP Data Structures

The generalized suffix tree of $\mathcal{D}$ (equivalently, the suffix tree of T), denoted by GST, is a lexicographic arrangement of all suffixes of T as a compact trie [11,15]. The GST consists of $|\mathsf{T}|$ leaves, and at most $(|\mathsf{T}| - 1)$ internal nodes all of which have at least two child nodes each. The edges are labeled with substrings of T. Let path of u refer to the concatenation of edge labels on the path from root to node u, denoted by $\mathsf{path}(u)$. If u is a leaf node, then its path corresponds to a unique suffix of T (equivalently a unique suffix of a unique sequence in $\mathcal{D}$) and vice versa. For any node, node-depth is the number of its ancestors and string-depth is the length of its path.

The suffix array [10], SA, is such that $\mathsf{SA}[i]$ is the starting position of the suffix corresponding to the ith left most leaf in the suffix tree of T, i.e., the starting position of the ith lexicographically smallest suffix of T. The inverse suffix array ISA is such that $\mathsf{ISA}[j] = i$, if $\mathsf{SA}[i] = j$. The Longest Common Prefix array LCP is defined as, for $1 \leq i < |\mathsf{T}|$

$$\mathsf{LCP}[i] = |\mathsf{lcp}(\mathsf{T}_{\mathsf{SA}[i]}, \mathsf{T}_{\mathsf{SA}[i+1]})|$$

In other words, $\mathsf{LCP}[i]$ is the string depth of the lowest common ancestor of the ith and $(i + 1)$th leaves in the suffix tree. There exist optimal sequential algorithms for constructing all these data structures in $O(|\mathsf{T}|)$ space and time [6,8,9,14].

All operations on GST required for our purpose can be simulated using SA, ISA, LCP array, and a range minimum query (RMQ) data structure over the LCP array [2]. A node u in GST can be uniquely represented by an interval $[\mathsf{sp}(u), \mathsf{ep}(u)]$, the range corresponding to the leaves in its subtree. The string depth of u is the minimum value in $\mathsf{LCP}[\mathsf{sp}(u), \mathsf{ep}(u) - 1]$ (can be computed in constant time using an RMQ). Similarly, the longest common prefix of any two suffixes can also be computed in $O(1)$ time. Finally, the k-mismatch longest common prefix of any two suffixes can be computed in $O(k)$ time as follows: let $l = |\mathsf{lcp}(\mathsf{T}[x..], \mathsf{T}[y..])|$, then for any $k \geq 1$, $|\mathsf{lcp}_k(\mathsf{T}[x..], \mathsf{T}[y..])|$ is also l if either of $\mathsf{T}[x+l]$, $\mathsf{T}[y+l]$ is a $\$_i$ symbol, else it is $l+1+|\mathsf{lcp}_{k-1}(\mathsf{T}[(x+l+1)..], \mathsf{T}[(y+l+1)..])|$.

2.2   Linear Time Sorting of Integers and Strings

Our algorithm relies heavily on sorting of integers and strings (specifically, suffixes). The integers are always within the range $[1, N]$. Therefore, linear time sorting (via radix sort) is possible when the number of integers to be sorted is sufficiently large (say $\geq N^\epsilon$ for some constant $\epsilon > 0$). In order to achieve linear sorting complexity for even smaller input sizes, we use the following strategy: combine multiple small collections such that the total size becomes sufficiently large. Assign a unique integer id in $[1, N]$ to each small collection, and with each element e within a small collection with id i, associate the key $i \cdot N + e$. We can now apply radix sort and sort all elements within the combined collection w.r.t. their associated key. An additional scanning step suffices to separate all small collections with their elements sorted. The strings to be sorted in our algorithms are always suffixes of T. Using GST, each suffix can be mapped to an integer in $[1, N]$ that corresponds to its lexicographic rank in constant time. Therefore, any further suffix sorting (or sparse suffix sorting) can also be reduced to linear time integer sorting.

3   Hardness Result

3.1   Boolean Matrix Multiplication (BMM)

The input to BMM problem consists of two $n \times n$ boolean matrices A and B with all entries either 0 or 1. The task is to compute their boolean product C. Specifically, let $A_{i,j}$ denotes the entry corresponding to ith row and jth column of matrix A, where $0 \leq i, j < n$, then for all i, j, compute

$$C_{i,j} = \bigvee_{k=0}^{n-1} (A_{i,k} \wedge B_{k,j}). \tag{1}$$

Here $\vee$ represents the logical OR and $\wedge$ represents the logical AND operations. The time complexity of the fastest matrix multiplication algorithm is $O(n^{2.372873})$ [16]. The result is obtained via algebraic techniques like Strassen's matrix multiplication algorithm. However, the fastest combinatorial algorithm is better than the standard cubic algorithm by only a poly-logarithmic factor [17]. As matrix multiplication is a highly studied problem over decades, any truly sub-cubic algorithm for matrix multiplication which is purely combinatorial is highly unlikely, and will be a breakthrough result.

3.2   Reduction

In this section, we prove the following result.

Theorem 1. *If there exists an $f(n, L)$ time algorithm for the existential version of Problem 1, then there exists an $O(f(2n, O(n \log n))) + O(n^2)$ time algorithm for the BMM problem.*

This implies BMM can be solved in sub-cubic time via purely combinatorial techniques if $f(n, L)$ is $O(n^{2-\epsilon}L)$ or $O(n^2 L^{1-\epsilon})$ for any constant $\epsilon > 0$. As the former is less likely, the later is also less likely. In other words, $O(n^2 L)$ bound is tight under BMM hardness assumption. We now proceed to show the reduction.

We create n sequences $X_1, X_2, \ldots, X_n$ from A and n sequences $Y_1, Y_2, \ldots, Y_n$ from B. Strings are of length at most $L = n\lceil \log n \rceil + n - 1$ over an alphabet $\Sigma = \{0, 1, \$, \#\}$. The construction is the following.

- For $1 \le i \le n$, let $U_i = \{k \mid A_{i,k} = 1\}$. Then, for each U_i, create a corresponding string X_i as follows: encode each element in U_i in binary in $\lceil \log n \rceil$ bits, append it with $\$$ symbol, then concatenate all of them. For example, if $n = 4$ and $U_i = \{0, 2\}$, then $X_i = 00\$10\$$.
- For $1 \le i \le n$, let $V_j = \{k \mid B_{k,j} = 1\}$. Then, for each V_j, create a corresponding string Y_j as follows: encode each element in V_j in binary in $\lceil \log n \rceil$ bits, append it with $\#$ symbol, then concatenate all of them. For example, if $n = 4$ and $V_j = \{1, 2\}$, then $Y_j = 01\#10\#$.
- The database $\mathcal{D}$ is the set of all X_i's and Y_j's.

Lemma 1. *The entry $C_{i,j} = 1$, iff $U_i \cap V_j$ is not empty. The set $U_i \cap V_j$ is not empty iff X_i and Y_j have a common substring of length $\lceil \log n \rceil$.*

Using this result, we can compute the boolean product of A and B from the output of the existential version of Problem 1 plus additional $O(n^2)$ time. Therefore, BMM can be solved in $f(2n, O(n \log n)) + O(n^2)$ time.

4 Our Algorithm for k-Mismatch Maximal Common Substrings

We present an $O(N \log^k N + \text{occ})$ expected time algorithm. Recall that each position x in the concatenated text T corresponds to a unique position in a unique sequence $S_d \in \mathcal{D}$, therefore we denote the sequence identifier d by $\mathsf{seq}(x)$. Each output can be represented as a pair (x, y) of positions in T, where

1. $\mathsf{seq}(x) \ne \mathsf{seq}(y)$
2. $\mathsf{T}[x - 1] \ne \mathsf{T}[y - 1]$
3. $|\mathsf{lcp}_k(\mathsf{T}[x..], \mathsf{T}[y..])| \ge \phi$.

4.1 The Exact Match Case

Without mismatches, the problem can be easily solved in optimal $O(N + \text{occ})$ worst case time. First create the GST, then identify all nodes whose string depth is at least ϕ, such that the string depth of their parent nodes is at most $(\phi - 1)$. Such nodes are termed as marked nodes. Clearly, a pair of suffixes satisfies condition (3) iff their corresponding leaves are under the same marked node. This allows us to process the suffixes under each marked node w independently as follows: let Suff_w denotes the set of starting positions of the suffixes of T corresponding to the leaves in the subtree of w. That is, $\mathsf{Suff}_w = \{\mathsf{SA}[j] \mid \mathsf{sp}(w) \le j \le \mathsf{ep}(w)\}$. Then,

1. Partition Suff_w into (at most) $\Sigma + 1$ buckets, such that for each $\sigma \in \Sigma$, there is a unique bucket containing all suffixes with previous character σ. All suffixes with their previous character is a $ symbol are put together in a special bucket. Note that a pair (x, y) is an answer only if both x are y are not in the same bucket, or if both of them are in the special bucket.
2. Sort all suffixes w.r.t the identifier of the corresponding sequence (i.e., $\mathsf{seq}(\cdot)$). Therefore, within each bucket, all suffixes from the same sequence appear contiguously.
3. For each x, report all answers of the from $(x, \cdot)$ as follows: scan every bucket, except the bucket in which x belongs to, unless x is in the special bucket. Then output (x, y) as an answer, where y is not an entry in the contiguous chunk of all suffixes from $\mathsf{seq}(x)$.

The construction of GST and Step (1) takes $O(N)$ time. Step (2) over all marked nodes can also be implemented in $O(N)$ time via integer sorting (refer to Sect. 2.2). By noting down the sizes of each chunk during Step (2), we can implement Step (3) in time proportional to the sizes of input and output. By combining all, the total time complexity is $O(N \cdot |\Sigma| + \mathsf{occ})$, i.e., $O(N + \mathsf{occ})$ under constant alphabet size assumption.

4.2   The k-Mismatch Case

Approximate matching problems are generally harder, because the standard data structures such as suffix trees and suffix arrays may not apply directly. Therefore, we follow a novel approach, where we transform the approximate matching problem over exact strings into an exact matching problem over inexact copies of strings, which we call as modified suffixes.

Definition 1. *Let # be a special symbol not in Σ. A k-modified suffix is a suffix of* T *with its k characters replaced by #.*

Let Δ be a set of positions. Then, $\mathsf{T}^\Delta[x..]$ denotes the $|\Delta|$-modified suffix obtained by replacing $|\Delta|$ positions in the suffix $\mathsf{T}[x..]$ as specified by Δ. For example, let $\mathsf{T} = aaccgattcaa$, $\Delta = \{2, 4\}$, then $\mathsf{T}[5..] = gattcaa$ and $\mathsf{T}^\Delta[5..] = g\#t\#caa$.

Our algorithm consists of two main phases. In the first phase, we create a collection of sets of k-modified suffixes. In the second phase, we independently process each set constructed in the first phase and extract the answers. The first phase takes $O(NH^k)$ time, where as the second phase takes $O(NH^k + \mathsf{occ})$ time. Here H is the height of GST. It is known that the expected value of H is $O(\log N)$ [3]. Therefore, by combining the time complexities of both phases with H replaced by $O(\log N)$, we obtain the expected run time as claimed. We now describe these phases in detail.

Details of Phase-1. This phase is recursive (with levels of recursion starting from 0 up to k), such that at each level $h > 0$, we create a collection of sets of

h-modified suffixes (denoted by $C_1^h, C_2^h, \ldots$) from the sets in the previous level. At level 0, we have only a single set C_0^1, the set of all suffixes of T. See Fig. 1 for an illustration. To generate the sets at level h, we take each set C_g^{h-1} at level $(h-1)$ and do the following:

- Create a compact trie of all strings in C_g^{h-1}
- For each internal node w in the trie, create a set consisting of the strings corresponding to the leaves in the subtree of w, but with their $(l+1)$th character replaced by #. Here l is the string depth of w. Those strings with their $(l+1)$-th character is a $\$_i$ symbol are not included.

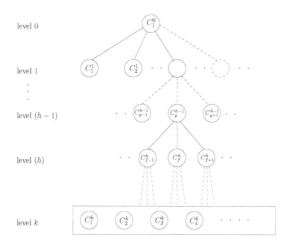

Fig. 1. The sets $\ldots, C_{f-1}^h, C_f^h, C_{f+1}^h \ldots$ of h-modified suffixes are generated from the set C_g^{h-1}.

From our construction procedure, the following properties can be easily verified.

Property 1. *All modified suffixes within the same set (at any level) have # symbols at the same positions and share a common prefix at least until the last occurrence of #.*

Property 2. *For any pair (x, y) of positions, there will be **exactly one** set at level k, such that it contain k-modified suffixes of $T[x..]$ and $T[y..]$ with # symbols at the first k positions in which they differ. Therefore, the lcp of those k-modified suffixes is equal to the lcp_k of $T[x..]$ and $T[y..]$.*

We have the following result about the sizes of these sets.

Lemma 2. *No set is of size more than N and the sum of sizes of all sets at a particular level h is $\leq N \times H^k$, where H is the height of GST.*

Proof. The first statement follows easily from our construction procedure and the second statement can be proved via induction. Let $\mathcal{S}_h$ be the sum of sizes of all sets at level h. Clearly, the base case, $\mathcal{S}_0 = |C_1^0| = N$, is true. The sum of sizes of sets at level h generated from C_g^{h-1} is at most $|C_g^{h-1}| \times$ the height of the compact trie over the strings in C_g^{h-1}. The height of the compact trie is $\leq H$, because if we remove the common prefix of all strings in C_g^{h-1}, they are essentially suffixes of T. By putting these together, we have $\mathcal{S}_h \leq \mathcal{S}_{h-1} \cdot H \leq \mathcal{S}_{h-2} \cdot H^2 \leq \cdots \leq NH^k$.

Space and Time Analysis: We now show that Phase-1 can be implemented in $O(N)$ space and $O(N \log^k N)$ time in expectation. Consider the step where we generate sets from C_g^{h-1}. The lexicographic ordering between any two $(h-1)$-modified suffixes in C_g^{h-1} can be determined in constant time. i.e., by simply checking the lexicographic ordering between those suffixes obtained by deleting their common prefix up to the last occurrence of $\#$. Therefore, suffixes can be sorted using any comparison based sorting algorithm. After sorting, the lcp between two successive strings can be computed in constant time. Using the sorted order and lcp values, we can construct the compact trie using standard techniques in the construction of suffix tree [4]. Finally, the new sets can be generated in time proportional to their size. In summary, the time complexity for a particular C_g^{h-1} is $O(|C_g^{h-1}|(\log N + H))$. Overall time complexity is

$$\sum_{h \leq k} \sum_f C_f^k + (H + \log n) \sum_{h < k} \sum_f C_f^h = O(N(\log N + H)^{k-1} H)$$

By replacing H by $O(\log N)$, we bound the expect run time of Phase-1 by $O(N \log^k N)$.

We generate the sets in pre-order of the corresponding node in the recursion tree. As soon as a set (at level k) is generated, we immediately pass it to Phase-2, extract the necessary information and discard it from the working space. Also, any set at level $h < k$ is also deleted after all k-level sets in its subtree are processed. This way, at any point of time in the execution of the algorithm, we need to maintain only k sets, corresponding to the sets in a root to leaf path in the recursion tree. Since the size of each set is at most N (Lemma 2), we can bound the working space also by $O(N)$, assuming $k = O(1)$.

Details of Phase-2. In this phase, we seek to process each set C_f^k created by Phase-1 independently and generate the answers in time linear to the total size of all sets and output. i.e., $O(NH^k + \text{occ})$. We first present a simple $O((N + \text{occ})H^k)$ time approach. Following are the key steps.

1. Create a compact trie over all k-modified suffixes in C_f^k. Then identify the marked nodes as before. Recall that a marked node has a string depth $\geq \phi$, where as the string depth of its parent is $< \phi$.
2. Let Δ be the set of k positions corresponding to modifications in the k-modified suffixes in C_f^k. Clearly, if the leaves corresponding to two modified suffixes (say $\mathsf{T}^\Delta[x..]$ and $\mathsf{T}^\Delta[y..]$) are in the same subtree of a marked node,

then their lcp_k is $\geq \phi$. If $\mathsf{seq}(x) \neq \mathsf{seq}(y)$ and $\mathsf{T}[x-1] \neq \mathsf{T}[y-1]$, then report (x, y) as an answer.

The trie can be created in time linear to the size of C_f^k. Note that the key step in the creation of a trie is the sorting of k-modified suffixes. To do it efficiently, we map each k-modified suffix to the lexicographic rank of the suffix obtained by removing all characters (from left) until its last $\#$ symbol. Using this as the key, the k-modified suffixes can be sorted via integer sorting (refer to Sect. 2.2). The second step of extracting answers can also be implemented using the exact same procedure described in Sect. 4.1. However, the problem with this approach is that, a pair (x, y) can get reported more than once, although only once per set. In the worst case, an answer can get reported H^k times. The resulting time complexity is therefore $O((N + \mathsf{occ})H^k)$.

Improving the Run Time Complexity. To achieve the claimed $O(NH^k + \mathsf{occ})$ run time, we need to ensure that each output (x, y) is reported exactly once. For this, we explore Property 2 as follows: while processing a pair of two k-modified suffixes $\mathsf{T}^\Delta[x..]$ and $\mathsf{T}^\Delta[y..]$ under the subtree of some marked node, report (x, y) as an answer iff

1. $\mathsf{lcp}(\mathsf{T}^\Delta[x..], \mathsf{T}^\Delta[y..]) = \mathsf{lcp}_k(\mathsf{T}[x..], \mathsf{T}[y..])$.
2. $\mathsf{seq}(x) \neq \mathsf{seq}(y)$
3. $\mathsf{T}[x-1] \neq \mathsf{T}[y-1]$

From Property 2, for a pair (x, y), there will be only one pair of k-modified suffixes satisfying this condition (1). The following is unique to that pair: $\mathsf{T}[x + l - 1] \neq \mathsf{T}[y + l - 1]$ for all $l \in \Delta$. Therefore, the task can be executed efficiently by processing the set of k-modified suffixes in the subtree of each marked node w as follows:

1. Partition them into (at most) $\Sigma + 1$ buckets based on the previous character as in Sect. 4.1.
2. Partition the k-modified suffixes in each bucket into (at most) $|\Sigma|^k$ sub-buckets based on the sequence of k characters that were originally at the positions in Δ. Each sub-bucket is therefore associated with a unique string of length $(1 + k)$: the (previous) character corresponding to the bucket in which it belongs to, followed by the sequence of k characters at the positions in Δ.
3. Within each sub-bucket, sort the k-modified suffixes based on the identifier of the sequence to which it belongs.
4. Finally, for each $\mathsf{T}^\Delta[x..]$, we visit each sub-bucket and find answers of the form $(x, \cdot)$ as follows: let $c_0, c_1, c_2, \ldots, c_k$ be the sequence of $(1+k)$ characters corresponding to a sub-bucket. If $c_0 \neq \mathsf{T}[x-1]$ or $\mathsf{T}[x-1]$ is some $\$_i$ symbol and $c_t \neq \mathsf{T}[x+t-1]$ for $t = 1, 2, \ldots k$, then for all entries $\mathsf{T}^\Delta[y..]$ in the sub-bucket with $\mathsf{seq}(x) \neq \mathsf{seq}(y)$, report (x, y) as an answer. Notice that the entries within a sub-bucket are sorted according to the sequence identifier. Therefore, all entries with $\mathsf{seq}(x) = \mathsf{seq}(y)$ comes together as a contiguous chunk, which can be easily skipped.

Analysis: The overall time for implementing the first three steps is $O(NH^k)$ and final step takes $O(NH^k|\Sigma|^k + \text{occ})$ time. Therefore, total time complexity is $O(NH^k + \text{occ})$, assuming k and $|\Sigma|$ are constants.

Theorem 2. *Problem 1 can be solved in $O(N)$ space and $O(N \log^k N + \text{occ})$ expected time, assuming k and $|\Sigma|$ are constants.*

Note: If the hamming distance between two reads is $< k$, then our algorithm will not output them as an answer. To capture such answers, we shall run the algorithm for all numbers of mismatches starting from 0 up to k. The run-time remains the same.

5   Preliminary Experiments

We have implemented our algorithm using the C++11 standard. As noted earlier, we use SA, LCP array, and RMQ data structures to simulate all the operations on the suffix trees. We construct the suffix array for the set of the sequences $\mathcal{D}$ using the `libdivsufsort` library [12]. We build ISA corresponding to SA by a single pass over SA. Construction of LCP array and RMQ tables are based on the implementations in the SDSL library [5]. For the construction of the LCP and RMQ tables, we use Kasai et al.'s algorithm [7] and Bender-Farach's algorithm [2], respectively. Although the SDSL library supports bit compression techniques to reduce the size of the tables and arrays in exchange for relatively longer time to answer queries, we do not compress these data structures. Instead, we use 32-bit integers both for indices and prefix lengths.

Note that our algorithm does not require the internal nodes to be processed in any specific order. Therefore, we do not need to construct parent-child links or suffix links, which are typically present in a suffix tree. We only need a list of all the internal nodes, their string depths and the corresponding suffixes. We represent the internal node u by the tuple $(\mathsf{sp}(u), \mathsf{ep}(u), string\text{-}depth)$, where $\mathsf{sp}(u)$ and $\mathsf{ep}(u)$ are the left and right indices in SA anchoring the range of suffixes having $\mathsf{path}(u)$ as their prefix.

We conducted our preliminary experiments on a system with Intel Xeon E5-2660 CPU having 10 cores and 64 GB RAM. We created a 2,049,118 read input derived from the RNA-Seq dataset with accession number SRX011546, from the NCBI SRA repository. This dataset corresponds to an Illumina sequencing of Human CD4 T cells, with 45bp reads. Table 1 shows the run-time results for this dataset for different numbers of mismatches allowed k. The length threshold ϕ is chosen as 25.

The run-time complexity for generating all valid occurrences increases exponentially with k. As shown in Table 1, this behavior is observed for smaller values of k. However, for larger values of k, the run-time escalation is significantly smaller than the exponential dependence predicted by the theory. This is probably because as the matches extend towards the end of the sequences, the number of compact tries that need to be generated is limited.

Table 1. Time to generate k-mismatch pairs with $\phi = 25$.

Hamming distance (k)	Run time (seconds)
0	21.44
1	931.33
2	3244.99
3	4584.36
4	4801.55
5	4920.68

For large input sizes, the value of k for which the algorithm can be run in practice will become limited. Our method can be supplanted with seed and extend heuristics, where the seed itself can be based on approximate sequence matching with a limited value of k. Because our algorithm can process internal nodes of the GST in any order, it allows easy parallelization on multiple cores sharing the same shared memory. This can be used to further increase the scale of the datasets or the values of k for which the algorithm is practical.

Acknowledgment. This research is supported in part by the U.S. National Science Foundation under IIS-1416259.

References

1. Aluru, S., Apostolico, A., Thankachan, S.V.: Efficient alignment free sequence comparison with bounded mismatches. In: Przytycka, T.M. (ed.) RECOMB 2015. LNCS, vol. 9029, pp. 1–12. Springer, Heidelberg (2015)
2. Bender, M.A., Farach-Colton, M.: The LCA problem revisited. In: Gonnet, G.H., Viola, A. (eds.) LATIN 2000. LNCS, vol. 1776, pp. 88–94. Springer, Heidelberg (2000)
3. Devroye, L., Szpankowski, W., Rais, B.: A note on the height of suffix trees. SIAM J. Comput. **21**(1), 48–53 (1992)
4. Farach-Colton, M., Ferragina, P., Muthukrishnan, S.: On the sorting-complexity of suffix tree construction. J. ACM **47**(6), 987–1011 (2000)
5. Gog, S., Beller, T., Moffat, A., Petri, M.: From theory to practice: plug and play with succinct data structures. In: Gudmundsson, J., Katajainen, J. (eds.) SEA 2014. LNCS, vol. 8504, pp. 326–337. Springer, Heidelberg (2014)
6. Kärkkäinen, J., Sanders, P.: Simple linear work suffix array construction. In: Baeten, J.C.M., Lenstra, J.K., Parrow, J., Woeginger, G.J. (eds.) ICALP 2003. LNCS, vol. 2719, pp. 943–955. Springer, Heidelberg (2003)
7. Kasai, T., Lee, G.H., Arimura, H., Arikawa, S., Park, K.: Linear-time longest-common-prefix computation in suffix arrays and its applications. In: Amir, A., Landau, G.M. (eds.) CPM 2001. LNCS, vol. 2089, pp. 181–192. Springer, Heidelberg (2001)
8. Kim, D.-K., Sim, J.S., Park, H.-J., Park, K.: Linear-time construction of suffix arrays. In: Baeza-Yates, R., Chávez, E., Crochemore, M. (eds.) CPM 2003. LNCS, vol. 2676, pp. 186–199. Springer, Heidelberg (2003)

9. Ko, P., Aluru, S.: Space efficient linear time construction of suffix arrays. In: Baeza-Yates, R., Chávez, E., Crochemore, M. (eds.) CPM 2003. LNCS, vol. 2676, pp. 200–210. Springer, Heidelberg (2003)

10. Manber, U., Myers, G.: Suffix arrays: a new method for on-line string searches. SIAM J. Comput. **22**(5), 935–948 (1993)

11. Edward, M.: McCreight.: a space-economical suffix tree construction algorithm. J. ACM **23**(2), 262–272 (1976)

12. Mori, Y.: Libdivsufsort: a lightweight suffix array construction library, pp. 1–12 (2003). https://github.com/y-256/libdivsufsort

13. Peterlongo, P., Pisanti, N., Boyer, F., Lago, A.P.D., Sagot, M.-F.: Lossless filter for multiple repetitions with hamming distance. J. Discrete Algorithms **6**(3), 497–509 (2008)

14. Ukkonen, E.: On-line construction of suffix trees. Algorithmica **14**(3), 249–260 (1995)

15. Weiner, P.: Linear pattern matching algorithms. In: Switching and Automata Theory, pp. 1–11 (1973)

16. Williams, V.V.: Multiplying matrices faster than coppersmith-winograd. In: Proceedings of the 44th Symposium on Theory of Computing Conference (STOC), New York, NY, USA, pp. 887–898, 19–22 May 2012 (2012)

17. Yu, H.: An improved combinatorial algorithm for boolean matrix multiplication. In: Halldórsson, M.M., Iwama, K., Kobayashi, N., Speckmann, B. (eds.) ICALP 2015. LNCS, vol. 9134, pp. 1094–1105. Springer, Heidelberg (2015)

Poisson-Markov Mixture Model and Parallel Algorithm for Binning Massive and Heterogenous DNA Sequencing Reads

Lu Wang, Dongxiao Zhu$^{(\boxtimes)}$, Yan Li, and Ming Dong

Department of Computer Science, Wayne State University,
Detroit, MI 48202, USA
{lu.wang3,dzhu,rock_liyan,mdong}@wayne.edu

Abstract. A major computational challenge in analyzing metagenomics sequencing reads is to identify unknown sources of massive and heterogeneous short DNA reads. A promising approach is to efficiently and sufficiently extract and exploit sequence features, i.e., k-mers, to bin the reads according to their sources. Shorter k-mers may capture base composition information while longer k-mers may represent reads abundance information. We present a novel Poisson-Markov mixture Model (PMM) to systematically integrate the information in both long and short k-mers and develop a parallel algorithm for improving both reads binning performance and running time. We compare the performance and running time of our PMM approach with selected competing approaches using simulated data sets, and we also demonstrate the utility of our PMM approach using a time course metagenomics data set. The probabilistic modeling framework is sufficiently flexible and general to solve a wide range of supervised and unsupervised learning problems in metagenomics.

Keywords: Probabilistic clustering · Expectation-Maximization algorithm · Metagenomics · Next-generation sequencing (NGS) · Parallel algorithm

1 Introduction

Metagenomics sequencing reads are typically sequenced from a large number of heterogeneous sources with diverse abundances. There are two related yet distinct computational problems. The first is unsupervised binning of the reads to identify unknown sources. Reads from the same sources are more similar compared to the rest and the sources can later be labeled as Operational Taxonomic Units (OTU's). The other is supervised classification of the reads to assign each read to a labeled known source, such as a taxonomic or a patient treatment/risk group. Here we will focus on the more challenging reads binning problem.

Reads binning has posed the following unprecedented algorithmic and computational challenges, ranked by decreasing priority, to bioinformatics research

© Springer International Publishing Switzerland 2016
A. Bourgeois et al. (Eds.): ISBRA 2016, LNBI 9683, pp. 15–26, 2016.
DOI: 10.1007/978-3-319-38782-6_2

community: (1) How to sufficiently and robustly extract discriminating features from the reads? This is essentially a k-mers (sequencing feature) counting and selection problem; (2) How to account for the differential abundances across bins? Some sources may generate more reads whereas others may generate less; (3) How to filter out the inseparable reads? Some reads contain useful feature information, but others don't. The latter can come from the common sequences shared among the sources and was referred to as inseparable reads, and (4) How to efficiently process ultra-high throughput (hundreds of millions), very short (≈ 100 bp) reads?

A key to overcome the first challenge is to sufficiently and robustly extract sequence features, i.e., k-mers (substring of length k), from NGS reads since it is the only information available from DNA sequencing data. Earlier approaches usually align the entire reads to non-redundant coding sequences (nr) and/or functional groups based on sequence similarity, usually via a BLASTX search. In metagenomics, familiar examples include CARMA [4], MEGAN [6] and Phymm [1]. CARMA attempts to assign short reads to known Pfam domains (structural components conserved across multiple proteins) and protein families [4]. MEGAN classifies reads to the Lowest Common Ancestor (LCA) based on multiple BLASTX score hits [6]. These dynamic programming approaches use information in the long k-mers to construct optimal read sequence alignment result.

Other approaches used information in the shorter k-mers. Phymm used interpolated Markov models (IMMs) [18] to characterize variable-length short k-mers that are typical of a phylogenetic grouping. Short k-mers, such as oligonucleotide [14], dinucleotide [7] and tetranucleotide counts [16,19], were used as the discriminative features to capture the information on base composition heterogeneity, perhaps in deference to the long sequencing contigs generated from the earlier sequencing technology. In particular, our recent work [16] used short k-mers in a mixture of Markov chains to calculate the probability of each read assigned to each bin. Presumably, reads binning approaches using both short k-mers and long k-mers as features are more desirable.

An effective approach to overcome the second challenge is to explicitly capture abundance information. For example, AbundanceBin extracted and used feature information from long k-mers of the reads, which directly yield read abundance information [21], to fit a mixture of Poisson models. Each component models the abundance of an individual bin. Similarly, an effective approach to overcome the third challenge is to develop non-mutually exclusive probabilistic clustering methods, where each read can simultaneously fall into different clusters with different posterior probabilities. A read with similar posterior probabilities across all the bins can be considered as non-informative, thus inseparable reads.

Due to the increasing degree of problem complexity, recent works focused more on developing analytic workflows, which exploit the information in short and/or long k-mers and solve the problem in a heuristic manner, e.g., [9,20]. However, the short and long k-mer reads features were not used systematically, i.e., the performance can be compromised by the choices of user-defined cutoff's and the heuristic k-means type algorithms. Thus, it is subject to high variance. Moreover, the deterministic reads partitioning significantly undermines performance, especially for the inseparable reads that are sequenced from the common and/or low-complexity regions of the meta-genomes.

Therefore, it is desirable to develop a systematic approach to robustly and sufficiently integrate reads base composition information and reads abundance information into a single probability model to maximize the binning performance. By assuming these two pieces of information are captured by short k-mers and long k-mers, respectively, we propose a novel Poisson-Markov Model (PMM) approach to integrate reads feature information for binning and classifying short reads. Specifically, we extract reads feature information in both short k-mers and long k-mers to combat the outstanding issues of read heterogeneity and abundance variation in short DNA sequencing reads. We use probability models to accommodate the uncertainties and errors in reads assignment, and we develop a joint mixture model to systematically integrate sequencing feature information. Additionally, our joint mixture model overcomes the third challenge by adopting a soft reads binning, which enables a better performance by filtering out inseparable reads, e.g., those from orthologs or introns across genomes.

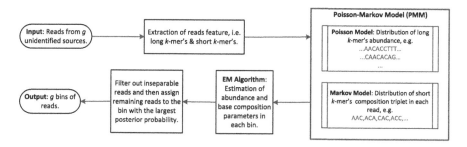

Fig. 1. A conceptual overview of the Poisson-Markov modeling approach for binning of DNA sequencing reads.

We claim that it is one-of-the-kind probabilistic modeling approaches to integrate feature information for binning and classifying short DNA sequencing reads. PMM has been applied in a number of different areas to solve a wide range of problems arising in biomedical science [12], animal science [11], agriculture science [8] and actuarial science [3]. By exploiting efficient data structures for counting k-mers and parallelizing likelihood calculations to multiple threads, we overcome the fourth challenge and make our binning approach more scalable to ever-increasing data volume. Figure 1 presents the main idea of this work.

2   Method

2.1   Poisson-Markov Model (PMM)

We assume a set of n DNA sequencing reads are sampled from g bins with N sequencing reads from each bins. A DNA sequence read is defined as S with discrete variables y_i from $\{A, T, C, G\}$. We also assume reads abundance in j^{th} bin follows a Poisson distribution with parameter λ_j and the reads base composition in the bin is calculated by a Markov model with parameter τ. Please refer to Table 1 for the list of mathematical symbols used in this paper. A joint probability model $f(y_i)$ is shown as:

$$f(y_i) = P(k_j|\lambda_j)P(y_i \mid \tau),\tag{1}$$

where i represents read index and j represents bin index. Assuming there are k_j sequences in j^{th} bin, so the abundance of jth bin can be shown in Poisson as:

$$P(k_j|\lambda_j) = \frac{\lambda_j^{k_j} e^{-\lambda_j}}{k_j!}.\tag{2}$$

In order to develop a probability model for binning and classification of DNA sequencing reads, we need to introduce another variable Z_{ij}, where $Z_{ij} = 1$ means the sequencing read S_i belongs to j^{th} bin, otherwise not. Z_{ij} is given (as the label)

Table 1. A list of mathematical symbols

Notations	Comments
n	number of DNA sequence reads
N	number of DNA sequence reads in each bin
S	a DNA sequencing read
i	index of the reads $\in [1, ..., n]$
S_i	i^{th} sequencing read in given dataset
y_i	discrete variables A, T, C, G
g	number of bins
j	index of the bins $\in [1, ..., g]$
τ	latent variable of Poisson-Markov Model
k_j	number of reads in j^{th} bin
λ_j	parameter of Poisson Model in j^{th} bin
Z_{ij}	indicator whether read S_i belongs to j^{th} bin
ϕ_j	4 by 4^m Transition Probability Matrix
m	tuple/order of TPM
π_j	proportion of jth tbin
c	G/C count
Θ	parameter of Poisson-Markov Model
τ_{ij}	posterior binning probability
l	index of the iterations
P_x	x^{th} partition in parallel computing of E-step
x	number of partitions in parallel computing of E-step

in supervised classification problems whereas it is a latent variable in unsupervised binning problem. Therefore, we focus on the more challenging read binning problem and applications to solve reads classification problem as follows.

In a Markov model, Transition Probability Matrix (TPM) is represented with parameter ϕ, and π_j is the initial proportion of jth bin. When $Z_{ij}=1$, the probability of a sequencing read S_i belongs to j^{th} bin is:

$$P(y_i \mid \tau) = P(Z_{ij} = 1 \mid S) = P(S_i \mid \phi_j), \tag{3}$$

and

$$P(S_i \mid \phi) = \sum_{j=1}^{g} \pi_j P(S_i \mid \phi_j). \tag{4}$$

$P(S_i \mid \phi_j)$ is the probability of observing read S_i which can be calculated using counts of the k-mers. ϕ_j is the TPM of 4 by 4^m calculated as:

$$\phi_j(c_{t-m} \cdots c_{t-1} c_t) = \frac{N(c_{t-m} \cdots c_{t-1} c_t)}{N(c_{t-m} \cdots c_{t-1})}, \tag{5}$$

where m is the tuple of TPM. $N(c_{t-m}, \ldots c_{t-1} c_t)$ is the count of the $(m+1)$-tuple, i.e., $c_{t-m} \ldots c_{t-1} c_t$, in S and $c_{t-m}, \ldots c_{t-1}$ is the count of the m-tuple $N(c_{t-m} \cdots c_{t-1})$ in S. For example, in a second-order Markov model, ϕ_j is the TPM using a 4 by 16 probability matrix, where m and t equal to 2 and 3 respectively, which can be calculated by counting the corresponding 3-mers. Please see [16] for further details in calculating $P(S_i \mid \phi_j)$.

The complete data log-likelihood of Poisson-Markov Model $L_c(\Theta)$ can be written as:

$$\begin{aligned}
\log L_c(\Theta) &= \log \left(\prod_{i=1}^{n} \sum_{j=1}^{g} Z_{ij} \frac{\lambda_j^{k_j} e^{-\lambda_j}}{k_j!} \pi_j P(S_i \mid \phi_j) \right) \\
&= \sum_{i=1}^{n} \sum_{j=1}^{g} Z_{ij} \{ \log \lambda_j^{k_j} - \lambda_j - \log k_j! + \log \pi_j \\
&\quad + \log P(S_i \mid \phi_j) \}.
\end{aligned} \tag{6}$$

The expected value of Z_{ij} is τ_{ij}, where Z_{ij} is a latent variable indicating whether the read i belongs to j^{th} bin:

$$\begin{aligned}
\tau_{ij} &= E[Z_{ij} = 1 \mid \pi_j S, \phi] = P(Z_{ij} = 1 \mid \pi_j S_i, \phi_j) \\
&= \frac{P(N = k_j) \pi_j P(S_i \mid \phi_j)}{\sum_{j=1}^{g} P(N = k_j) \pi_j P(S_i \mid \phi_j)}.
\end{aligned} \tag{7}$$

2.2   An Expectation-Maximization Algorithm

Here we develop an Expectation-Maximization (EM) algorithm to maximize the complete data log-likelihood function $\log L_c(\Theta)$. In the E-step, we calculate the expected values of the log-likelihood function $\log L_c(\Theta)$, i.e., $Q(\Theta \mid \Theta^{(l)})$, under the current estimate of the parameters $\Theta^{(l)}$ in l^{th} iteration, where $\Theta = (\lambda_j, \tau)$, the set of parameters in Poisson and Markov models.

$$Q(\Theta \mid \Theta^{(l)}) = \sum_{i=1}^{n} \sum_{j=1}^{g} \tau_{ij}^{(l+1)} \{ \log^{(l)} \lambda_j^{k_j} - \lambda_j^{(l)} - \log^{(l)} k_j!$$

$$+ \log^{(l)} \pi_j + \log P(S_i \mid \phi_j) \}.$$

In the M-step, we find the parameter values that maximize the $Q(\Theta \mid \Theta^{(l)})$. Specifically, τ_{ij} after $l+1$ iterations is calculated as:

$$\tau_{ij}^{(l+1)} = E[Z_{ij} = 1 \mid \pi_j^{(l)} S, \phi^{(l)}] = P(Z_{ij} = 1 \mid \pi_j^{(l)} S_i, \phi_j^{(l)})$$

$$= \frac{P(N = k_j)^{(l)} \pi_j^{(l)} P(S_i \mid \phi_j^{(l)})}{\sum_{j=1}^{g} P(N = k_j)^{(l)} \pi_j^{(l)} P(S_i \mid \phi_j^{(l)})}. \tag{8}$$

π_j is the proportion of jth bin, so that π_j is updated by summarizing the expected counts of reads as:

$$\pi_j^{(l+1)} = \sum_{i=1}^{n} \frac{\tau_{ij}^{(l+1)}}{n}. \tag{9}$$

$\phi_j^{(l+1)}$ is the second-order TPM which can be updated as in [16]:

$$\phi_j^{(l+1)}(c_{t-m} \ldots c_{t-1} c_t) = \frac{N_j^{(l+1)}(c_{t-m} \ldots c_{t-1} c_t)}{N_j^{(l+1)}(c_{t-m} \ldots c_{t-1})},$$

$$N_j^{(l+1)}(c_{t-m} \ldots c_{t-1} c_t) = \sum_{i=1}^{n} \tau_{ij}^{(l+1)} N_j(c_{t-m} \ldots c_{t-1} c_t),$$

$$N_j^{(l+1)}(c_{t-m} \ldots c_{t-1}) = \sum_{i=1}^{n} \tau_{ij}^{(l+1)} N_j(c_{t-m} \ldots c_{t-1}).$$

λ_j is estimated by calculating the first derivative of $Q(\Theta \mid \Theta^{(l)})$ as:

$$\frac{dQ(\Theta \mid \Theta^{(l)})}{d\lambda_j} = 0. \tag{10}$$

Thus we have:

$$\lambda_j^{(l+1)} = k_j^{(l+1)}. \tag{11}$$

The E and M steps alternates until convergence.

2.3   A Parallel Implementation of the PMM Algorithm

The E-step calculates the expected values of complete data log-likelihood which can be calculated using multiple threads in parallel where each thread calculates

---

Algorithm 1. The Parallelized PMM Algorithm

Input: n DNA sequencing reads $S = S_1, ..., S_i, ..., S_n$, Number of clusters g.

1 **for** $j = 1$ to g **do**
2   Initialize $\Theta^{(0)}$:
3          $\pi_j = \frac{1}{g}$, $k_j = \frac{n}{g}$, $\phi_j(c_{t-m} \cdots c_{t-1} c_t) = \frac{N(c_{t-m} \cdots c_{t-1} c_t)}{N(c_{t-m} \cdots c_{t-1})}$ and $\lambda_j = \frac{n}{g}$;
4 **end**
5 **repeat**
6   **E-step:** Compute the responsibilities at l^{th} iteration
7   Distribute the log-likelihood table $(n \times g)$ into x partitions for parallel computation;
8          $\hat{\tau}_{ij} = E[Z_{ij} = 1 \mid \pi_j, S, \phi] = p(Z_{ij} = 1 \mid \pi_j, S_i, \phi_j)$ by Eq. (7);
9   **M-step:** Update the corresponding parameters
10          $\tau^{(l+1)} = E[Z_{ij} = 1 \mid \pi_j^{(l)} S, \phi^{(l)}] = P(Z_{ij} = 1 \mid \pi_j^{(l)} S_i, \phi_j^{(l)})$ by Eq. (8);
11          $\pi_j^{(l+1)} = \sum_{i=1}^{n} \frac{\tau_{ij}^{(l+1)}}{n}$ by Eq. (9), $\phi_j^{(l+1)}$ by Eq. (2.2);
12          $\lambda_j^{(l+1)} = n_{k_j}^{(l+1)}$ by Eq. (11) ;
13 **until** $|\tau^{(l+1)} - \tau^{(l)}| < \epsilon$;

---

a fraction of Q function values. The M-step then sums up all these values and update the parameters. We use a $n \times g$ table storing the log-likelihood for each read calculated in E-step. The table has been randomly separated into x partitions, where each partition contains n/x reads. The latter is computed in x threads in parallel by using "IntStream" technique in Java. We summarize our workflow as shown in Fig. 2.

3 Results

We developed a PMM model and a Parallel algorithm (hence thereafter referred as PMMBin, Algorithm 1), to capture both long k-mer and short k-mer information in the DNA sequencing reads. We compared our methods to the competing methods that use long k-mers (i.e., AbundanceBin) only and short k-mers (i.e., MarkovBin) only.

3.1 Simulation Data Analysis

We used MetaSim [17], an open-source DNA sequencing reads simulation system, to generate six data sets, each with 10 million reads with 100 bases in length, which are "sequenced" from 10 randomly selected source species. We assigned the abundances of those species in the taxon profiles of MataSim following a normal distribution and used the empirical error model that was recommended for simulating Illumina reads. The ground truth of the reads abundances are shown in Fig. 3.

We compared the performance and running time of PMMBin and fPMM-Bin (derived from PMMBin by filtering out the inseparable DNA reads where

Table 2. The Accuracy (Acc.), Precision (Pre.), and adjusted Rand index (ARI) of PMMBin, fPMMBin, MarkovBin and AbundanceBin. The best performance results (excluding fPMMBin due to the added filtering procedure) are in bold face.

	PMMBin			AbundanceBin			MarkovBin			fPMMBin		
Data	Acc	Pre	ARI	Acc	Pre	ARI	Acc	Pre	ARI	Acc	Pre	ARI
1	0.77	0.85	**0.75**	0.56	0.59	0.14	0.59	0.81	0.56	0.96	0.86	0.95
2	0.70	0.76	**0.73**	0.42	0.65	0.12	0.44	0.74	0.44	0.93	0.78	0.92
3	0.84	0.85	**0.82**	0.52	0.63	0.15	0.55	0.82	0.53	0.92	0.86	0.91
4	0.75	0.80	**0.73**	0.50	0.66	0.24	0.68	0.79	0.66	0.90	0.82	0.90
5	0.63	0.91	0.54	0.43	0.57	0.13	0.90	0.74	**0.65**	0.98	0.88	0.81
6	0.66	0.84	0.51	0.56	0.67	0.22	0.99	0.63	**0.58**	0.91	0.87	0.76

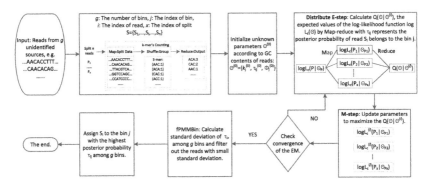

Fig. 2. A flowchart of the Parallel PMM algorithm implementation, where the dotted boxes represent a more efficient k-mer counting step to further speed up the algorithm.

standard deviation of $\tau_{ij}^{(l+1)}$ among clusters is less than 0.25) with that of AbundanceBin (long k-mers) [21] (version 1.01, February 2013) and MarkovBin (short k-mers) [16] (version 1.01, July 2013) in terms of accuracy, precision and adjusted Rand index (ARI) [5]. When calculating accuracy and precision, we consider a pair of reads to be positive if they are from the same source, negative otherwise. Let us denote N_P as the total number of the positive pairs, N_N as the total number of the negative pairs, N_{TP} (true positive) as the number of positive pairs that were assigned to the same bin, N_{TN} (true negative) as the number of negative pairs that were assigned to different bins. We define Accuracy as $\frac{N_{TP}}{N_P}$ and Precision as $\frac{N_{TN}}{N_N}$. To highlight the unique advantage of PMMBin in recovering the bin abundances with high variance, we designed a set of case-control experiments. Specifically, we used a bin size distribution with high variance to generate the data sets 1 to 4, and a true bin size distribution with low variance to generate the data sets 5 and 6.

From Table 2, PMMBin performs the best in the first 4 simulated data sets when compared with MarkovBin and AbundanceBin, but not in the last 2 data

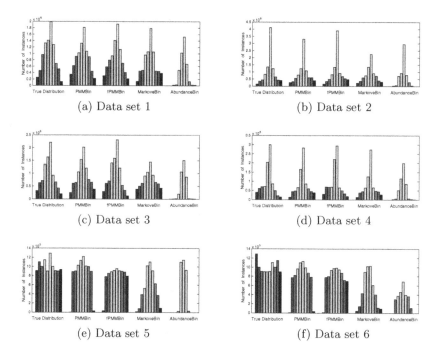

Fig. 3. Comparison of reads binning performance in terms of recovering the true bin size distribution. Each panel corresponds to one data set and from left to right, bar plots represents: the True Distribution of bin size, the one estimated by our proposed PMM and fPMM approaches, the one estimated by MarkovBin approach and the one estimated by AbundanceBin approach.

sets, highlighting the unique capability of PMMBin in detecting bins of diverse sizes. Compared to PMMBin, fPMMBin enjoys much higher accuracy and ARI due to the removal of inseparable reads. Thus, the abundance variation information is duly captured by Poisson mixtures through extracting long k-mers while the base composition information is sufficiently captured by the mixture Markov models by extracting short k-mers. Therefore, our simulation studies strongly support the notion that short k-mers and long k-mers capture uncorrelated yet complementary feature information in the reads.

Figure 3 gives a more visually compelling comparison of the binning performance. PMMBin and fPMMBin successfully identified each of the 10 reads sources (species), represented by a "peak" for each source with negligible surrounding noises. Both MarkovBin and AbundanceBin miss a number of sources (peaks) albeit the former identities more sources than the latter. In data sets 1-4 when the bin sizes are truly diverse, the bin size distributions recovered by PMMBin and fPMMBin are much closer to the true distribution compared with MarkovBin and AbundanceBin. In data sets 5 and 6 when the bin sizes are more uniform, MarkovBin performs best whereas AbundanceBin capturing bin size variation performs the worst.

Table 3. Comparison of running time per iteration (100 million reads).

Data set	No Partition	4 Partitions	10 Partitions	40 Partitions
1	92.9 mins	41.7 mins	35.8 mins	6.2 mins
2	92.7 mins	41.3 mins	35.5 mins	5.8 mins
3	92.9 mins	41.6 mins	35.8 mins	6.1 mins
4	92.5 mins	41.4 mins	35.6 mins	5.9 mins
5	92.3 mins	41.1 mins	35.7 mins	5.8 mins
6	93.1 mins	41.7 mins	35.9 mins	6.2 mins

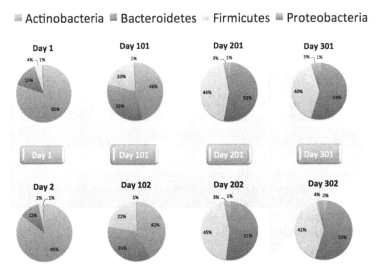

Fig. 4. The temporal changes of the individual's microbiome composition from Day 1 to Day 302.

We also compared running time of the parallelized PMM algorithm with the non-parallelized version. As shown in Algorithm 1, we split the calculation of expected log-likelihood into different number of partitions so that we calculate all partitions in parallel. We ran the parallel PMM algorithm on the 6 data sets (one hundred million reads) on a server (4x Twelve-Core AMD Opteron 2.6 GHz, 256 GB RAM). We compare the running time per iteration since different numbers of iterations are needed for different data sets. From Table 3, we observe a markedly faster running time of the parallelized PMM algorithm compared with the non-parallelized version without sacrificing the accuracy and precision.

Ideally, the running time of the parallelized algorithm per iteration can be reduced to $\frac{1}{x}$ of that of the non-parallelized algorithm, where x is the number of partitions. But it is not the case in reality as shown in Table 3. The reason is that we only parallelized the E-step since the E-step calculation dominates the entire computational complexity whereas M-step calculation is relatively trivial.

3.2   Real-World Data Analysis

We analyzed a human microbiome time course data set in which an individual's microbiome was sequenced daily over a period of one year [2]. We looked at the individual's microbiome data at eight days: day 1, day 2, day 101, day 102, day 201, day 202, day 301, and day 302, and we partitioned the reads from each metagenomic sample into four bins, i.e., Actinobacteria, Bacteroidetes, Firmicuts and Proteobacteria. From Fig. 4, it is evident that the individual's microbiomes are similarly between two consecutive days (per columns) whereas are radically different among distant days (per rows). It was also noted in [2] that the drastically changed microbiome at days 101–102 is due to the individual's trip abroad.

4   Conclusion

In this paper, we presented a novel probability model and a parallel algorithm to bin short DNA sequencing reads. Our original contributions lie in the systematic extraction and integration of both short and long k-mers information into the same probability model, and the parallel implementation of the optimization algorithm, which results in a vastly improved performance in terms of accuracy, precision and running time. Albeit the joint probability model was presented in the context of unsupervised reads binning, it is sufficiently flexible to be extended to solving supervised reads classification problems.

To further improve the running time, we will leverage efficient data structures for counting k-mers. Specifically, longer k-mers are sparse meaning that a majority of the k-mers are unique [15]. Thus the k-mers counting and hashing can be significantly accelerated by filtering out infrequent k-mers using a Bloom filter, and store frequent k-mers using a suffix tree in both memory and hard disk [10]. There are other existing k-mer counting approaches as well, such as in [13,22]. To this end, we will develop a versatile and scalable toolbox for facilitating data mining and machine learning of short DNA sequencing reads.

Acknowledgment. This research is partially supported by NSF grant CCF: 1451316 to D.Z.

References

1. Brady, A., Salzberg, S.L.: Phymm and PhymmBL: metagenomic phylogenetic classification with interpolated Markov models. Nat. Methods **6**(9), 673–676 (2009)
2. David, L.A., Materna, A.C., Friedman, J., Campos-Baptista, M.I., Blackburn, M.C., Perrotta, A., Erdman, S.E., Alm, E.J.: Host lifestyle affects human microbiota on daily timescales. Genome Biol. **15**(7), R89 (2014)
3. di Milano, U.C.S.: Poisson hidden markov models for time series of overdispersed insurance counts

4. Gerlach, W., Stoye, J.: Taxonomic classification of metagenomic shotgun sequences with CARMA3. Nucleic Acids Res. **39**(14), e91 (2011)
5. Hubert, L., Arabie, P.: Comparing partitions. J. Classif. **2**(1), 193–218 (1985)
6. Huson, D.H., Mitra, S., Ruscheweyh, H.-J., Weber, N., Schuster, S.C.: Integrative analysis of environmental sequences using MEGAN4. Genome Res. **21**(9), 1552–1560 (2011)
7. Kariin, S., Burge, C.: Dinucleotide relative abundance extremes: a genomic signature. Trends Genet. **11**(7), 283–290 (1995)
8. Karunanayake, C.: Multivariate Poisson Hidden Markov Models for Analysis of Spatial Counts. Canadian theses. University of Saskatchewan (Canada) (2007)
9. Kelley, D., Salzberg, S.: Clustering metagenomic sequences with interpolated Markov models. BMC Bioinform. **11**(1), 544 (2010)
10. Kurtz, S., Narechania, A., Stein, J.C., Ware, D.: A New Method to compute K-mer frequencies and its application to annotate large repetitive plant genomes. BMC Genomics **9**(1), 517 (2008)
11. Leroux, B.G., Puterman, M.L.: Maximum-Penalized-Likelihood estimation for independent and Markov-Dependent mixture models. Biometric **48**, 545–558 (1992)
12. Lu, J., Bushel, P.R.: Dynamic expression of 3' UTRs revealed by poisson hidden Markov modeling of RNA-Seq: implications in gene expression profiling. Gene **527**(2), 616–623 (2013)
13. Marçais, G., Kingsford, C.: A fast, lock-free approach for efficient parallel counting of occurrences of K-mers. Bioinform. **27**(6), 764–770 (2011)
14. Meinicke, P., Asshauer, K.P., Lingner, T.: Mixture models for analysis of the taxonomic composition of metagenomes. Bioinform. **27**(12), 1618–1624 (2011)
15. Melsted, P., Pritchard, J.K.: Efficient counting of K-mers in dna sequences using a bloom filter. BMC Bioinform. **12**(1), 333 (2011)
16. Nguyen, T.C., Zhu, D.: MarkovBin : an algorithm to cluster metagenomic reads using a mixture modeling of hierarchical distributions. In: Proceedings of the International Conference on Bioinformatics, Computational Biology and Biomedical Informatics, p. 115. ACM (2013)
17. Richter, D.C., Ott, F., Auch, A.F., Schmid, R., Huson, D.H.: Metasim - a sequencing simulator for genomics and metagenomics. PLoS ONE **3**(10), e3373 (2008)
18. Salzberg, S.L., Delcher, A.L., Kasif, S., White, O.: Microbial gene identification using interpolated Markov models. Nucleic Acids Res. **26**(2), 544–548 (1998)
19. Wang, Y., Leung, H.C., Yiu, S.M., Chin, F.Y.: MetaCluster 4.0: a novel binning algorithm for NGS reads and huge number of species. J. Comput. Biol. J. Comput. Mol. Cell Biol. **19**(2), 241–249 (2012)
20. Wang, Y., Leung, H.C., Yiu, S.-M., Chin, F.Y.: Metacluster 5.0: a two-round binning approach for metagenomic data for low-abundance species in a noisy sample. Bioinform. **28**(18), i356–i362 (2012)
21. Wu, Y.-W., Ye, Y.: A novel abundance-based algorithm for binning metagenomic sequences using l-tuples. J. Comput. Biol. **18**(3), 523–534 (2010)
22. Zhang, Q., Pell, J., Canino-Koning, R., Howe, A.C., Brown, C.T.: These are not the K-mers you are looking for: efficient online K-mer counting using a probabilistic data structure. PloS one **9**(7), e101271 (2014)

FSG: Fast String Graph Construction for De Novo Assembly of Reads Data

Paola Bonizzoni, Gianluca Della Vedova, Yuri Pirola, Marco Previtali$^{(\boxtimes)}$, and Raffaella Rizzi

DISCo, University of Milano-Bicocca, Milan, Italy
{bonizzoni,yuri.pirola,marco.previtali,rizzi}@disco.unimib.it,
gianluca.dellavedova@unimib.it

Abstract. The string graph for a collection of next-generation reads is a lossless data representation that is fundamental for de novo assemblers based on the overlap-layout-consensus paradigm. In this paper, we explore a novel approach to compute the string graph, based on the FM-index and Burrows-Wheeler Transform (BWT). We describe a simple algorithm that uses only the FM-index representation of the collection of reads to construct the string graph, without accessing the input reads. Our algorithm has been integrated into the SGA assembler as a stand-alone module to construct the string graph.

The new integrated assembler has been assessed on a standard benchmark, showing that FSG is significantly faster than SGA while maintaining a moderate use of main memory, and showing practical advantages in running FSG on multiple threads.

1   Introduction

De novo sequence assembly continues to be one of the most fundamental problems in Bioinformatics. Most of the available assemblers [1,12,13,19,20,25] are based on the notions of de Bruijn graphs and of k-mers (short k-long substrings of input data). Currently, biological data are produced by different Next-Generation Sequencing (NGS) technologies which routinely and cheaply produce a large number of reads whose length varies according to the specific technology. For example, reads obtained by Illumina technology (which is the most used) have length between 50 and 150 bases [21].

To analyze datasets coming from different technologies, hence with a large variation of read lengths, an approach based on same-length strings is likely to be limiting, as witnessed by the recent introduction of variable-length de Bruijn graphs [9]. The *string graph* [18] representation is an alternative approach that does not need to break the reads into k-mers (as in the de Bruijn graphs), and has the advantage of immediately distinguishing the repeats that result in different arcs. The string graph is the main data representation used by assemblers based on the overlap-layout-consensus paradigm. Indeed, in a string graph, the vertices are the input reads and the arcs corresponds to overlapping reads, with the property that contigs are paths of the string graph. An immediate advantage of string graphs is that they can disambiguate some repeats that methods based on

© Springer International Publishing Switzerland 2016
A. Bourgeois et al. (Eds.): ISBRA 2016, LNBI 9683, pp. 27–39, 2016.
DOI: 10.1007/978-3-319-38782-6_3

de Bruijn graphs might resolve only at later stages—for example, the repeats that are longer than $k/2$ but contained in a read. Even without repetitions, analyzing only k-mers instead of the longer reads can result in some information loss, since bases of a read that are more than k positions apart are not part of the same k-mer, but might be part of the same read. Indeed, differently from de Brujin graphs, any path of a string graph is a valid assembly of reads. On the other hand, string graphs are more computationally intensive to compute [24], justifying our search for faster algorithms. From an algorithmic point of view, the most used string graph assembler is SGA [23], which first constructs the BWT [11] and the FM-index of a set of strings, and then uses those data structures to efficiently compute the arcs of the string graph (connecting overlapping reads). Another string graph assembler is Fermi [17] which implements a variant of the original SGA algorithm [23] that is tailored for SNP and variant calling. A number of recent works face the problem of designing efficient algorithmic strategies or data structures for building string graphs. Among those works we can find a string graph assembler [4], based on a careful use of hashing and Bloom filters, with performance comparable with the first SGA implementation [23]. Another important alternative approach to SGA is Readjoiner [15] which is based on an efficient computation of a subset of exact suffix-prefix matches, and by subsequent rounds of suffix sorting, scanning, and filtering outputs the non-redundant arcs of the graph.

All assemblers based on string graphs (such as SGA) need to both (1) query an indexing data structures (such as an FM-index), and (2) access the original reads set to detect prefix-suffix overlaps between the elements. Since the self-indexing data structures, such as FM-index, represent the whole information of the original dataset, an interesting problem is to design efficient algorithms for the construction of string graphs that only require to keep the index and do not need to access the read set together with the index. Improvements in this direction have both theoretical and practical motivations. Indeed, detecting prefix-suffix overlaps only by analyzing the (compressed) index is an almost unexplored problem, and managing such data structure is usually more efficient.

Following this research direction, we propose a new algorithm, called FSG, to compute the string graph of a set R of reads, whose $O(nm)$ time complexity matches that of SGA—n is the number of reads in R and m is the maximum read length. To the best of our knowledge, it is the first algorithm that computes a string graph using only the FM-index of the input reads. The vast literature on BWT and FM-index hints that this approach is amenable to further research. An important observation is that SGA computes the string graph basically performing, for each read r, a query to the FM-index for each character of r, to compute the arcs outgoing from r. While this approach works in $O(nm)$ time, it can perform several redundant queries, most notably when the reads share common suffixes (a very common case). Our algorithm queries the FM-index in a specific order, so that each string is processed only once, while SGA might process more than once each repeated string. It is important to notice that our novel algorithm uses a characterization of a string graph that is different, but equivalent, to the one in [18] stated in [7] and which is quite useful when processing reads with their FM-index. Moreover, since we have integrated our algorithm into SGA, the read correction and the assembly phases of SGA can be applied without any modification. These facts guarantees that the assemblies produced by our approach and SGA are the same. In a previous paper, we have tackled the problem of constructing the string graph in external memory [8] by taking advantages of some recent results on the external memory implementation of the FM-index [2]. Experimental results [8] have revealed that computing the FM-index and LCP (Longest Common Prefix) array are the two main limiting factors towards an efficient (in terms of running time and main memory requirements) external memory algorithm to construct the string graph. In fact, even the best known algorithms for these steps do not have an optimal I/O complexity [2,3].

The FSG algorithm provides an approach to build a string graph that could be used for different read assembly purposes. We have implemented FSG and integrated it with the SGA assembler, by replacing in SGA the step related to the string graph construction. Our implementation follows the SGA guidelines, *i.e.*, we use the correction step of SGA before computing the overlaps without allowing mismatches (which is also SGA's default). Notice that SGA is a finely tuned implementation that has performed very nicely in the latest Assemblathon competition [10]. We have compared FSG with SGA, where we have used the latter's default parameter (that is, we compute overlaps without errors). Our experimental evaluation on a standard benchmark dataset shows that our approach is 2.3–4.8 times faster than SGA in terms of wall clock time.

2   Preliminaries

We briefly recall some standard definitions that will be used in the following. Let Σ be a constant-sized alphabet and let S be a string over Σ. We denote by $S[i]$ the i-th symbol of S, by $\ell = |S|$ the length of S, and by $S[i : j]$ the substring $S[i]S[i+1]\cdots S[j]$ of S. The *suffix* and *prefix* of S of length k are the substrings $S[\ell - k + 1 : \ell]$ (denoted by $S[\ell - k + 1 :]$) and $S[1 : k]$ (denoted by $S[: k]$) respectively. Given two strings (S_i, S_j), we say that S_i *overlaps* S_j iff a nonempty suffix β of S_i is also a prefix of S_j, that is $S_i = \alpha\beta$ and $S_j = \beta\gamma$. In this paper we consider a set R of n strings over Σ that are terminated by the sentinel \$, which is the smallest character. To simplify the exposition, we will assume that all input strings have exactly m characters, excluding the \$. The *overlap graph* of a set R of strings is the directed graph $G_O = (R, A)$ whose vertices are the strings in R, and each two overlapping strings $r_i = \alpha\beta$ and $r_j = \beta\gamma$ form the arc $(r_i, r_j) \in A$ labeled by α. In this case β is called the *overlap* of the arc and α is called the *extension* of the arc. Observe that the notion of overlap graph originally given by [18] is defined by labeling with γ the arc $(r_i, r_j) \in A$.

The notion of a string graph derives from the observation that in a overlap graph the label of an arc (r, s) may be obtained by concatenating the labels of a pair of arcs (r, t) and (t, s), thus arc (r, s) can be removed from the overlap graph without loss of information, since removing all such arcs, called *redundant* arcs, does not changet the set of valid paths. In [18] redundant arcs are those arcs (r, s) labeled by $\alpha\beta$, for α the prefix of an arc (r, t). An equivalent definition of string graphs is below. An arc $e_1 = (r_i, r_j)$ of G_O labeled by α is *transitive* (or *reducible*) if there exists another arc $e_2 = (r_k, r_j)$ labeled by δ where δ is a suffix of α [7]. Therefore, we say that e_1 is *non-transitive* (or *irreducible*) if no such arc e_2 exists. The string graph of R is obtained from G_O by removing all reducible arcs. This definition allows to use the FM-index to compute the labels of the string graph via backward extensions on the index.

The *Generalized Suffix Array (GSA)* [22] of R is the array SA where each element $SA[i]$ is equal to (k, j) iff the k-long suffix $r_j[|r_j| - k + 1 :]$ of the string r_j is the i-th smallest element in the lexicographic ordered set of all suffixes of the strings in R. The *Burrows-Wheeler Transform (BWT)* of R is the sequence B such that $B[i] = r_j[|r_j| - k]$, if $SA[i] = (k, j)$ and $k > 1$, or $B[i] = \$$, otherwise. Informally, $B[i]$ is the symbol that precedes the k-long suffix of a string r_j where such suffix is the i-th smallest suffix in the ordering given by SA. For any string ω, all suffixes of (the lexicographically sorted) SA whose prefix is ω appear consecutively in SA. Consequently, we define the ω-interval [2], denoted by $q(\omega)$, as the maximal interval $[b, e]$ such that $b \leq e$, $SA[b]$ and $SA[e]$ both have prefix ω. Notice that the width $e - b + 1$ of the ω-interval is equal to the number of occurrences of ω in some read of R. Since the BWT B and SA are closely related, we also say that $[b, e]$ is a ω-interval on B. Given a ω-interval and a character c, the *backward c-extension* of the ω-interval is the $c\omega$-interval.

3   The Algorithm

Our algorithm is based on two steps: the first is to compute the overlap graph, the second is to remove all transitive arcs. Given a string ω and R a set of strings (reads), let $R^S(\omega)$ and $R^P(\omega)$ be respectively the subset of R with suffix (resp. prefix) ω. As usual in string graph construction algorithms, we will assume that the set R is *substring free*, *i.e.*, no string is a substring of another. A fundamental observation is that the list of all nonempty overlaps β is a compact representation of the overlap graph, since all pairs in $R^S(\beta) \times R^P(\beta)$ are arcs of the overlap graph. Our approach to compute all overlaps between pairs of strings is based on the notion of *potential overlap*, which is a nonempty string $\beta^* \in \Sigma^+$, s.t. there exists at least one input string $r_i = \alpha\beta^*$ ($\alpha \neq \epsilon$) with suffix β^*, and there exists at least one input string $r_j = \gamma\beta^*\delta$ ($\delta \neq \epsilon$) with β^* as a substring (possibly a prefix). The first part of Algorithm 1 (lines 3–11) computes all potential overlaps, starting from those of length 1 and extending the potential overlaps by adding a new leading character. For each potential overlap, we check if it is an actual overlap. Lemma 1 is a direct consequence of the definition of potential overlap.

Lemma 1. *Let β be an overlap. Then all suffixes of β are potential overlaps.*

The second part of our algorithm, that is to detect all transitive arcs, can be sped up if we cluster together and examine some sets of arcs. We start considering the set of all arcs sharing the same overlap and a suffix of their extensions, as stated in the following definition.

Definition 2. *Assume that* $\alpha, \beta \in \Sigma^*, \beta \neq \epsilon$ *and* $X \subseteq R^P(\beta)$. *The* arc-set $ARC(\alpha, \alpha\beta, X)$ *is the set* $\{(r_1, r_2) : \alpha\beta$ *is a suffix of* r_1, β *is a prefix of* r_2, *and* $r_1 \in R, r_2 \in X\}$. *The strings* α *and* β *are called the* extension *and the* overlap *of the arc-set. The set* X *is called the* destination set *of the arc-set.*

In other words, an arc-set contains the arcs with overlap β and extension α. An arc-set is *terminal* if there exists $r \in R$ s.t. $r = \alpha\beta$, while an arc-set is *basic* if $\alpha = \epsilon$ (the empty string). Since the arc-set $ARC(\alpha, \alpha\beta, X)$ is uniquely determined by strings α, $\alpha\beta$, and X, the triple $(\alpha, \alpha\beta, X)$ encodes the arc-set $ARC(\alpha, \alpha\beta, X)$. Moreover, the arc-set $ARC(\alpha, \alpha\beta, X)$ is *correct* if X includes all irreducible arcs that have overlap β and extension with suffix α, that is $X \supseteq \{r_2 \in R^P(\beta) : r_1 \in R^S(\alpha\beta)$ and (r_1, r_2) is irreducible$\}$. Observe that our algorithm computes only correct arc-sets. Moreover, terminal arc-sets only contain irreducible arcs (Lemma 5). Lemma 3 shows the use of arc-sets to detect transitive arcs. Due to space constraints, all proofs are omitted.

Lemma 3. *Let* (r_1, r_2) *be an arc with overlap* β. *Then* (r_1, r_2) *is transitive iff (i) there exist* $\alpha, \gamma, \delta, \eta \in \Sigma^*$, $\gamma, \eta \neq \epsilon$ *such that* $r_1 = \gamma\alpha\beta$, $r_2 = \beta\delta\eta$, *(ii) there exists an input read* $r_3 = \alpha\beta\delta$ *such that* (r_3, r_2) *is an irreducible arc of a nonempty arc-set* $ARC(\alpha, \alpha\beta\delta, X)$.

A direct consequence of Lemma 3 is that a nonempty correct terminal arc-set $ARC(\alpha, \alpha\beta\delta, X)$ implies that all arcs of the form $(\gamma\alpha\beta, \beta\delta\eta)$, with $\gamma, \eta \neq \epsilon$ are transitive. Another consequence of Lemma 3 is that an irreducible arc $(\alpha\beta\delta, \beta\delta\eta)$ with extension α and overlap $\beta\delta$ reduces all arcs with overlap β and extension $\gamma\alpha$, with $\gamma \neq \epsilon$. Lemma 3 is the main ingredient used in our algorithm. More precisely, it computes terminal correct arc-sets of the form $ARC(\alpha, \alpha\beta\delta, X)$ for extensions α of increasing length. By Lemma 3, $ARC(\alpha, \alpha\beta\delta, X)$ contains arcs that reduce all the arcs contained in $ARC(\alpha, \alpha\beta, X')$ which have a destination in X. Since the transitivity of an arc is related to the extension α of the arc that is used to reduce it, and our algorithm considers extensions of increasing length, a main consequence of Lemma 3 is that it computes terminal arc-sets that are correct, that is they contain only irreducible arcs. We will further speed up the computation by clustering together the arc-sets sharing the same extension.

Definition 4. *Let* T *be a set of arc-sets, and let* α *be a string. The* cluster *of* α, *denoted by* $C(\alpha)$, *is the union of all arc-sets of* T *whose extension is* α.

We sketch Algorithm 1 which consists of two phases: the first phase to compute the overlap graph, and the second phase to remove all transitive arcs. In our description, we assume that, given a string ω, we can compute in constant time (1) the number $\mathsf{suff}(\omega)$ of input strings whose suffix is ω, (2) the number $\mathsf{pref}(\omega)$ of input strings whose prefix is ω, (3) the number $\mathsf{substr}(\omega)$ of occurrences of ω in the input strings. Moreover, we assume to be able to list the set $\mathsf{listpref}(\omega)$ of input strings with prefix ω in $O(\|\mathsf{listpref}(\omega)\|)$ time. In Sect. 4 we will describe such a data structure. The first phase (lines 3–11) exploits Lemma 1 to compute all overlaps. Potential overlaps are defined inductively. The empty string ϵ is a potential overlap of length 0; given an i-long potential overlap β^*, the $(i+1)$-long string $c\beta^*$, for $c \in \Sigma$, is a potential overlap iff $\mathsf{suff}(c\beta^*) > 0$ and $\mathsf{substr}(c\beta^*) > \mathsf{suff}(c\beta^*)$. Our algorithm uses this definition to build potential overlaps of increasing length, starting from those with length 1, *i.e.*, symbols of Σ (line 2). The lists *Last* and *New* store the potential overlaps computed at the previous and current iteration respectively. Observe that a potential overlap β^* is an overlap iff $\mathsf{pref}(\beta^*) > 0$. Since a potential overlap is a suffix of some input string, there are at most nm distinct suffixes, where m and n are the length and the number of input strings, respectively. Each query $\mathsf{suff}(\cdot), \mathsf{pref}(\cdot), \mathsf{substr}(\cdot)$ requires $O(1)$ time, thus the time complexity related to the total number of such queries is $O(nm)$. Given two strings β_1 and β_2, when $|\beta_1| = |\beta_2|$ no input string can be in both $\mathsf{listpref}(\beta_1)$ and $\mathsf{listpref}(\beta_2)$. Since each overlap is at most m long, the overall time spent in the $\mathsf{listpref}(\cdot)$ queries is $O(nm)$. The first phase produces (line 7) the set of disjoint *basic* arc-sets $ARC(\epsilon, \beta, R^p(\beta))$ for each overlap β, whose union is the set of arcs of the overlap graph. Recall that $\mathsf{listpref}(\beta)$ gives the set of reads with prefix β, which has been denoted by $R^p(\beta)$.

The second phase (lines 13–25) classifies the arcs of the overlap graph into reducible or irreducible by computing arc-sets of increasing extension length, starting from the basic arc-sets $ARC(\epsilon, \epsilon\beta, R^p(\beta))$ obtained in the previous phase. By Lemma 3, we compute all correct terminal arc-sets $ARC(\alpha, \alpha\beta, X)$ and remove all arcs that are reduced by $ARC(\alpha, \alpha\beta, X)$. The set Rdc is used to store the destination set X of the computed terminal arc-sets. Notice that if $ARC(\alpha, \alpha\beta, X)$ is terminal, then all of its arcs have the same origin $r = \alpha\beta$, *i.e.*, $ARC(\alpha, \alpha\beta, X) = \{(r, x) : x \in X\}$. By Lemma 3 all arcs in the cluster $C(\alpha)$ with a destination in X and with an origin different from r are transitive and can be removed, simply by removing X from all destination sets in the arc-sets of $C(\alpha)$. Another application of Lemma 3 is that when we find a terminal arc-set all of its arcs are irreducible, *i.e.*, it is also correct. In fact, Lemma 3 classifies an arc as transitive according to the existence of a read $r = \alpha\beta$ with extension α. Since the algorithm considers extensions α of increasing length, all arcs whose extensions is shorter than α have been reduced in a previous step, thus all terminal arc-set of previous iterations are irreducible. More precisely, the test at line 18 is true iff the current arc-set is terminal. In that case, at line 19 all arcs of the arc-set are output as arcs of the string graph, and at line 20 the destination set X is added to the set Rdc that contains the destinations of $C(\alpha)$ that must be removed. For each cluster $C(\alpha)$, we read twice all arc-sets that are included in $C(\alpha)$. The first time to determine which arc-sets are terminal and, in that case, to determine the set Rdc of reads that must be removed from all destinations of the arc-sets included in $C(\alpha)$. The second time to compute the clusters $C(c\alpha)$ that contain the nonempty arc-sets with extension $c\alpha$ consisting of the arcs that we still have to check if they are transitive or not (that is the arcs with destination set $X \setminus Rdc$). In Algorithm 1, the cluster $C(\alpha)$ that is currently analyzed is stored in *CurrentCluster*, that is a list of the arc-sets included in the cluster. Moreover, the clusters that still have to be analyzed are stored in the stack *Clusters*. We use a stack to guarantee that the clusters are analyzed in the correct order, that is the cluster $C(\alpha)$ is analyzed after all clusters $C(\alpha[i :])$—$\alpha[i :]$ is a generic suffix

Algorithm 1. Compute the string graph

Input : The set R of input strings
Output: The string graph of R, given as a list of arcs
1 Cluster ← empty list;
2 Last ← $\{c \in \Sigma \mid \mathsf{suff}(c) > 0 \text{ and } \mathsf{substr}(c) > \mathsf{suff}(c)\}$;
3 **while** *Last is not empty* **do**
4   New ← ∅;
5   **foreach** $\beta^* \in Last$ **do**
6   **if** $\mathsf{pref}(\beta^*) > 0$ **then**
7   Append $(\epsilon, \beta^*, \mathsf{listpref}(\beta^*))$ to Cluster;
8   **for** $c \in \Sigma$ **do**
9   **if** $\mathsf{suff}(c\beta^*) > 0$ *and* $\mathsf{substr}(c\beta^*) > \mathsf{suff}(c\beta^*)$ **then**
10   Add $c\beta^*$ to New;
11   Last ← New;
12 Clusters ← the stack with Cluster as its only element;
13 **while** *Clusters is not empty* **do**
14   CurrentCluster ← Pop(Clusters);
15   Rdc ← ∅;
16   Let ExtendedClusters be an array of $|\Sigma|$ empty clusters;
17   **foreach** $(\alpha, \alpha\beta, X) \in CurrentCluster$ **do**
18   **if** $\mathsf{substr}(\alpha\beta) = \mathsf{pref}(\alpha\beta) = \mathsf{suff}(\alpha\beta) > 0$ **then**
19   Output the arcs $(\alpha\beta, x)$ with label α for each $x \in X$;
20   Rdc ← Rdc $\cup X$;
21   **foreach** $(\alpha, \alpha\beta, X) \in CurrentCluster$ **do**
22   **if** $X \not\subseteq Rdc$ **then**
23   **for** $c \in \Sigma$ **do**
24   **if** $\mathsf{suff}(c\alpha\beta) > 0$ **then**
25   Append $(c\alpha, c\alpha\beta, X \setminus Rdc)$ to ExtendedClusters[c];
26   Push each non-empty cluster of ExtendedClusters to Clusters;

---

of α. We can prove that a generic irreducible arc (r_1, r_2) with extension α and overlap β belongs exactly to the clusters $C(\epsilon), \ldots, C(\alpha[2:]), C(\alpha)$. Moreover, r_2 does not belong to the set Rdc when considering $C(\epsilon), \ldots, C(\alpha[2:])$, hence the arc (r_1, r_2) is correctly output when considering the cluster $C(\alpha)$. The lemmas leading to the correctness of the algorithm follow.

Lemma 5. *Let $ARC(\alpha, \alpha\beta, X)$ be an arc-set inserted into a cluster by Algorithm 1. Then such arc-set is correct.*

Lemma 6. *Let e_1 be a transitive arc (r_1, r_2) with overlap β. Then the algorithm does not output e_1.*

Theorem 7. *Given as input a set of strings R, Algorithm 1 computes exactly the arcs of the string graph.*

We can now sketch the time complexity of the second phase. Previously, we have shown that the first phase produces at most $O(nm)$ arc-sets, one for each distinct overlap β. Since each string $\alpha\beta$ considered in the second phase is a suffix of an input string, and there are at most nm such suffixes, at most nm arc-sets are considered in the second phase. In the second phase, for each cluster a set Rdc is computed. If Rdc is empty, then each arc-set of the cluster can be examined in constant time, since all unions at line 20 are trivially empty and at line 25 the set $X \setminus Rdc$ is equal to X, therefore no operation must be computed. The interesting case is when $X \neq \varnothing$ for some arc-set. In that case the union at line 20 and the difference $X \setminus Rdc$ at line 25 are computed. Let $d(n)$ be the time complexity of those two operations on n-element sets (the actual time complexity depends on the data structure used). Notice that X is not empty only if we have found an irreducible arc, that is an arc of the string graph. Overall, there can be at most $|E|$ nonempty such sets X, where E is the set of arcs of the string graph. Hence, the time complexity of the entire algorithm is $O(nm + |E|d(n))$.

4   Data Representation

Our algorithm entirely operates on the (potentially compressed) FM-index of the collection of input reads. Indeed, each processed string ω (both in the first and in the second phase) can be represented in constant space by the ω-interval $[b_\omega, e_\omega]$ on the BWT (*i.e.*, $\mathsf{q}(\omega)$), instead of using the naïve representation with $O(|\omega|)$ space. Notice that in the first phase, the i-long potential overlaps, for a given iteration, are obtained by prepending a symbol $c \in \Sigma$ to the $(i-1)$-long potential overlaps of the previous iteration (lines 8–10). In the same way the arc-sets of increasing extension length are computed in the second phase. In other words, our algorithm needs in general to obtain string $c\omega$ from string ω, and, since we represent strings as intervals on the BWT, this operation can be performed in $O(1)$ time via backward c-extension of the interval $\mathsf{q}(\omega)$ [14].

Moreover, both queries $\mathsf{pref}(\omega)$ and $\mathsf{substr}(\omega)$ can be answered in $O(1)$ time. In fact, given $\mathsf{q}(\omega) = [b_\omega, e_\omega]$, then $\mathsf{substr}(\omega) = e_\omega - b_\omega + 1$ and $\mathsf{pref}(\omega) = e_{\$\omega} - b_{\$\omega} + 1$ where $\mathsf{q}(\$\omega) = [b_{\$\omega}, e_{\$\omega}]$ is the result of the backward $\$$-extension of $\mathsf{q}(\omega)$. Similarly, it is easy to compute $\mathsf{listpref}(\omega)$ as it corresponds to the set of reads that have a suffix in the interval $\mathsf{q}(\$\omega)$ of the GSA. The interval $\mathsf{q}(\omega\$) = [b_{\omega\$}, e_{\omega\$}]$ allows to answer to the query $\mathsf{suff}(\omega)$ which is computed as $e_{\omega\$} - b_{\omega\$} + 1$. The interval $\mathsf{q}(\omega\$)$ is maintained along with $\mathsf{q}(\omega)$. Moreover, since $\mathsf{q}(\omega\$)$ and $\mathsf{q}(\omega)$ share the lower extreme $b_\omega = b_{\omega\$}$ (recall that $\$$ is the smallest symbol), each string ω can be compactly represented by the three integers $b_\omega, e_{\omega\$}, e_\omega$. While in our algorithm a substring ω of some input read can be represented by those three integers, we exploited the following representation for greater efficiency. In the first phase of the algorithm we mainly have to represent the set of potential overlaps. At each iteration, the potential overlaps in $Last$ (New, resp.) have the same length, hence their corresponding intervals on the BWT are disjoint. Hence we can store those intervals using a pair of $n(m + 1)$-long bitvectors. For each potential overlap $\beta \in Last$ (New, resp.) represented by the β-interval $[b_\beta, e_\beta]$, the first bitvector has 1 in position b_β and the second bitvector has 1 in positions $e_{\beta\$}$ and e_β. Recall that we want also to maintain the interval $q(\beta\$) = [b_\beta, e_{\beta\$}]$. Since substr($\beta$) > suff($\beta$), then $e_{\beta\$} \neq e_\beta$ and can be stored in the same bitvector. In the second phase of the algorithm, we mainly represent clusters. A cluster groups together arc-sets whose overlaps are pairwise different or one is the prefix of the other. Thus, the corresponding intervals on the BWT are disjoint or nested. Moreover, also the destination set of the *basic* arc-sets can be represented by a set of pairwise disjoint or nested intervals on the BWT (since listpref(β) of line 7 correspond to the interval q($\$\beta$)). Moreover, the loop at lines 13–25 preserves the following invariant: let $ARC(\alpha, \alpha\beta_1, X_1)$ and $ARC(\alpha, \alpha\beta_2, X_2)$ be two arc-sets of the same cluster $C(\alpha)$ with β_1 prefix of β_2, then $X_2 \subseteq X_1$. Hence, each subset of arc-sets whose extensions plus overlaps share a common nonempty prefix γ is represented by means of the following three vectors: two integers vectors V_b, V_e of length $e_\gamma - b_\gamma + 1$ and a bitvector B_x of length $e_{\$\gamma} - b_{\$\gamma} + 1$, where $[b_\gamma, e_\gamma]$ is the γ-interval and $[b_{\$\gamma}, e_{\$\gamma}]$ is the $\$\gamma$-interval. More specifically, $V_b[i]$ ($V_e[i]$, resp.) is the number of arc-sets whose representation (BWT interval) of the overlap starts (ends, resp.) at $b_\gamma + i$, while $B_x[i]$ is 1 iff the read at position $b_{\$\gamma} + i$, in the lexicographic order of the GSA, belongs to the destination set of all the arc-sets. As a consequence, the number of backward extensions performed by Algorithm 1 is at most $O(nm)$, while SGA performs $\Theta(nm)$ extensions.

5 Experimental Analysis

A C++ implementation of our approach, called FSG (short for Fast String Graph), has been integrated in the SGA suite and is available at http://fsg. algolab.eu under the GPLv3 license. We have evaluated the performance of FSG on a standard benchmark of 875 million 101 bp-long reads sequenced from the NA12878 individual of the International HapMap and 1000 genomes project and comparing the running time of FSG with SGA. We have run SGA with its default parameters, that is SGA has compute exact overlaps after having corrected the input reads. Since the string graphs computed by FSG and SGA are the same, we have not compared the entire pipeline, but only the string graph construction phase. We could not compare FSG with Fermi, since Fermi does not split its steps in a way that allows to isolate the running time of the string graph construction—most notably, it includes reads correction and scaffolding.

Especially on the DNA alphabet, short overlaps between reads may happen by chance. Hence, for genome assembly purposes, only overlaps whose length is larger than a user-defined threshold are considered. The value of the minimum overlap length threshold that empirically showed the best results in terms of genome assembly quality is around the 75 % of the read length [24]. To assess how graph size affects performance, different values of minimum overlap length (called τ) between reads have been used (clearly, the lower this value, the larger the graph). The minimum overlap lengths used in this experimental assessment are 55, 65, 75, and 85, hence the chosen values test the approaches also on larger-than-normal ($\tau = 55$) and smaller-than-normal ($\tau = 85$) string graphs. Another aspect that we have wanted to measure is the scalability of FSG. We have run the programs with 1, 4, 8, 16, and 32 threads. In all cases, we have measured

Table 1. Comparison of FSG and SGA, for different minimum overlap lengths and numbers of threads. The wall-clock time is the time used to compute the string graph. The CPU time is the overall execution time over all CPUs actually used.

Min. overlap	No. of threads	Wall time [min]			Work time [min]		
		FSG	SGA	$\frac{FSG}{SGA}$	FSG	SGA	$\frac{FSG}{SGA}$
55	1	1,485	4,486	0.331	1,483	4,480	0.331
	4	474	1,961	0.242	1,828	4,673	0.391
	8	318	1,527	0.209	2,203	4,936	0.446
	16	278	1,295	0.215	3,430	5,915	0.580
	32	328	1,007	0.326	7,094	5,881	1.206
65	1	1,174	3,238	0.363	1,171	3,234	0.363
	4	416	1,165	0.358	1,606	3,392	0.473
	8	271	863	0.315	1,842	3,596	0.512
	16	255	729	0.351	3,091	4,469	0.692
	32	316	579	0.546	6,690	4,444	1.505
75	1	1,065	2,877	0.37	1,063	2,868	0.371
	4	379	915	0.415	1,473	2,903	0.507
	8	251	748	0.336	1,708	3,232	0.528
	16	246	561	0.439	2,890	3,975	0.727
	32	306	455	0.674	6,368	4,062	1.568
85	1	1,000	2,592	0.386	999	2,588	0.386
	4	360	833	0.432	1,392	2,715	0.513
	8	238	623	0.383	1,595	3,053	0.523
	16	229	502	0.457	2,686	3,653	0.735
	32	298	407	0.733	6,117	3,735	1.638

the elapsed (wall-clock) time and the total CPU time (the time a CPU has been working). All experiments have been performed on an Ubuntu 14.04 server with four 8-core Intel® Xeon E5-4610v2 2.30 GHz CPUs. The server has a NUMA architecture with 64 GiB of RAM for each node (256 GiB in total).

Table 1 summarizes the running times of both approaches on the different configurations of the parameters. Notice that LSG approach is from 2.3 to 4.8 times faster than SGA in terms of wall-clock time and from 1.9 to 3 times in terms of CPU time. On the other hand, FSG uses approximately 2.2 times the memory used by SGA—on the executions with at most 8 threads. On a larger number of threads, and in particular the fact that the elapsed time of FSG on 32 threads is larger than that on 16 threads suggests that, in its current form, FSG might not be suitable for a large number of threads. However, since the current implementation of FSG is almost a proof of concept, future improvements to its codebase and a better analysis of the race conditions of our tool will likely lead to better performances with a large number of threads. Furthermore, notice that also the SGA algorithm, which is (almost) embarrassingly parallel and has a stable implementation, does not achieve a speed-up better than 6.4 with 32 threads. As such, a factor that likely contributes to a poor scaling behaviour of both FSG and SGA could be also the NUMA architecture of the server used for the experimental analysis, which makes different-unit memory accesses more expensive (in our case, the processors in each unit can manage at most 16 logical threads, and only 8 on physical cores). FSG uses more memory than SGA since genome assemblers must correctly manage reads extracted from both strands of the genome. In our case, this fact has been addressed by adding each reverse-and-complement read to the set of strings on which the FM-index has been built, hence immediately doubling the size of the FM-index. Moreover, FSG needs some additional data structures to correctly maintain potential overlaps and arc-sets: two pairs of $n(m + 1)$-long bitvectors and the combination of two (usually) small integer vectors and a bitvector of the same size. Our experimental evaluation shows that the memory required by the latter is usually negligible, hence a better implementation of the four bitvectors could decrease the memory use. The main goal of FSG is to improve the running time, not the memory use.

The combined analysis of the CPU time and the wall-clock time on at most 8 threads (which is the number of physical cores of each CPU on our server) suggests that FSG is more CPU efficient than SGA and is able to better distribute the workload across the threads. In our opinion, our greater efficiency is achieved by operating only on the FM-index of the input reads and by the order on which extension operations (*i.e.*, considering a new string $c\alpha$ after α has been processed) are performed. These two characteristics of our algorithm allow to eliminate the redundant queries to the index which, instead, are performed by SGA. In fact, FSG considers each string that is longer than the threshold at most once, while SGA potentially reconsiders the same string once for each read in which the string occurs. Indeed, FSG uses 2.3–3 times less user time than SGA when $\tau = 55$ (hence, when such sufficiently-long substrings occur more frequently) and "only" 2–2.6 times less user time when $\tau = 85$ (hence, when such sufficiently-long substrings are more rare).

6 Conclusions and Future Work

We present FSG: a tool implementing a new algorithm for constructing a string graph that works directly querying a FM-index representing a collection of reads, instead of processing the input reads. Our main goal is to provide a simpler and fast algorithm to construct string graphs, so that its implementation can be easily integrated into an assembly pipeline that analyzes the paths of the string graph to produce the final assembly. Indeed, FSG could be used for related purposes, such as transcriptome assembly [5,16], and haplotype assembly [6]. These topics are some of the research directions that we plan to investigate.

References

1. Bankevich, A., Nurk, S., et al.: SPAdes: a new genome assembly algorithm and its applications to single-cell sequencing. J. Comput. Biol. **19**(5), 455–477 (2012)
2. Bauer, M., Cox, A., Rosone, G.: Lightweight algorithms for constructing and inverting the BWT of string collections. Theoret. Comput. Sci. **483**, 134–148 (2013)
3. Bauer, M.J., Cox, A.J., Rosone, G., Sciortino, M.: Lightweight LCP construction for next-generation sequencing datasets. In: Raphael, B., Tang, J. (eds.) WABI 2012. LNCS, vol. 7534, pp. 326–337. Springer, Heidelberg (2012)
4. Ben-Bassat, I., Chor, B.: String graph construction using incremental hashing. Bioinformatics **30**(24), 3515–3523 (2014)
5. Beretta, S., Bonizzoni, P., Della Vedova, G., Pirola, Y., Rizzi, R.: Modeling alternative splicing variants from RNA-Seq data with isoform graphs. J. Comput. Biol. **16**(1), 16–40 (2014)
6. Bonizzoni, P., Della Vedova, G., Dondi, R., Li, J.: The haplotyping problem: an overview of computational models and solutions. J. Comput. Sci. Technol. **18**(6), 675–688 (2003)
7. Bonizzoni, P., Della Vedova, G., Pirola, Y., Previtali, M., Rizzi, R.: Constructing string graphs in external memory. In: Brown, D., Morgenstern, B. (eds.) WABI 2014. LNCS, vol. 8701, pp. 311–325. Springer, Heidelberg (2014)
8. Bonizzoni, P., Della Vedova, G., Pirola, Y., Previtali, M., Rizzi, R.: LSG: an external-memory tool to compute string graphs for NGS data assembly. J. Comp. Biol. **23**(3), 137–149 (2016)
9. Boucher, C., Bowe, A., Gagie, T., et al.: Variable-order de bruijn graphs. In: 2015 Data Compression Conference (DCC), pp. 383–392. IEEE (2015)
10. Bradnam, K.R., Fass, J.N., Alexandrov, A., et al.: Assemblathon 2: evaluating de novo methods of genome assembly in three vertebrate species. GigaScience **2**(1), 1–31 (2013)
11. Burrows, M., Wheeler, D.J.: A block-sorting lossless data compression algorithm. Technical report, Digital Systems Research Center (1994)
12. Chikhi, R., Limasset, A., Jackman, S., Simpson, J.T., Medvedev, P.: On the representation of de bruijn graphs. J. Comp. Biol. **22**(5), 336–352 (2015)
13. Chikhi, R., Rizk, G.: Space-efficient and exact de Bruijn graph representation based on a Bloom filter. Alg. Mol. Biol. **8**(22), 1–9 (2013)
14. Ferragina, P., Manzini, G.: Indexing compressed text. J. ACM **52**(4), 552–581 (2005)
15. Gonnella, G., Kurtz, S.: Readjoiner: a fast and memory efficient string graph-based sequence assembler. BMC Bioinform. **13**(1), 82 (2012)
16. Lacroix, V., Sammeth, M., Guigo, R., Bergeron, A.: Exact transcriptome reconstruction from short sequence reads. In: Crandall, K.A., Lagergren, J. (eds.) WABI 2008. LNCS (LNBI), vol. 5251, pp. 50–63. Springer, Heidelberg (2008)
17. Li, H.: Exploring single-sample SNP and INDEL calling with whole-genome de novo assembly. Bioinformatics **28**(14), 1838–1844 (2012)
18. Myers, E.: The fragment assembly string graph. Bioinformatics **21**(s2), 79–85 (2005)
19. Peng, Y., Leung, H.C., Yiu, S.-M., Chin, F.: IDBA-UD: a de novo assembler for single-cell and metagenomic sequencing data with highly uneven depth. Bioinformatics **28**(11), 1420–1428 (2012)
20. Salikhov, K., Sacomoto, G., Kucherov, G.: Using cascading bloom filters to improve the memory usage for de brujin graphs. Alg. Mol. Biol. **9**(1), 2 (2014)

21. Salzberg, S.L., et al.: GAGE: a critical evaluation of genome assemblies and assembly algorithms. Genome Res. **22**(3), 557–567 (2012)

22. Shi, F.: Suffix arrays for multiple strings: a method for on-line multiple string searches. In: Jaffar, J., Yap, R.H.C. (eds.) ASIAN 1996. LNCS, vol. 1179, pp. 11–22. Springer, Heidelberg (1996)

23. Simpson, J., Durbin, R.: Efficient construction of an assembly string graph using the FM-index. Bioinformatics **26**(12), i367–i373 (2010)

24. Simpson, J., Durbin, R.: Efficient de novo assembly of large genomes using compressed data structures. Genome Res. **22**, 549–556 (2012)

25. Simpson, J., Wong, K., Jackman, S., et al.: ABySS: a parallel assembler for short read sequence data. Genome Res. **19**(6), 1117–1123 (2009)

OVarCall: Bayesian Mutation Calling Method Utilizing Overlapping Paired-End Reads

Takuya Moriyama[1], Yuichi Shiraishi[1], Kenichi Chiba[1], Rui Yamaguchi[1],
Seiya Imoto[2], and Satoru Miyano[1,2(✉)]

[1] Human Genome Center, Institute of Medical Science,
The University of Tokyo, Tokyo, Japan
{moriyama,yshira,kchiba}@hgc.jp,ruiy@ims.u-tokyo.ac.jp
[2] Health Intelligence Center, Institute of Medical Science,
The University of Tokyo, Tokyo, Japan
{imoto,miyano}@ims.u-tokyo.ac.jp

Abstract. Detection of somatic mutations from tumor and matched normal sequencing data has become a standard approach in cancer research. Although a number of mutation callers are developed, it is still difficult to detect mutations with low allele frequency even in exome sequencing. We expect that overlapping paired-end read information is effective for this purpose, but no mutation caller has modeled overlapping information statistically in a proper form in exome sequence data. Here, we develop a Bayesian hierarchical method, OVarCall, where overlapping paired-end read information improves the accuracy of low allele frequency mutation detection. Firstly, we construct two generative models: one is for reads with somatic variants generated from tumor cells and the other is for reads that does not have somatic variants but potentially includes sequence errors. Secondly, we calculate marginal likelihood for each model using a variational Bayesian algorithm to compute Bayes factor for the detection of somatic mutations. We empirically evaluated the performance of OVarCall and confirmed its better performance than other existing methods.

Keywords: Somatic mutation detection · Next-generation sequencing data · Overlapping paired-end reads · Bayesian hierarchical model

1 Introduction

Cancer is driven by genomic alterations. Acquired somatic mutations, together with individual germ line haplotypes, are definitive factors for cancer evolution. Together with decreasing massively parallel sequencing costs, mutation calling from tumor and matched normal sequence data has been a fundamental analysis in cancer research [11].

There are several important points in somatic mutation detection for achieving high accuracy. For example, a somatic mutation caller should distinguish somatic mutations from sequence errors, alignment errors, and germ line

© Springer International Publishing Switzerland 2016
A. Bourgeois et al. (Eds.): ISBRA 2016, LNBI 9683, pp. 40–51, 2016.
DOI: 10.1007/978-3-319-38782-6_4

variants. In this paper, we focus on the development of somatic mutation caller for mutations with low allele frequencies because of low tumor contents in sample, copy number alterations and tumor heterogeneity. The detection of low allele frequency mutations is important for capturing minor subclones which are thought as a cause of therapeutic resistance or recurrence of cancer [17].

Although there are many methods for somatic mutation detection; VarScan2 [7], Genomon [21], SomaticSniper [8], JointSNVMix [14], Strelka [16], MuTect [4] and HapMuC [20], it is still difficult to detect low allele frequency mutations of 1 % to 5 % in exome sequence data. These mutations are detectable in PCR-targeted sequence data. On the other hand, Chen-Harris [3] reported the efficacy of overlapping paired-end reads, and a mutation caller was developed for PCR-targeted sequence data. However, this mutation caller uses only overlapping paired-end reads and was not designed for usual exome sequence data that include overlapping and non-overlapping paired-end reads.

Here, we constructed a Bayesian hierarchical method, OVarCall, for the detection of somatic mutations with low allele frequencies from exome sequence data. In this method, we construct two generative models: one is for reads with somatic variants generated from tumor cells and the other is for reads with sequence errors, and then calculate marginal likelihood for each model using a variational Bayesian algorithm. Under the assumption that the sequence error occurs randomly, it is low probability that sequence errors occur in the same position. Our generative model realizes this idea and determines the statistical significance of the candidate somatic mutation by using Bayes factor.

Firstly, we evaluated our assumption and the validity of the use of overlapping paired-end reads by observing that the sequence errors occur mostly on one side of the overlapping paired-end reads and the validated mutations appear mostly on both sides of the overlapping paired-end reads. Secondly, we empirically showed the effectiveness of our method using simulation data. Finally, in 20 ccRCC patients' exome sequence data, we confirmed that our method outperforms existing methods for detection of low allele frequency somatic mutations.

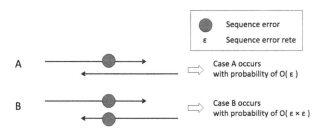

Fig. 1. Assumptions of sequence error for overlapping paired-end reads. (**A**) Sequence error occurs on one side of the reads. This type of error is considered to be common for overlapping paired-end reads, since sequence error needs to occur independently on both sides of the reads. (**B**) Sequence errors occur on both sides of the reads. This type of error is considered to be rare since sequence error should occur twice.

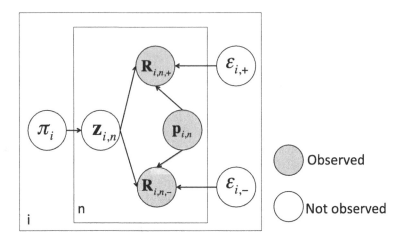

Fig. 2. Graphical representation of OVarCall. Notation i stands for the index of the candidate mutation positions, and n stands for the index of the paired-end reads. In this model, strand bias of sequence error is considered, and sequence error of each strand on the i-th position occurs independently.

2   Methods

2.1   Graphical Model of OVarCall

We define the 1-of-K expression vectors, $\boldsymbol{R}_{i,n,+}$ and $\boldsymbol{R}_{i,n,-}$, which represent the base or ins/del of the n-th read pair on the i-th reference position on each strand. We also define the 1-of-K expression vector, $\boldsymbol{p}_{i,n}$, which indicates the strand information of reads covering the candidate mutation positions, i.e., covering by only plus strand reads ($\boldsymbol{p}_{i,n} = [1,0,0]$), only minus strand reads ($\boldsymbol{p}_{i,n} = [0,1,0]$), or both strands ($\boldsymbol{p}_{i,n} = [0,0,1]$). The vector of latent variables of the pair-end reads is denoted by $\boldsymbol{z}_{i,n}$, which explains whether the read pair is sequenced from a DNA fragment with true somatic mutation or not. For example, if the original DNA fragment is the same as the reference sequence, $\boldsymbol{z}_{i,n}$ is equal to $[1,0]$, and if the original DNA fragment contains somatic mutation, $\boldsymbol{z}_{i,n}$ is equal to $[0,1]$. Allele frequency for the candidate mutation position is denoted by π_i; $(1-\pi_i)$ represents the proportion of variant supporting reads. Sequence error probability on the i-th position of each strand is denoted by $\epsilon_{i,+}, \epsilon_{i,-}$. We set the prior distributions of the parameters and the latent variables as follows:

$$\pi_i \sim p_{beta}(\pi_i | \alpha_{ref}, \beta_{obs})$$
$$\epsilon_{i,+} \sim p_{beta}(\epsilon_{i,+} | \delta_{+,1}, \delta_{+,2})$$
$$\epsilon_{i,-} \sim p_{beta}(\epsilon_{i,-} | \delta_{-,1}, \delta_{-,2})$$
$$p(z_{i,n,0}) = \pi_i$$
$$p(z_{i,n,1}) = 1 - \pi_i$$

where $p_{beta}(\cdot|a, b)$ is the probability density of the beta distribution with the parameters a and b, i.e., hyperparameters of beta distribution for $\pi_i, \epsilon_{i,+}$, and $\epsilon_{i,-}$ are denoted as $(\alpha_{ref}, \beta_{obs}), (\delta_{+,1}, \delta_{+,2})$, and $(\delta_{-,1}, \delta_{-,2})$, respectively.

2.2   Modeling of Sequence Errors on Paired-End Reads

We assumed that sequence errors occur independently in the overlapping paired-end reads as expressed in Fig. 1. For simplicity, here we denote only the modeling of sequence error when candidate mutation is single nucleotide variant (SNV). If there is no mismatch between the latent variables $z_{i,n}$ and observed read bases $R_{i,n,+}$ or $R_{i,n,-}$, sequence error is supposed to occur with probability of $(\frac{1}{3}\epsilon_{i,+})$ or $(\frac{1}{3}\epsilon_{i,-})$. The other case, in which there is no sequence error, is supposed to occur with probability of $(1 - \epsilon_{i,+})$ or $(1 - \epsilon_{i,-})$. We introduce an indicator variable $m_{i,+}$ which is 1 if the latent variables $z_{i,n}$ is the same as the observed bases $R_{i,n,+}$, and which is 0 otherwise. The indicator variable $m_{i,-}$ is defined in the same way. We define the likelihood of the observed reads as follows:

$$p(R_{i,n,+}, R_{i,n,-}|z_{i,n}, \epsilon_{i,+}, \epsilon_{i,-}, \pi_{i,n}, p_{i,n})$$

$$= \begin{cases} (1 - \epsilon_{i,+})^{m_{i,+}} \cdot (\frac{1}{3}\epsilon_{i,+})^{1-m_{i,+}} & (p_{i,n} = [1, 0, 0]) \\ (1 - \epsilon_{i,-})^{m_{i,-}} \cdot (\frac{1}{3}\epsilon_{i,-})^{1-m_{i,-}} & (p_{i,n} = [0, 1, 0]) \\ (1 - \epsilon_{i,+})^{m_{i,+}} \cdot (\frac{1}{3}\epsilon_{i,+})^{1-m_{i,+}} & \\ \quad \cdot (1 - \epsilon_{i,-})^{m_{i,-}} \cdot (\frac{1}{3}\epsilon_{i,-})^{1-m_{i,-}} & (p_{i,n} = [0, 0, 1]) \end{cases}$$

From the probabilistic distributions above and the dependencies of variables denoted in Fig. 2, the joint distribution is given in the following form:

$$\begin{aligned} p(\mathcal{R}_i, \mathcal{Z}_i|\gamma_i, \alpha_{i,+}, \alpha_{i,-}) &= p(\pi_i|\gamma_i)p(\epsilon_{i,+}|\alpha_{i,+})p(\epsilon_{i,-}|\alpha_{i,-}) \\ &\quad \cdot \prod_n p(R_{i,n,+}, R_{i,n,-}|z_{i,n}, \epsilon_{\pm,i}, \pi_{i,n}, p_{i,n})p(z_{i,n}|\pi_i) \\ &= p(\pi_i|\gamma_i)p(\epsilon_{i,+}|\alpha_{i,+})p(\epsilon_{i,-}|\alpha_{i,-}) \\ &\quad \cdot p(R_{i,+}, Z_{i,+}|\epsilon_{i,+}, \pi_i) \cdot p(R_{i,-}, Z_{i,-}|\epsilon_{i,-}, \pi_i) \\ &\quad \cdot p(R_{i,\pm}, Z_{i,\pm}|\epsilon_{i,+}, \epsilon_{i,-}, \pi_i) \end{aligned}$$

where we denote all paired-end read information or corresponding latent variables as $(R_{i,+}, Z_{i,+}), (R_{i,-}, Z_{i,-})$ or $(R_{i,\pm}, Z_{i,\pm})$, depending on the value of $p_{i,n}$, and $\mathcal{R}_i = (R_{i,+}, R_{i,-}, R_{i,\pm})$, $\mathcal{Z}_i = (Z_{i,+}, Z_{i,-}, Z_{i,\pm}, \pi_i, \epsilon_{i,+}, \epsilon_{i,-})$. In this equation, the hyperparameters are denoted as $\gamma_i = (\alpha_{ref}, \beta_{obs})$, $\alpha_{i,+} = (\delta_{+,1}, \delta_{+,2})$, $\alpha_{i,-} = (\delta_{-,1}, \delta_{-,2})$.

2.3   Difference of the Tumor Model and the Error Model

In the tumor model, we set $(\alpha_{ref}, \beta_{obs}) = (100.0, 1.0)$ because the number of true somatic mutation is small, compared to error prone sites. In the error model, there should be no mutations except for contaminations in a sample, and we set

$(\alpha_{ref}, \beta_{obs}) = (1000000.0, 10.0)$. In both models, we used the hyperparameters of the posterior beta distributions in the error model given the normal sample data, where the hyperparameters for sequence errors are $(\delta_{+,1}, \delta_{+,2}) = (\delta_{-,1}, \delta_{-,2}) = (2.0, 30.0)$, which is same as the error probabilities in simulation model. In the real data analysis, we set $(\delta_{+,1}, \delta_{+,2}) = (\delta_{-,1}, \delta_{-,2}) = (1.0, 100.0)$, $(\alpha_{ref}, \beta_{obs}) = (100.0, 1.0)$, since positions with over 100 depth coverage are selected in the minimum criteria.

2.4   Bayes Factor for Detecting Somatic Mutations

For the detection of the somatic mutations, Bayes factor (BF) is defined as:

$$BF = \frac{p_T(\mathcal{R}_i|\gamma_i, \alpha_{i,+}, \alpha_{i,-})}{p_E(\mathcal{R}_i|\gamma_i, \alpha_{i,+}, \alpha_{i,-})}$$

where p_T and p_E represent the marginal likelihoods calculated by the tumor and error models, respectively. In order to compute BF, marginal likelihoods is calculated for both models as:

$$p_T(\mathcal{R}_i|\gamma_i, \alpha_{i,+}, \alpha_{i,-}) = \sum_{\mathcal{Z}_i} \int p_T(\mathcal{R}_i\mathcal{Z}_i|\gamma_i, \alpha_{i,+}, \alpha_{i,-})d\pi_i d\epsilon_{i,+} d\epsilon_{i,-}$$

However, calculation of the marginal likelihoods needs high-dimensional integrals and sums, and the closed form cannot be obtained. We use variational Bayes approach for approximating the values of the marginal likelihoods.

2.5   Variational Bayes Procedure

Here we represent the computation of $p(\mathcal{R}_i|\gamma_i, \alpha_{i,+}, \alpha_{i,-})$, or $\ln p(\mathcal{R}_i|\gamma_i, \alpha_{i,+}, \alpha_{i,-})$. From the fact that log function is concave, we can derive the following inequality [6].

$$\ln p(\mathcal{R}_i|\gamma_i, \alpha_{i,+}, \alpha_{i,-}) \geq E_q \left[\frac{p(\mathcal{R}_i, \mathcal{Z}_i|\gamma_i, \alpha_{i,+}, \alpha_{i,-})}{q(\mathcal{Z}_i)} \right] \qquad (1)$$

where $q(\mathcal{Z}_i)$ is the variational distribution for $\mathcal{Z}_i$, and the equality holds when $q(\mathcal{Z}_i)$ is equal to the posterior distribution of $p(\mathcal{Z}_i|\mathcal{R}_i, \gamma_i, \alpha_{i,+}, \alpha_{i,-})$. In this variational Bayes approach, we approximately decompose $q(\mathcal{Z}_i)$ in the following form, and update $q(\mathcal{Z}_i)$ iteratively to maximize the right hand side of the Eq. (1).

$$q(\mathcal{Z}_i) = q(Z_+)q(Z_-)q(Z_\pm)q(\pi_i)q(\epsilon_{i,+})q(\epsilon_{i,-})$$

The update formula of q is expressed as:

$$q^*(Z_{i,+}) = \prod_{n_+}^{N_+} \prod_{j=0}^{1} (\zeta_{i,n_+,+,j}^*)^{Z_{i,n_+,j}}$$

$$q^*(Z_{i,-}) = \prod_{n_-}^{N_-} \prod_{j=0}^{1} (\zeta^*_{i,n_-,-,j})^{Z_{i,n_-,j}}$$

$$q^*(Z_{i,\pm}) = \prod_{n_\pm}^{N_\pm} \prod_{j=0}^{1} (\zeta^*_{i,n_\pm,\pm,j})^{Z_{i,n_\pm,j}}$$

$$q^*(\pi_i) = p_{beta}(\pi_i|\boldsymbol{\gamma_i^*})$$

$$q^*(\epsilon_{i,+}) = p_{beta}(\epsilon_{i,+}|\boldsymbol{\alpha_{i,+}^*})$$

$$q^*(\epsilon_{i,-}) = p_{beta}(\epsilon_{i,-}|\boldsymbol{\alpha_{i,-}^*})$$

Updated parameters are obtained in the following equations.

$$\zeta^*_{i,n_+,j} = \frac{\rho^*_{i,n_+,j}}{\rho^*_{i,n_+,0} + \rho^*_{i,n_+,1}}$$

$$\zeta^*_{i,n_-,j} = \frac{\rho^*_{i,n_-,j}}{\rho^*_{i,n_-,0} + \rho^*_{i,n_-,1}}$$

$$\zeta^*_{i,n_\pm,j} = \frac{\rho^*_{i,n_\pm,j}}{\rho^*_{i,n_\pm,0} + \rho^*_{i,n_\pm,1}}$$

$$\alpha^*_{i,+,0} = \delta_{+,1} + N_+ + N_\pm + 2 - s_+$$

$$\alpha^*_{i,+,1} = \delta_{+,2} + s_+$$

$$\alpha^*_{i,-,0} = \delta_{-,1} + N_- + N_\pm + 2 - s_+$$

$$\alpha^*_{i,-,1} = \delta_{-,2} + s_-$$

$$\gamma^*_{i,j} = \sum_{n_+}^{N_+} \zeta^*_{i,n_+,j} + \sum_{n_-}^{N_-} \zeta^*_{i,n_-,j} + \sum_{n_\pm}^{N_\pm} \zeta^*_{i,n_\pm,j} + \gamma_{i,j}$$

$$s_+ = \sum_{n_+}^{N_+} \sum_{j=0}^{1} \zeta^*_{i,n_+,j} R_{i,+,n_+,j} + \sum_{n_\pm}^{N_\pm} \sum_{j=0}^{1} \zeta^*_{i,n_\pm,j} R_{i,+,n_\pm,j}$$

$$s_- = \sum_{n_-}^{N_-} \sum_{j=0}^{1} \zeta^*_{i,n_-,j} R_{i,-,n_-,j} + \sum_{n_\pm}^{N_\pm} \sum_{j=0}^{1} \zeta^*_{i,n_\pm,j} R_{i,-,n_\pm,j}$$

$$\ln \rho^*_{i,n_+,j} = E_q\left[\ln \pi_{i,j}\right] + E_q\left[R_{i,n_+,+,j}\left\{-\ln\left(\frac{\epsilon_{i,+}}{3}\right) + \ln(1 - \epsilon_{i,+})\right\}\right]$$

$$\ln \rho^*_{i,n_-,j} = E_q\left[\ln \pi_{i,j}\right] + E_q\left[R_{i,n_-,-,j}\left\{-\ln\left(\frac{\epsilon_{i,-}}{3}\right) + \ln(1 - \epsilon_{i,-})\right\}\right]$$

$$\ln \rho^*_{i,n_\pm,j} = E_q\left[\ln \pi_{i,j}\right]$$
$$+ E_q\left[R_{i,n_\pm,+,j}\left\{-\ln\left(\frac{\epsilon_{i,+}}{3}\right) + \ln(1 - \epsilon_{i,+})\right\}\right]$$
$$+ E_q\left[R_{i,n_\pm,-,j}\left\{-\ln\left(\frac{\epsilon_{i,-}}{3}\right) + \ln(1 - \epsilon_{i,-})\right\}\right]$$

2.6   Criteria for Selecting the Error Prone Site

To examine the correlation between sequence error and overlapping paired-end reads, we collected the error prone sites using ccRCC patients' 20 normal samples, whose sequencing reads were aligned to hg19 Reference Genome using Burrows-Wheeler Aligner [9], with default parameter settings. We firstly excluded the germ line SNP positions, as listed in the following conditions.

1. The read coverage is ≥ 20
2. The non-reference allele frequency is >0.2
3. At least one variant read is observed in both positive and negative strands.

We counted the candidate error positions which satisfy the following conditions:

1. The read coverage is ≥ 20
2. The non-reference allele frequency is >0.03
3. The read coverage of plus strand read ≥ 1
4. The read coverage of minus strand read ≥ 1

If the number of samples satisfying the above four conditions is greater than or equal to 5, then we decide the position as an error prone site. The above conditions is based on [19], but to consider low depth positions, we ignore the condition that read coverages between all samples are ≥ 20.

2.7   Minimum Criteria for the Simulation Study

We retained the candidate positions if they met with the following criteria:

1. The tumor allele frequency is $>thres$.
2. The normal allele frequency is <0.1.
3. The number of variant-supporting reads in tumor is ≥ 4.
4. The read coverage is ≥ 12.

Where we set *thres* as 0.07 if tumor allele frequency is 10 %, and we set *thres* as 0.005 if tumor allele frequency is 1 %, in order to collect as many true somatic mutations as possible.

2.8   Minimum Criteria for the Real Data Study

We set several filter conditions to exclude false positive SNV. We filtered reads if they met with at least one filtering conditions described below:

1. Mapping quality is <30.
2. Number of insertion or deletion is ≥ 3.
3. Number of SNV is ≥ 3.
4. Number of insertion or deletion or SNV is ≥ 4.
5. The proportion of soft-clipping is >0.25.
6. Read is not mapped in proper pair.
7. Intersections with sam flag 3840 is not 0.

If the proportions of filtered reads that cover a candidate position is ≥ 0.3 or indels are present within a 25 bp distance, we excluded the position. After applying the above filters, we collected candidate positions if they met with the following criteria:

1. The read coverage is ≥ 100.
2. The tumor allele frequency is ≥ 0.01 and ≤ 0.07.
3. The number of variant-supporting reads in a tumor sample is ≥ 3.
4. The normal allele frequency is ≤ 0.01.
5. The number of variant-supporting reads in a normal sample is ≤ 1.

We should also note that potential mapping errors are excluded by using genomic super duplications, simple repeats [1], and dbSNP138 [18], and we selected positions within exome regions.

2.9   Parameters for Alternative Methods

Fisher's exact test: Two-sided Fisher's exact test is executed. We simply used the library function from scipy.stats.fisher_exact

VarScan2 (v2.3.9): –min-var-freq 0.01 –min-coverage 10 –min-coverage-normal 10 –min-coverage-tumor 10 –somatic-p-value 0.5.

Strelka (v1.0.14): isSkipDepthFilters = 1 is set on the default setting.

MuTect (v1.1.4): –minimum_mutation_cell_fraction 0.01

3   Numerical Examples

3.1   Validation of the Sequence Error Assumption in ccRCC Patients' Data

We investigated the independence of the sequence error on the overlapping paired-end reads, using exome tumor and matched normal sequences of 20 ccRCC patients [15, 19]. Sequence errors in Illumina sequencer data do not occur uniformly, i.e., errors occur in a sequence specific manner [5,10,12]. Shiraish [19] examined such errors, by collecting errors which occur repeatedly in multiple samples, and he defined these sites as error prone sites. Firstly, we extracted error prone sites using the normal sequence samples and the somatic mutation positions which were validated by PCR-targeted sequence. Secondly, for each position of error prone sites or true somatic mutations, we checked the numbers of two types of overlapping paired-end reads: (i) VX, overlapping paired-end reads where at least one read is a variant supporting read. (ii) VV, overlapping paired-end reads where both reads are variant supporting read and have same variant in the candidate position. For each candidate SNV position, we count the numbers of VV and VX reads, denoted by #VV and #VX, and show the scatter plot of (#VV, #VX) in Fig. 3. We should note that potential mapping errors are excluded by using genomic super duplications, simple repeats [1], and dbSNP138 [18]. For almost all of the validated somatic mutations, the numbers of VX and VV are almost same. However, for sequence errors, almost no VV read is observed. This observation indicates that sequence error occurs randomly on the overlapping paired-end reads as expressed in Fig. 1.

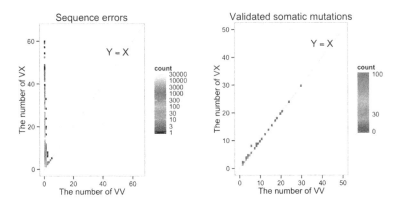

Fig. 3. Consistency within the overlapping paired-end reads from 20 ccRCC patients' sequence data. For each error prone site or validated SNV position, we counted #VV and #VX. For each point of (#VV, #VX) in this plot, color gradient represents the number of positions. (**A**) For almost all of the position error prone sites, we could hardly observe the VV paired-end reads. (**B**) However, for almost all of the validated SNV positions, we can observe that the number of VX and VV paired-end reads are same.

3.2   Simulation Study

We evaluated OVarCall using simulation data. In this simulation, we used two types of tumor allele frequency (10 % and 1 %), four pairs of average and variance of DNA fragment size, and three pairs of average and variance of depth around the positive mutations or negative error prone sites. We generated positive mutations and negative error prone sites as follows:

(1) Generate a random reference DNA sequence.
(2) Generate somatic mutations and error prone sites randomly, and generate paired-end reads around them.
    (2-a) Determine the number of paired-end reads covering the position, by generating a random value d from a norm distribution of $N(\mu_d, \sigma_d)$, and round d to the nearest integer value.
    (2-b) Determine randomly whether the original DNA fragment is from tumor cell DNA or normal cell DNA, according to the tumor allele frequency.
    (2-c) For each paired-end reads, determine the DNA fragment size, by generating a random value l from $N(\mu_l, \sigma_l)$, and round l to the nearest integer value.
    (2-d) Generate the 100 bp length read sequence on plus strand. Each observed base flips with sequence error probability of p_{error}. If the position of each observed base is the error prone site, p_{error} is generated from a beta distribution of $Beta(2, 30)$. If the position of each observed base is not the error prone site, p_{error} is generated from $Beta(10, 1000)$.
    (2-e) Generate the read sequence on minus strand like (2-d).

We used only those positions that passed the minimum criteria for allele frequencies, read coverage, the number of variant supporting reads, etc., as listed in Method section. As a counterpart method, we prepared a simple Fisher's exact test-based method, which uses a 2 by 2 contingency table of read counts; tumor and normal samples / variant and reference alleles. We calculated the area under curve (AUC) values [2], plotted the ROC curve for each simulation conditions and summarized the results in Table 1. In the simulation data of which tumor allele frequency is 10 %, OVarCall outperforms Fisher's exact test methods, when overlapping paired-end reads are available or enough depth is obtained, and even if we cannot use the enough overlapping paired-end reads, OVarCall performs competitively. In the simulation data of which tumor allele frequency is 1 %, OVarCall succeeded in detecting low allele frequency mutations accurately.

Table 1. Simulation results summary

Allele frequency	μ_d	σ_d	μ_l	σ_l	Fisher AUC	OVarCall AUC	#SNV	#False positive
10 %	50	2	180	40	0.867	0.941	667	2442
			200	40	0.868	0.912	707	3152
			300	40	0.870	0.863	738	4697
	100	2	180	40	0.983	0.994	788	2254
			200	40	0.976	0.988	794	2858
			300	40	0.973	0.981	796	5203
1 %	300	2	150	20	0.745	0.917	455	14394

3.3 Mutations of Low Allele Frequency in cRCC Patients' Data

By using the error prone sites and validated somatic SNVs in 20 ccRCC patients' whole exome sequence data, we also evaluated the performance of OVarCall and compared with the existing mutation callers, i.e., VarScan2 [7] and MuTect [4] and Strelka [16]. From each ccRCC patients' sample, we prepared two data sets as normal and tumor samples, following these rules below.

Error Prone site in normal sample: Output all reads in original normal sample, covering error prone sites, to the new tumor sample, and output all reads in original normal sample, except for variant supporting reads, to the new normal sample.

Validated SNV in tumor sample: Output all reads in original tumor sample, covering error prone sites, to the new tumor sample, and output all reads in original normal sample to the new normal sample.

After applying the minimum criteria in Method section for collection of candidate mutations, we collected 44 validated somatic mutations and 1884 false positive positions. We applied OVarCall, VarScan2, MuTect and Strelka to the data sets and drew the ROC curves shown in Fig. 4. By calculating AUC value for each method, it is clear that OVarCall outperformed to the other methods.

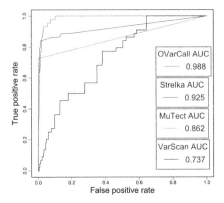

Fig. 4. Performance evaluation of OVarCall using real data which contains validated mutations of low allele frequency and error prone sites.

4   Conclusions

In this paper, we focused on the problem of accurate detection of low allele frequency somatic mutations from tumor and matched normal sequence data. In 20 ccRCC patients' exome sequence data, we observed the consistent overlapping paired-end reads on the validated somatic mutations, however, we observed the inconsistent overlapping paired-end reads on the error prone sites. From these observations and existing observations in PCR-targeted sequence [3,13], we developed a novel Bayesian statistical method, OVarCall, which can use the information of overlapping paired-end reads for detecting low allele frequency somatic mutations.

In our simulation study, we prepared multiple simulation data set, based on the conditions of tumor allele frequency, DNA fragment size, and depth around the positive mutations or negative error prone sites. As the results, we succeeded in showing the superiority of our method for low allele frequency somatic mutation when overlapping pair-end reads are available. Even if such reads are not available substantially, our method can show comparable performance with Fisher's exact test based method. Furthermore, in 20 ccRCC patients' exome sequence data, we confirmed that our method could detect validated low allele frequency somatic mutations and outperformed existing methods for detection of mutations of low allele frequency.

Also, in our methods, the hyperparameters for tumor and sequence error model were chosen empirically, and we should develop the learning methods from other sequence data. In the learning of the hyperparameters, a problem is expected. When the distribution of tumor allele frequency follows a multimodal form [19], accurate learning of the tumor allele frequencies is expected to be difficult in our methods. The same problem is also expected for the learning of the sequence error. One possible solution for this problem is to use the framework of nonparametric Bayes approach for learning the multimodal distributions from the training data set. We will investigate this problem in our future papers.

Acknowledgments. The super-computing resource was provided by Human Genome Center, the Institute of Medical Science, the University of Tokyo.

References

1. Benson, G.: Tandem repeats finder: a program to analyze DNA sequences. Nucleic Acids Res. **27**(2), 573–580 (1999)
2. Bradley, A.P.: The use of the area under the ROC curve in the evaluation of machine learning algorithms. Pattern Recogn. **30**(7), 1145–1159 (1997)
3. Chen-Harris, H., et al.: Ultra-deep mutant spectrum profiling: improving sequencing accuracy using overlapping read pairs. BMC Genomics **14**(1), 96 (2013)
4. Cibulskis, K., et al.: Sensitive detection of somatic point mutations in impure and heterogeneous cancer samples. Nat. Biotechnol. **31**(3), 213–219 (2013)
5. Dohm, J.C., et al.: Substantial biases in ultra-short read data sets from high-throughput DNA sequencing. Nucleic Acids Res. **36**(16), e105 (2008)
6. Jensen, J.L.W.V.: Sur les fonctions convexes et les inégalités entre les valeurs moyennes. Acta Math. **30**(1), 175–193 (1906)
7. Koboldt, D.C., et al.: VarScan 2: somatic mutation and copy number alteration discovery in cancer by exome sequencing. Genome Res. **22**(3), 568–576 (2012)
8. Larson, D.E., et al.: SomaticSniper: identification of somatic point mutations in whole genome sequencing data. Bioinformatics **28**(3), 311–317 (2012)
9. Li, H., et al.: Fast and accurate short read alignment with Burrows-Wheeler transform. Bioinformatics **25**(14), 1754–1760 (2009). Oxford, England
10. Li, M., Stoneking, M.: A new approach for detecting low-level mutations in next-generation sequence data. Genome Biol. **13**(5), R34 (2012)
11. Meyerson, M., et al.: Advances in understanding cancer genomes through second-generation sequencing. Nat. Reviews. Genet. **11**(10), 685–696 (2010)
12. Nakamura, K., et al.: Sequence-specific error profile of Illumina sequencers. Nucleic Acids Res. **39**(13), e90 (2011)
13. Pope, B.J., et al.: ROVER variant caller: read-pair overlap considerate variant-calling software applied to PCR-based massively parallel sequencing datasets. Source Code Biol. Med. **9**(1), 3 (2014)
14. Roth, A., et al.: JointSNVMix: a probabilistic model for accurate detection of somatic mutations in normal/tumour paired next-generation sequencing data. Bioinformatics **28**(7), 907–913 (2012)
15. Sato, Y., et al.: Integrated molecular analysis of clear-cell renal cell carcinoma. Nat. Genet. **45**(8), 860–867 (2013)
16. Saunders, C.T., et al.: Strelka: accurate somatic small-variant calling from sequenced tumor-normal sample pairs. Bioinformatics **28**(14), 1811–1817 (2012)
17. Shah, S.P., et al.: Mutational evolution in a lobular breast tumour profiled at single nucleotide resolution. Nature **461**(7265), 809–813 (2009)
18. Sherry, S.T.: dbSNP: the NCBI database of genetic variation. Nucleic Acids Res. **29**(1), 308–311 (2001)
19. Shiraishi, Y., et al.: An empirical Bayesian framework for somatic mutation detection from cancer genome sequencing data. Nucleic Acids Res. **41**(7), e89 (2013)
20. Usuyama, N., et al.: HapMuC: somatic mutation calling using heterozygous germ line variants near candidate mutations. Bioinformatics **30**(23), 3302–3309 (2014)
21. Yoshida, K., et al.: Frequent pathway mutations of splicing machinery in myelodysplasia. Nature **478**(7367), 64–69 (2011)

High-Performance Sensing
of DNA Hybridization on Surface
# of Self-organized MWCNT-Arrays Decorated
by Organometallic Complexes

V.P. Egorova[1], H.V. Grushevskaya[1(✉)], N.G. Krylova[1], I.V. Lipnevich[1],
T.I. Orekhovskaja[1], B.G. Shulitski[2], and V.I. Krot[1]

[1] Physics Department, Belarusan State University,
4 Nezavisimasti Ave., 220030 Minsk, Belarus
{grushevskaja,nina-kr,lipnevich,taisa-o}@bsu.by
[2] Belarusan State University of Informatics and Radioelectronica,
6 P. Brovki Str., 220013 Minsk, Belarus
shulitski@bsuir.by

Abstract. In this paper a high-sensitive capacitive DNA-nanosensor based on spin-dependent polarization effects has been proposed. We demonstrate that the polarization effects of charge-carriers transport in multi-walled carbon nanotubes (MWCNT) decorated by organometallic complexes lead to the surface-resonance-enhanced signals of DNA-hybridization sensor. According to obtained experimental data, such DNA-sensor allows to discover the forming duplex with DNA targets, including single-base-pair-mismatched DNA.

1 Introduction

At present, development of nanostructures for sensor devices which detect biomacromolecular binding is an actual problem. Utilizing of nanostructures allows to register reagents in molecular recognition reactions at ultra-low concentrations without using of any type of labels, so called "label-free"-detection via a surface plasmon resonance, for example, in oligonucleotide-capped gold nanoparticle [1]. Since the intensity of signal, which is transduced by single nanostructure, is not high enough, arrays of nanostructures have to be used. These arrays have to be previously arranged to decrease experimental data dispersion stipulated by multipoles interaction between nanostructures themselves. As for example, an "antigen-antibody"-interaction signal transducer, which consists of a layer of silver nanowires with diameter of about 60 nm and a layer of silver nanoribbons with a width of about 500 nm, allows to detect specific binding up to limiting antigen concentrations about 300 nM under conditions of "label-free"-detection [2]. An effective medical diagnostics needs DNA-nanosensors for specific detection of molecular recognition process *in vitro* and *in vivo* with sensitivity to analyzed molecules at ultra-low concentrations about pico- and femtoM and less [1]. To detect such ultra-low concentrations, superlattices with characteristic size

© Springer International Publishing Switzerland 2016
A. Bourgeois et al. (Eds.): ISBRA 2016, LNBI 9683, pp. 52–66, 2016.
DOI: 10.1007/978-3-319-38782-6_5

$\leq 2\,\text{nm}$ have to be fabricated. Because of weak interactions between components of such a nanosystem, for example, Van der Waals or stacking $\pi - \pi$-electron interactions, the system emerges and gains stability due to cooperative effects in a self-organizing process. Therefore, nanotechnologies, by means of which such superlattices can be produced, are the Langmuir-Blodgett (LB) technique and the method of molecules deposition on a coating, on which these molecules self-assemble into a monolayer [3]. The sensitivity of DNA sensors functioning on the base of π-stacking interaction of fluorescein-dye-labelled DNA-probes with MWCNT and graphene is of about 300 nM [4–6], whereas though usage of single-walled non-labelled CNT increases sensitivity to complimentary ssDNA [7], but restricts the recognition of base-pair nucleotide mismatch to five pairs [8]. Hydration of nucleic acids, including deoxyribonucleic acids (DNA), occurs in aqueous mediums [9]. At that, DNA electron density distribution is high sensitive to pH of the medium. Because of this, structural parameters of hydrated nucleic acids are hardly to be established [10].

Besides this, heat γ-radiation (irradiation by quanta of electro-magnetic field with the energy of about 10^{-13} eV/molecule and higher) can lead to γ-radiolysis of water with reactive oxygen species (ROS) formation, namely hydroxyl radicals $HO^{\cdot}$ and hydrogen peroxide H_2O_2 et al. [11,12].

Detection of DNA hybridization in living systems is realized in the presence of hydrogen peroxide H_2O_2 and hypo(pseudo)halous acids (for example, hypochlorous acid HOCl). HOCl reacts with water to form hydrochloric acid and H_2O_2. H_2O_2 produces hydroxyl radical in the Fenton reaction also [13]. Hydroxyl radicals damage DNA [14]. Electrochemical H_2O_2 biosensors are based on heme-proteins such as horseradish peroxidase [15]. However, since their low stability, at present, one utilizes fluorescence and nuclear or electron magnetic resonance (NMR or EPR) to detect ROS on base of chemoselective probes (see [16,17] and refs. therein).

Elecrtochemical sensors promise to provide simple, low cost, sensitive, and miniaturized platforms for DNA-based diagnostics. To date, most of the electrochemical DNA sensors demand a single-stranded (ss) nucleic acid probe sequence immobilization on electrode surface and utilizing of special redox-active labels to enhance their sensitivity and improve the signal-to-noise ratio [18,19]. Immobilization of ssDNA within recognition layer usually decreases the electrochemical DNA-sensor sensitivity due to the steric hindrance effect of the electrode surface and the loss of configurational freedom of the immobilized probe DNA. To avoid this, label-free and immobilization-free electrochemical sensors have been actively proposing [20–22].

Moreover, because of hydroxyl radicals neutrality, DNA hybridization detection faces difficulties at determination of hydroxyl radicals contribution in transducer signal. To get over the difficulties, one can design DNA-sensor based on magneto-electric effects in graphene-like materials [23].

Thus, the absence of DNA-sensors with control of $HO^{\cdot}$ low-concentration and, respectively, of DNA-oxidation level is a principal problem at recognition of DNA-markers including single nucleotide polymorphism (SNP) at ultra-low DNA-concentrations.

The nanosensors based on graphene and graphene-like materials are considered as the most perspective ones due to their high electro-conductivity in comparison with ordinary tree-dimensional metals [6,8,24,25]. Nanobiosensors formed from graphene and CNT in self-organizing processes are able to recognize a specific antigen at ultra-low concentrations [6]. Minimal conductivity $\frac{4e^2}{h}$ (conductivity quantum) of graphene and carbon nanotubes is about $154\,\mu S$ (respectively, resistivity is equal to 6 kOm for graphene and 6.5 kOm for CNT, that is less than resistivity of silver), whereas conductivity of a contact is much less, about 65 nS [26–28]. Here e is the electron charge, h is the Plank constant. However, utilizing of graphene and graphene-like materials is impeded by linearity of charge-carrier energy dispersion law which consists in the absence of charge carriers on the Fermi level for non-doped samples and in a shift of the Fermi level into hole or electron bands at addition of infinitely small amount of electrically charged impurities in material.

The goal of the paper is to show that it is possible to control and enhance signals of biomolecular binding on interphase boundary between a water subphase and self-organized LB-arrays of carbon nanotubes decorated by organometallic complexes, and then to propose a DNA-detection via a surface plasmon resonance combined with DNA-covered LB-MWCNT-array signal enhancement. This enhancement phenomenon will be used to determine reactive oxygen species at ultra-low concentrations and for nanosensorics of DNA-markers including single-base-mismatched DNA-target.

2   Materials and Methods

MWCNTs with diameters ranging from 2.0 to 5 nm and length of $\sim$2.5 μm were obtained by the method of chemical vapor deposition (CVD-method) [29]. Salts $Fe(NO_3)_3 \cdot 9H_2O$, $Ce_2(SO_4)_3$ (Sigma, USA), hydrochloric acid, deionized water were used to preparate subphases. Iron-containing films were fabricated from an amphiphilic oligomer of thiophene derivatives with chemically bounded hydrophobic 16-link hydrocarbon chain: 3-hexadecyl-2,5-di(thiophen-2-yl)-1H-pyrrole (H-DTP, H-dithiopyrrole). H-dithionilepyrrole was synthesized by a method proposed in [30]. Working solution of H-dithionilepyrrole, $1.0 \cdot 10^{-3}$ M, was prepared by dissolving precisely weighed substances in hexane. All salt solutions have been prepared with deionized water with resistivity $18.2\,M\Omega\cdot cm$. Complexes primer ssDNA/MWCNT were obtained by means of ultrasonic treatment of alcoholic solution of ssDNA with MWCNT [31,32]. All used materials belong to class of analytical pure reagents.

Langmuir - Blodgett monolayer formation was carried out on an automated hand-made Langmuir trough with controlled deposition on a substrate, and with computer user interface working under Microsoft Windows operational system. Control of the surface tension has been performed by a highly sensitive resonant inductive sensor. The Y-type transposition of monolayers on supports was performed by their vertical dipping. The complexes $Fe(II)DTP_3$ of high-spin Fe(II) with the dithionilepyrrole ligands were synthesized by LB-method at compression of H-dithionilepyrrole molecules on the surface of subphase with salts of three-valence Fe [33]. Horizontally and vertically arranged LB-MWCNT-bundles can be fabricated from the multi-walled CNTs [34]. We use the Langmuir–Blodgett technique to fabricate new layered nanoheterostructures consisting of two MWCNT/ssDNA LB-monolayers which are deposited on five-monolayer LB-film of the organometallic complexes.

For electrochemical studies, we use a planar capacitive sensor of interdigital-type on pyroceramics support. N pairs, $N = 20$, of aluminum electrodes of the sensors are arranged in an Archimedes-type spiral configuration. Every such pair is an "open type" capacitor. A dielectric coating of the electrodes represents itself nanoporous anodic alumina layer (AOA) with a pore diameter of 10 nm. The obtained LB-nanoheterostructures were suspended on the interdigital electrode system covered by dielectric nanoporous anodic alumina. To excite harmonic auto-oscillations of electric current (charge-discharge processes in the capacitors), the sensor was connected as the capacitance C into the relaxation RC-generator (self-excited oscillator) [35]. Operating of such RC-generator is based on the principle of self-excitation of an amplifier with a positive feedback on the quasi-resonance frequency. The capacitance C of the sensor entered in measuring RC-oscillating circuit has been calculated by the formula $C = 1/(2\pi R f)$, where R is the measuring resistance, f is the frequency of quasi-resonance.

3 Plasmon Resonance Phenomenon in Decorated Graphene and Graphene-Like Materials

Let consider the screening in true two-dimensional (2D) semimetals to which graphene and graphene-like materials belong. As known [36–38], free charge carriers at the Fermi level in these materials are absent. In single-pole approximation for small momenta p the condition of emerging of plasmon resonance can be found from a following p dependency of dielectric permeability $\epsilon(p,\omega)$ at a frequency ω [39]:

$$\epsilon(p,\omega) \approx 1 - \frac{1}{(\omega^2/\omega_{plasmon}^2) - p^2 r_D^2}, \quad p \ll 1 \qquad (1)$$

where $r_D = \left(\sum_a \frac{e_a^2 n_a}{\epsilon_0 k_B T}\right)^{-1/2}$ is the Debye screening radius, $\omega_{plasmon} = \left(\sum_a \frac{e_a^2 n_a}{\epsilon_0 m_a}\right)^{1/2}$ is a plasmon frequency, a, $a = 1, 2$ is a type of the charge carriers; n_a, m_a, and e_a are a density, a mass and a charge of a-th type, respectively; ϵ_0 is the dielectric constant, T is the temperature, k_B is the Boltzmann constant. Pseudofermion charge carriers in bipolar graphene-like two-dimensional material are massless: $m_a = 0$, and their charge density is vanishing: $n_a \to 0$ in Dirac point $K(K')$ of a Brillouin zone. Due to the fact that $r_D \to \infty$, and plasmon frequency gets a finite value $\omega_{plasmon}^2 = 2e^2\epsilon_0$, the right side of the condition (1) is always positive and, respectively, plasmon resonance in 2D system is absent. Now, one can examine two-dimensional semimetals decorated by impurity adsorbed metal atoms (adatoms). In Coulomb field of adatoms a band

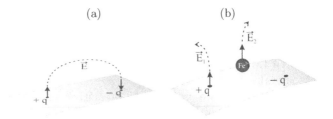

Fig. 1. Origin of screening effect in 2D Dirac metal (a) after decorating (b)

structure of the monolayer acquires the energy gap E_g, the mass of the charge carriers becomes non-zero: $m_a = E_g$, and on Fermi level there appears non-zero charge-carrier density $n_a = \delta n_F < 1$. Therefore, the dielectric permeability $\epsilon_{LB}(p, \omega)$ of the decorated system is determined by the following expression:

$$\epsilon_{LB}(p, \omega) \approx 1 - \frac{1}{(\epsilon_0 \omega^2 E_g/(2e^2 \delta n_F)) - 2p^2 \frac{2e^2 \delta n_F}{\epsilon_0 k_B T}}, \quad p \ll 1. \tag{2}$$

The expression (2) can acquire zero and negative values, corresponding to plasmon resonance. Figure 1 demonstrates the absence of screening and, respectively, plasmon oscillations for non-decorated 2D Dirac metals.

In what follows we will show that decorating of CNT by organometallic compounds provides such values of δn_F, at which CNTs effectively screen the Coulomb field.

4 Screening Effects

Near-electrode double electrically charged layer of interdigital capacity transducer, shown in Fig. 2, is similar to a plane capacitor [35, 40]. Due to the fact that the distance between plates of such capacitor is small, the field strength is high. Let us consider the screening of the near-electrode double electrically charged layer of such sensor by metal- and CNT-containing LB-monolayers. Metal atoms add impurity charge density into graphene-like monolayers of the LB-film. Therefore, in accord with the results of the previous section, free charge carriers with non-zero mass appear at the Fermi level. Addition of free charge carriers of conducting Ce- and/or Fe-containing LB-films into the electric density of near-electrode layer reveals as an observable screening effect (curves "0" and "1" in Fig. 2b,c).

Two LB-monolayers of carboxylated MWCNT have been deposited on transducer insulating layer, which has been previously modified by metal-containing LB-film of conducting thiophene-pyrrole series oligomer (dithionilepyrrole – 3-hexadecyl-2,5-di(thiophen-2-yl)-1H-pyrrole) [33, 41]. Note that CNTs decorated by the metallic complexes of dithionilepyrrole possess non-zero electron density at the Fermi level. Due to this fact the CTNs screen the electrical field of electrodes, as one can see from comparison of curves "1" and "2" in Fig. 2b,c.

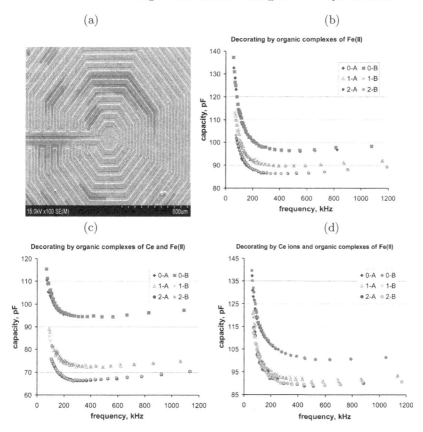

Fig. 2. Interdigital structure of electrodes (a) and cyclic frequency dependencies of capacity (b–d) of such sensors without (0) and with dithionilepyrrole LB-films (1) and MWCNTs (2) in deionized water. A and B denote direct and reverse branches, respectively.

A high value of low-frequency dielectric permeability of water is due to the ionization of water molecules and impurities in water with following formation of hydrated complexes of ions and of impurity ions. The existence of such complexes impedes the ion recombination. According to frequency dependencies of capacity C (curves "0" in Fig. 2), the sensor capacity is minimal at a frequency of oscillating alternative electric field (a plasmon resonance frequency) of the near-electrode layer. This frequency coincides with an eigenfrequency of hydrated-complex oscillations (ion vibrations). The plasmon resonance enhances a process of decay of hydrated complexes and, respectively, subsequent recombination of ions into neutral molecules H_2O and impurity molecules that results in decreasing of the dielectric permeability of the medium. The density of charge carriers in the double electrically charged layer, which does not modified by ultra-thin LB-film, is moderate because this resonance is observed as a wide band in dielectric spectrum with maximum in frequency range from 200 to 600 kHz.

Because of these screening effects are stipulated by sharp increasing of the double layer electric field strength, the probability of hydrated complexes decay with subsequent recombination into neutral molecules increases sharply. It leads to narrowing of the dielectric band and appearing of an explicit extremum at frequency 220 kHz in dielectric spectrum of CNTs decorated by Fe(II) complexes and at 270 kHz for CNTs decorated by complexes of Fe(II) and Ce atoms (curves "2" in Fig. 2a,b). The well-exhibited screening effect is due to decoration of all surface of MWCNTs by the complexes of Fe(II) and Ce atoms. An addition of impurity ions Ce^{3+} into water decreases the screening effect sharply, according the comparison of displacements of curves "1" and "2" respect to each other in Fig. 2d. It occurs because of reducing (up to full suppression) of the charge carrier mobility in graphene devices at decoration with small and large doses of adatoms, respectively [23].

Further, let us study magnetoelectric effects of spin-dependent dielectric polarization in CNTs decorated by complexes of dithionilepyrrole with metal atoms.

5   Spin-Dependent Polarization of Metal- and CNT-Containing LB-Coatings

As is known [23], the distribution of adatom electron density near the graphene sheet causes a local spin-orbit field, and, respectively, the scattering of current components with positive (negative) angular momentum is enhanced (suppressed) for charge carriers with a spin projection $s_z = +1/2$ ($s_z = -1/2$). Furthermore, a spin-orbit splitting of the band dispersion occurs by bringing heavy metal atoms in close contact to graphene.

Thus, the charge carriers scattering in LB-monolayer of CNT decorated by complexes of Ce and/or high-spin Fe(II) can be spin-dependent. Vector of spin-dependent polarization of LB-coating will precess in alternative magnetic field of the sensor. Spin precession is a quantum phenomenon consisting in quantization of levels of a system in a magnetic field. Dependencies of capacity upon ions concentration for the sensor, electrodes of which have been modified by LB-coating or persist unmodified, are represented in Fig. 3. According to experimental data, the LB-coatings screen an electric field of the near-electrode double layer, significantly decreasing the capacitance value and, therefore, not allowing appearance of break-down voltage. These dependencies have at least three inflection points. The stepwise dependence of the sensor capacity is due to a quantized low-frequency Maxwell-Wagner polarization of the spin-polarized LB-coating.

If atoms of reagents with lone-electron pairs have unpaired electrons, a magnetic field produced by their non-zero magnetic moments causes a precession of polarization vector coupled with spin polarization of a sample. This magneto-electric effect will be higher, the more non-paired electrons are and, respectively, the higher reagent concentration. Because of this, in subsequent we will study the enhancement of the concentration signal of a sensor for substances, decay of which leads to formation of neutral $HO^{\cdot}$.

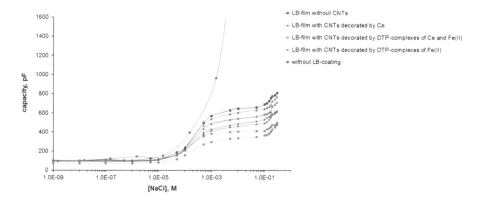

Fig. 3. Concentration dependencies of capacity of sensors without and with Fe-containing dithionilepyrrolle LB-film and MWCNTs at frequency 150 kHz.

6 Reactive Oxygen Species Influence on Sensor Response

Due to large spin relaxation time and spin polarization of charge carrier current components in graphene and graphene-like materials, electronic graphene devices can be utilized to detect magnetotransport [42]. Further we show that large spin relaxation time for graphene-like materials allows to detect ROS at ultra-low concentrations. To do this, we study sensor response at different concentrations of H_2O_2 and NaOCl, hydrolysis of the last leads to generation of HOCl [43]. Addition of ROS results in increase of the capacity in low-frequency range, as one can see in Fig. 4a. At that, Garnett law holds for Maxwell-Wagner dielectric polarization of a mixture [44,45] (Fig. 4b, c and d).

Using experimental data, we estimate sensitivity of the LB-array of CNTs decorated by organometallic complexes. Let C and C_0 denote capacities of the sensor in media with and without ROS. Then, estimation can be obtained by intersection of the straight-line dependence of a difference $(C - C_0)$ logarithm and logarithmic axis corresponding to ROS concentration. The sensor sensitivity depends on a measurement frequency. Maximal sensitivity of a sensor to ROS must be at a frequency of the plasmon resonance. At frequencies near the plasmon resonance, which have been obtained by using measuring resistance $R = 1.96$ kOm, the sensitivity to H_2O_2 has been observed of about 10^{-15} M and less (Fig. 4b) for sensor put in salt solution (Earle medium). Far from the resonance, that corresponds to measuring resistances $R = 20$ and 10 kOm, the sensitivity decreases up to 10^{-12} M (Fig. 4b). For deionized water, the limiting concentration sensitivity to H_2O_2 has an order of 10^{-14} M, according to Fig. 4d. One can suppose that salt ions in water reduce free charge carriers density δn_F due to partial electric neutralizing of metallocenters. Therefore, according the expression (2), dielectric permeability decreases and sensor sensitivity increases. Moreover, the experimental data for neutral molecules NaOCl allow to conclude that NaOCl decay and, respectively, appearing of hydroxyl radicals $HO^\cdot$ occurs at large enough concentration of this reagent.

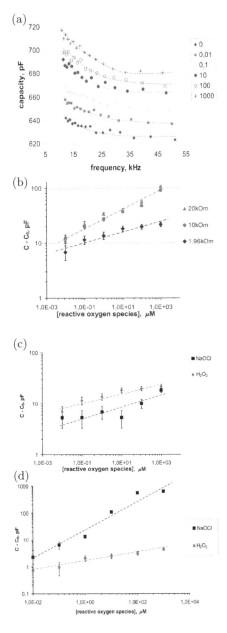

Fig. 4. Frequency-capacity dependencies for sensor in Earle solution (0) at addition of hydrogen peroxide in concentrations 0.01, 0.1, 10, 10^2, and 10^3 μM (a). Concentration dependence of capacity increment for sensor in Earle solution (b,c) and in deionized water (d) at addition of hydrogen peroxide (b–d) and sodium hypochlorite (c,d). Measurements have been carried out at measuring-resistor values $R = 20$ (b, d), 10 (b), 1.96 (b and c, H_2O_2), 1.89 (c, NaOCl) kOm. C and C_0 are sensor capacities in presence and absence of ROS, respectively.

7   DNA-Hybridization Signal Surface-Enhancement Phenomenon

LB-monolayers consisting of ssDNA/MWCNT complexes have been formed by Langmuir-Blodgett technique from hydrophobic micelles of stearic acid with ssDNA/MWCNT complexes inside (Fig. 5). Two of such monolayers were deposited on near-electrode insulating layers modified by metal-containing dithionilepyrrole LB-film. Micelles were obtained preliminary by ultrasonic irradiating of the mixture from ssDNA-probe and carboxylated MWCNT (Fig. 5) in hexane. DNA, which is compactified due to stacking π-electron interaction between nitrogenous bases of DNA and carboxylated MWCNT, represents itself an electron-dense layer of oligonucleotide (primer ssDNA), on which the epitaxial growth of DNA-crystals can occur [31, 46]. The obtained data of ssDNA/MWCNT-complex structural and diffraction analysis allow to propose a model of complementary or non-complementary binding of ssDNA-probe with ssDNA-target, which is represented in Fig. 6.

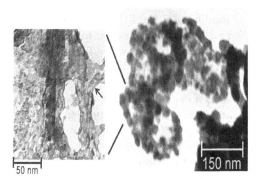 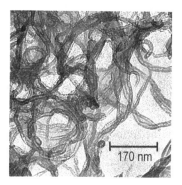

Fig. 5. Transmission electron microscopy (TEM) images of hydrophobic stearic acid micelles with DNA compactified on MWCNT surface, inside; arrow on insert directs on one of CNTs. Micella is located on the edge of etching nanoporous alumina membrane (left). TEM-image of original MWCNTs (right).

Carbon nanotubes quench the fluorescence of FAM-type dyes (FAM - fluorescein phosphoramidites) [7, 32]. This property of CNT can be used to study the process of DNA-hybridization on the DNA/CNT-complex surface. Electrophoretic data of FAM-labeled primer ssDNA binding with different ssDNA-targets are shown in Fig. 7a. They demonstrate that utilizing of carboxylated MWCNT makes it possible to recognize single nucleotide polymorphism (SNP).

Now, we apply this ssDNA/MWCNT-complex to recognize DNA-targets by means of the electrophysical method. The proposed capacitive DNA-sensor detects forming homo- and heteroduplexes including heteroduplexes probe ssDNA/single-base-pair-mismatched oligonucleotide. The determination is due

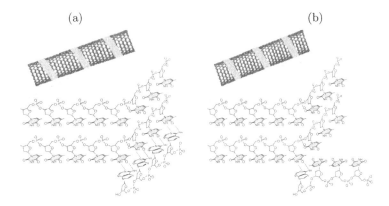

Fig. 6. Model of complementary (a) or non-complementary (b) binding of ssDNA-probe with ssDNA-target.

to above considered effect of resonance-enhanced spin-dependent Maxwell-Wagner depolarization/polarization of interphase boundary in electric field of electrodes, insulating coating of which is modified by the LB-films of MWCNT decorated by the organometallic complexes. Further we discuss results of the electrochemical analysis concerning the degree of binding between different-types unlabeled dsDNA and MWCNTs on the surface of the LB-film.

In deionized water the dielectric properties of MWCNT-LB-films and oligonucleotide/MWCNT-complexes reveal as a capacity change of the charged double Helmholtz layer formed on the interface – insulating barrier layer of anodic alumina – water. Two frequency ranges ($\sim$100 kHz – low-frequency one being in the vicinity of plasmon resonance, and $\sim$400–800 kHz – high-frequency one being far from resonance) have been used for the electrochemical detection. In the low-frequency range the main contribution into film polarization is given by surface-enhanced Maxwell-Wagner polarization of conducting impurity inclusions. According to DNA-sensor model in Fig. 6, far from the resonance the number of hydrated nitrogenous bases and, respectively, the dipole polarization vector for the hydrate complex of heteroduplex DNA-probe/non-complementary DNA-target are larger than for the homoduplex DNA-probe/complementary DNA-target. Because of this, the dielectric response is higher in the case of non-complementary binding than in the case of complementary one.

Let us define a degree of heterogeneity s_h of biosensitive coating as a ratio of difference between capacities $C_{\text{ssDNA/CNT}}$ and $C_{\text{dsDNA/CNT}}$ for DNA-sensors without and with DNA marker, respectively, to difference between $C_{\text{ssDNA/CNT}}$ and C_{CNT} for the sensor without DNA probe:

$$s_h = \frac{C_{\text{ssDNA/CNT}} - C_{\text{dsDNA/CNT}}}{C_{\text{ssDNA/CNT}} - C_{\text{CNT}}}. \tag{3}$$

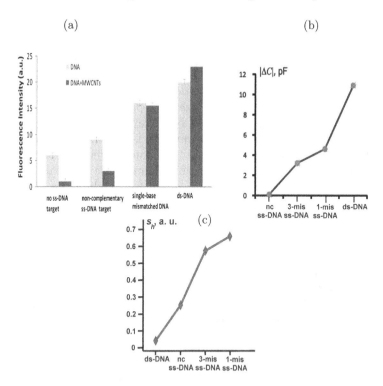

Fig. 7. (a) Intensity of FAM-ssDNA (light bars) and complexes FAM-ssDNA/MWCNT (dark bars) fluorescence in the absence of ssDNA-target, at the presence non-complementary ssDNA-target and single-base mismatched target, and at the presence of perfect complementary ssDNA (left diagram). (b) Absolute value of low-frequency sensor capacity change $|\Delta C|$, $\Delta C < 0$ at interaction of DNA probe with perfect complementary one, and non-complementary (nc ssDNA), three-base mismatched, and single-base mismatched DNA markers (diagram in center). (c) Degree s_h of sensor coating heterogeneity for different types of ssDNA-target at frequency 700 kHz. ds-DNA denotes homoduplex, non-complementary, three-base mismatched, and single-base mismatched DNA markers are denoted by symbols "nc ss-DNA", "3-mis ss-DNA", and "1-mis ss-DNA", respectively (right diagram).

The experimental data of s_h are shown in Fig. 7c. As one can see, the probability of heteroduplex formation decreases with the increase of mismatched-nucleotide numbers. Due to this, the capacity increment $|\Delta C_m|$ of DNA-sensor placed in solution of m-base-pair-mismatched DNA-marker is less than ΔC_n for n-base-pair-mismatched DNA-marker at $n < m$. It is known [47], spin-dependent dielectric polarization is absent at the helicoidal structure in a sample. Because of this, a sharp decrease of low-frequency sensor capacity value in Fig. 7c at homoduplex formation is due to the helicoidal structure of DNA-probe/complementary DNA-marker.

8   Conclusion

So, a capacitive DNA-nanosensor based on spin-dependent polarization effects for high-performance sensing of DNA hybridization has been proposed. The new method of hybridization assay allows to detect the interaction between DNA-probe and DNA marker, including single-base-pair-mismatched DNA-target occuring on surface of self-organized MWCNT-arrays decorated by organometallic complexes. The DNA-sensorics of formating homo- and heteroduplexes is based on the studied above phenomenon of resonance-enhanced spin-dependent Maxwell-Wagner polarization/depolarization on interphase boundary in the electric field of electrodes, insulating coating of which is modified by LB-film consisting of carboxilated MWCNT decorated by organometallic complexes.

References

1. Yao, X., Li, X., Toledo, F., Zurita-Lopez, C., Gutova, M., Momand, J., Zhou, F.M.: Sub-attomole oligonucleotide and p53 cDNA determinations via a high-resolution surface plasmon resonance combined with oligonucleotide-capped gold nanoparticle signal amplification. Anal. Biochem. **354**, 220–224 (2006)
2. Lee, K.-L., You, M.-L., Tsai, C.-L., Hung, C.-Y., Hsieh, S.-Y., Wei, P.-K.: Visualization of biosensors using enhanced surface plasmon resonances in capped silver nanostructures. Analyst **141**, 974–980 (2016)
3. Ozin, G.A., Hou, K., Lotsch, B.V., Cademartiri, L., Puzzo, D.P., Scotognella, F., Ghadimi, A., Thomson, J.: Nanofabrication by self-assembly. Mater. Today **12**, 12–23 (2009)
4. Li, H., Tian, J., Wang, L., Zhang, Y., Sun, X.: Multi-walled carbon nanotubes as an effective fluorescent sensing platform for nucleic acid detection. J. Mater. Chem. **21**, 824–828 (2011)
5. Lu, C.-H., Yang, H.-H., Zhu, C.-L., Chen, X., Chen, G.-N.: A graphene platform for sensing biomolecules. Angew. Chem. Int. Ed. **48**, 4785–4787 (2009)
6. Zhang, B., Cui, T.: An ultrasensitive and low-cost graphene sensor based on layer-by-layer nano self-assembly. Appl. Phys. Lett. **98**, 073116 (2011)
7. Zhu, Z., Yang, R., You, M., Zhang, X., Wu, Y., Tan, W.: Single-walled carbon nanotube as an effective quencher. Anal. Bioanal. Chem. **396**, 73–83 (2010)
8. Bansal, J., Singh, I., Bhatnagar, P.K., Mathur, P.C.: DNA sequence detection based on Raman spectroscopy using single walled carbon nanotube. J. Biosci. Bioeng. **115**, 438–441 (2013)
9. Vardevanyan, P.O., Karapetyan, R.A., Parsadanyan, M.A.: Pequliarities of interaction of ethidium bromide with different GC-content DNAs. Biol. J. Armenia **4**(62), 6–11 (2010)
10. Zhizhina, G.P., Oleynik, E.F.: Infrared spectroscopy of nucleic acid. Chem. Usp. **41**, 475–511 (1972)
11. Milinchuk, V.K.: Radiation chemistry. Soros Educ. J. Chem. **6**(4), 24–29 (2000)
12. Doshi, R., Day, P.J.R., Tirelli, N.: Dissolved oxygen alteration of the spectrophotometric analysis and quantification of nucleic acid solutions. Biochem. Soc. Trans. **37**(2), 466–470 (2009)
13. Prousek, J.: Fenton chemistry in biology and medicine. Pure Appl. Chem. **79**, 2325–2338 (2007)

14. Brooker, R.J.: Genetics: Analysis and Principles, 4th edn. McGraw-Hill Science, Columbus (2011)
15. Calas-Blanchard, C., Catanante, G., Noguer, T.: Electrochemical sensor and biosensor strategies for ROS/RNS detection in biological systems. Electroanalysis **26**, 1–10 (2014)
16. Burns, J., Cooper, W., Ferry, J., King, D.W., DiMento, B., McNeill, K., Miller, C., Miller, W., Peake, B., Rusak, S., Rose, A., Waite, T.: Methods for reactive oxygen species (ROS) detection. Aquat. Sci. **74**, 683–734 (2012)
17. Lo Conte, M., Carroll, K.S.: The chemistry of thiol oxidation and detection. In: Jakob, U., Reichmann, D. (eds.) Oxidative Stress and Redox Regulation, pp. 1–42. Springer Science+Business Media Dordrecht, Dordrecht (2013)
18. Drummond, T.G., Hill, M.G., Barton, J.K.: Electrochemical DNA sensors. Nat. Biotechnol. **21**, 1192–1199 (2003)
19. Kelley, S., Boon, E., Barton, J., Jackson, N., Hill, M.: Single-base mismatch detection based on charge transduction through DNA. Nucleic Acids Res. **27**, 4830–4837 (1999)
20. Kafka, J., Panke, O., Abendroth, B., Lisdat, F.: A label-free DNA sensor based on impedance spectroscopy. Electrochim. Acta **53**, 7467–7474 (2008)
21. Kongpeth, J., Jampasa, S., Chaumpluk, P., Chailapakul, O., Vilaivan, T.: Immobilization-free electrochemical DNA detection with anthraquinone-labeled pyrrolidinyl peptide nucleic acid probe. Talanta **146**, 318–325 (2016)
22. Luo, X., Hsing, I.-M.: Immobilization-free multiplex electrochemical DNA and SNP detection. Biosens. Bioelectron. **25**, 803–808 (2009)
23. Ferreira, A., Rappoport, T.G., Cazalilla, M.A., Neto, A.H.C.: Extrinsic spin hall effect induced by resonant skew scattering in graphene. Phys. Rev. Lett. **112**, 066601 (2014)
24. Fortina, P., Wang, J., Surrey, S., Park, J.Y., Kricka, L.J.: Beyond microtechnology–nanotechnology in molecular diagnosis. In: Liu, R.H., Lee, A.P. (eds.) Integrated Biochips for DNA Analysis, pp. 187–197. Landes Bioscience and Springer, New York (2007). Chapter 13
25. Rahman, M.M., Li, X.-B., Lopa, N.S., Ahn, S.J., Lee, J.-J.: Electrochemical DNA hybridization sensors based on conducting polymers. Sensors **15**, 3801–3829 (2015)
26. Suzdalev, I.P.: Nanothechnology: Physical Chemistry of Nanoclusters, Nanostructures, and Nanomaterials. KomKniga, Moscow, p. 383 (2006)
27. Blake, P., Yang, R., Morozova, S.V., Schedin, F., Ponomarenko, L.A., Zhukov, A.A., Nair, R.R., Grigorieva, I.V., Novoselov, K.S., Geim, A.K.: Influence of metal contacts and charge inhomogeneity on transport properties of graphene near the neutrality point. Solid State Commun. **149**, 1068–1071 (2009)
28. Anantram, M.P., Leonard, F.: Physics of carbon nanotube electronic devices. Rep. Prog. Phys. **69**, 507–561 (2006)
29. Labunov, V., Shulitski, B., Prudnikava, A., Shaman, Y.P., Basaev, A.S.: Composite nanostructure of vertically aligned carbon nanotube array and planar graphite layer obtained by the injection CVD method. Quantum Electron. Optoelectron. **13**, 137–141 (2010)
30. Kel'in, A., Kulinkovich, O.: Folia pharm. Univ. Carol. (supplementum), 18, 96 (1995)
31. Egorov, A.S., Egorova, V.P., Krot, V.I., Krylova, H.V., Lakhvich, F.F., Lipnevich, I.V., Orekhovskaya, T.I., Veligura, A.A., Govorov, M.I., Shulitski, B.G., Ulashchik, V.S.: Electrochemical and electrophoretic detection of hybridization on DNA/carbon nanotubes: SNP genetic typing. Herald Found. Basic Res. **3**(14), 63–87 (2014)

32. Veligura, A.A., Egorov, A.S., Egorova, V.P., Krylova, H.V., Lipnevich, I.V., Shulitsky, B.G.: Fluiorofor quenching in oligonucleotide layers self-organized on carbon nanotubes. Uzhhorod Univ. Sci. Herald Ser. Phys. **34**, 206–210 (2013)

33. Grushevskaya, H.V., Lipnevich, I.V., Orekhovskaya, T.I.: Coordination interaction between rare earth and/or transition metal centers and thiophene series oligomer derivatives in ultrathin langmuir-blodgett films. J. Mod. Phys. **4**, 7–17 (2013)

34. Egorov, A.S., Krylova, H.V., Lipnevich, I.V., Shulitsky, B.G., Baran, L.V., Gusakova, S.V., Govorov, M.I.: A Structure of modified multi-walled carbon nanotube clusters on conducting organometallic langmuir - blodgett films. J. Nonlinear Phenom. Complex Sys. **15**, 121 (2012)

35. Abramov, I.I., Hrushevski, V.V., Krylov, G.G., Krylova, H.V., Lipnevich, I.V., Orekhovskaya, T.I.: Method to calculate impedance of interdigital capacity sensor. St. Petersburg J. Electron. **4**(73), 59–67 (2012)

36. Morozov, S.V., Novoselov, K.S., Geim, A.K.: Electron transport in graphene. Phys. Usp. **178**, 776–780 (2008)

37. Grushevskaya, H.V., Krylov, G.: Partially breaking pseudo-dirac band symmetry in graphene. J. Nonlinear Phenom. Complex Sys. **17**, 86–96 (2014)

38. Grushevskaya, H.V., Krylov, G.G.: Beyond the massless dirac's fermion approach. In: Bonĉa, J., Kruchinin, S. (eds.) Nanotechnology in the Security Systems. NATO Science for Peace and Security Series C: Environmental Security, pp. 21–31. Springer Science+Business Media, Dordrecht (2015)

39. Kraeft, V.D., Kremp, D., Ebeling, W., Röpke, G.: Quantum statistics of charged particle systems. Akademie-Verlag, Berlin (1986)

40. Krylova, H.V., Drapeza, A.I., Khmelnitski, A.I.: Electrical field of open capacity. Devices Control Syst. **6**, 43–44 (1993)

41. Golubeva, E.N., Grushevskaya, H.V., Krylova, N.G., Lipnevich, I.V., Orekhovskaya, T.I.: Raman scattering in conducting metal-organic films deposited on nanoporous anodic alumina. J. Phys. CS. **541**, 012070 (2014)

42. Andrei, E.Y., Li, G., DuRep, X.: Electronic properties of graphene: a perspective from scanning tunneling microscopy and magnetotransport. Prog. Phys. **75**, 056501 (2012)

43. Estrela, C., Estrela, C.R.A., Barbin, E.L., Spano, J.C.E., Marchesan, M.A., Pecora, J.D.: Mechanism of action of sodium hypochlorite. Braz Dent J. **13**(2), 113–117 (2002)

44. Garnett, J.C.M.: Philos. Trans. R. Soc. Lond. 205, 237 (1906)

45. Vinogradov, A.P.: Electrodynamics of Composite Materials. URSS, Moscow (2001). p. 63

46. Egorova, V.P., Krylova, H.V., Lipnevich, I.V., Veligura, A.A., Shulitsky, B.G., Fedotenkova, L.Y.: Molecular beacon CNT-based detection of SNPs. J. Phys. CS. **643**, 012023 (2015)

47. Catalan, G., Scott, J.F.: Physics and applications of bismuth ferrite. Adv. Mater. **21**, 2463–2485 (2009)

Towards a More Accurate Error Model
for BioNano Optical Maps

Menglu Li[1], Angel C.Y. Mak[2], Ernest T. Lam[3], Pui-Yan Kwok[2], Ming Xiao[4],
Kevin Y. Yip[5], Ting-Fung Chan[6], and Siu-Ming Yiu[1(✉)]

[1] Department of Computer Science, The University of Hong Kong,
Hong Kong, China
smyiu@cs.hku.hk
[2] Cardiovascular Research Institute, University of California,
San Fransisco, USA
[3] BioNano Genomics, San Diego, USA
[4] School of Biomedical Engineering, Science and Health Systems,
Drexel University, Philadelphia, USA
[5] Department of Computer Science and Engineering,
The Chinese University of Hong Kong, Hong Kong, China
[6] School of Life Sciences, The Chinese Univeristy of Hong Kong,
Hong Kong, China

Abstract. Next-generation sequencing technologies has advanced our
knowledge in genomics by a tremendous step in the past years. On the
other hand, there are still critical questions left unanswered due to the
intrinsic limitations of short read length. To address this issue, several
new sequencing platforms came into view. However, a lack of comprehensive understanding of the sequencing error poses a primary challenge for
their optimal use. Here, we focus on optical mapping, a high-throughput
laboratory technique that provides long-range information of a genome.
Existing error model is based on OpGen maps. It is not clear if the model
is also good for BioNano maps. In this paper, we try to provide a more
accurate error model for BioNano optical maps based on real data. Due
to the limited availability of real datasets, as an indirect validation of
our model, we predict the regions that are difficult to cover and compare the predicted results with the empirical results (both simulated and
real data) on human chromosomes. The results are promising, with most
of the difficult regions correctly predicted. Tested on BioNano maps,
our model is more accurate than the most popular existing error model
developed based on OpGen maps. Although we may not have captured
all possible errors of the technology, our model should provide important
insights for the development of downstream analysis tools using BioNano
optical maps.

Keywords: Optical mapping · Error model · Difficult to cover regions

1 Introduction

In recent years, next-generation sequencing technologies (e.g. short reads from
Solexa) enabled researchers to discover many critical findings in genomics. On the

© Springer International Publishing Switzerland 2016
A. Bourgeois et al. (Eds.): ISBRA 2016, LNBI 9683, pp. 67–79, 2016.
DOI: 10.1007/978-3-319-38782-6_6

other hand, it is believed that the intrinsic limitations of these technologies, in particular, short read length, are a major obstacle for future research. Many sequencing platforms have been developed in recent years to tackle this issue. However, a lack of comprehensive understanding of the sequencing errors poses a primary challenge for their effective use. Among these new high-throughput technologies, optical mapping (OM) is one that produces high-resolution restriction maps at no risk of PCR artifacts. Each restriction map, known as an "optical map", represents a stained DNA molecule digested by a given restriction enzyme and imaged by optics. Consequently, an optical map can be regarded as an ordered sequence of lengths, each corresponding to the distance between two adjacent digested sites. On average, an optical map spans 200,000 bases, which is much longer than short reads of a few hundred bases.

Long-range information contributes significantly to a more comprehensive analysis of genomes. It is crucial to the connectivity of genome assembly, sequence alignments in repeat regions and the detection of long structural variants. With long spans, optical maps have been applied either alone or together with short reads to genome assembly [2,7,12,19,21] and sequence alignments [10,11,16,18,20], bringing in insights in explorations of bacteria [12,13], animal [4], plant [5,21,22] and human [8,14,17] genomes. To achieve better accuracy in these studies, it is important to understand the inherent errors during the production of optical maps.

The modern optical mapping system works as follows. Single molecules of DNA are placed onto a nanofluidic platform. With nanoconfinement, molecules are unraveled and stretched uniformly towards their full contour length. When they dock on a slide electrostatically, restriction enzymes are added to cleave the molecules at specific recognition sites, leaving gaps identifiable under the microscope. The resulting DNA fragments are stained by fluorescent dyes and are imaged through microscope. Lengths of these fragments are measured by the integrated fluorescence intensity, comprising optical maps. Details may differ from platform to platform. For example, the OpGen Argus System follows the above workflow, while the BioNano Genomics Irys System applies enzyme before the elongation of molecules. Furthermore, Irys System uses nicking enzymes that create single-strand cuts rather than cleaving enzymes that cut on both strands at its restriction sites. Repaired by fluorescently labelled nucleotides, the nicking sites appear as identifiable labels instead of gaps under microscope. Keeping the molecule intact, this strategy keeps short fragments from being flushed away.

Various types of errors are involved in different stages of the procedure. Lengths of fragments are estimated by comparing the observed fluorescent intensity with that of a molecule with known size and intensity. Because a fragment is not guaranteed to stretch perfectly to its full contour length,the observed fragment length may deviate from its true underlying length. Meng et al. [9] concluded from experiments on lambda bacteriophage genome that shorter fragments tend to have greater sizing error. The restriction enzyme is another source of inaccuracy. There are a small fraction of missing sites because some restriction sites may not be fully digested by the enzyme, while false sites appear when the enzyme cleaves at wrong sites.

Although errors in optical mapping were addressed by some previous studies, most of the studies, if not all, are based on OpGen maps and some were not validated using real data. For examples, Anatharaman el al. [1] discussed sizing error, false sites and missing sites in error modeling; however, the validity of their model was not examined with real data. Valouev et al. [18] proposed a popular representation of sizing error e, and showed the resemblance between e of the Kim strain of *Y. Pestis* and Normal distribution. Their error model was prevalently adopted in subsequent investigations and continues to be the most popular model. Taking it a step further, Sarkar et al. [15] derived a new error model based on the GM075535 data set. In their study, Sarkar found that the sizing error of optical maps has a heavier tail than a Normal distribution as what Anatharaman and Valouev assumed; however, Sarkar only presented a few fragment-length Q-Q plots of his simulated data and real data, which may not provide enough evidence to support the accuracy of his model[1]. Zhou el al. [21] had a similar hitch that they only validated their error model on rice genome optical maps with fragment sizes. We remark that despite the limitations of these studies (probably due to the limited availability of real optical map data at that moment), they laid down a good standing point for the study in this paper. In this paper, we want to focus on the raw maps produced from BioNano as more and more institutions are now using BioNano maps. The aim is to derive a more accurate model for BioNano maps.

Our study is based on a trio of samples (NA12878, NA12891, and NA12892) from the CEU collection. Long DNA molecules from the cell lines are digested by nicking endonuclease Nt.BspQI with BioNano IrysPrep Reagents. An *in silico* reference map for GRCh38 genome is generated by detecting the BspQI restriction sites (GCTCTTCN̂). The trio maps are aligned to the GRCh38 map using BioNano RefAligner. By studying this alignment result, we derive an error model. Using the trio samples and simulated data, we demonstrate that our error model fits optical maps generated by BioNano Genomics Irys System much better than the previous model. To further validate our result, based on our error model, we try to predict the regions in the human genome that are difficult to cover by optical maps and compare our results with the empirical alignment results on both simulated and real datasets. The results are consistent and promising. Most of the predicted regions overlap with the empirical results. The rest of the paper is organized as follows. Section 2 provides the details of the proposed error model. We also show how we can compute a probability that captures how likely a particular region may not be covered by the data. It is a difficult task to evaluate such an error model with limited availability of real data. Our evaluation is based on alignment results of RefAligner. With proper settings, the alignment results achieves an accuracy of 98.88 %, which should provide enough evidence for our evaluation. The details of evaluation are given in Sect. 3. Finally, we discuss the applicability and limitations of this error model in Sect. 4.

---

[1] Fragment size distribution mainly depends on the distribution of nicking sites, thus may not be a strong evidence to show the accuracy of an error model.

2   Methods

2.1   Error Model

An optical map can be represented by a vector of fragment lengths, i.e., $O = (o_1, o_2, \ldots, o_m)$, where a fragment is bounded by a pair of digested recognition sites visible under the microscope. This map originates from a region $R = (r_1, r_2, \ldots, r_n) = \text{Ref}[i \ldots j] = (\text{ref}_i, \text{ref}_{i+1}, \ldots, \text{ref}_j)$ on the whole reference map $\text{Ref} = (\text{ref}_1, \text{ref}_2, \ldots, \text{ref}_N)$, with some random errors introduced during the wet bench procedures. We will explore each type of error in more detail. Note that the following study is based on the CEU trio dataset, which consists of human optical maps of three individuals, namely NA12878, NA12891, NA12892. The parameters of the error model in this section are derived from NA12878 by maximum likelihood estimation. We also present how such an error model fits the optical maps of NA12891 (a similar result was also obtained on NA12892).

Sizing Error. The length of a fragment is observed from microscopy images. Due to limited resolution of microscopy and imprecision of image processing, o_k's ($1 \leq k \leq m$) may not be exactly their corresponding reference length. Assuming no site error, the expected value of $\frac{o_k}{r_k}$ is around 1.

Valouev el al. [18] concluded that $o_k \sim N(r_k, \sigma^2 r_k)$, or equivalently $e_k = \frac{o_k - r_k}{\sqrt{r_k}} \sim N(0, \sigma)$, where $\sigma = \frac{\mu}{\gamma}$, μ and γ are the mean and standard deviation of fluorescent intensity per nucleotide. This conclusion holds when the length of fragments is determined by $\frac{W}{U}$, where W is the accumulation of fluorescence intensity and U is the intensity per unit length. However, with the advances in nanochannel, microscopy and imaging technology, recent studies [3,6] have adopted various image processing techniques to calculate the length of fragments, which may deviate from $\frac{W}{U}$. As a result, Fig. 1, which illustrates how sizing error $e_k = \frac{o_k - r_k}{\sqrt{r_k}}$ of NA12878 optical maps fits a Normal distribution, shows that Valouev's model for sizing error is not very consistent with optical maps generated by Irys System.

For any molecule fragment that originates from a reference fragment of length r_k but is observed to be o_k in length, its sizing error is defined as $s_k = \frac{o_k}{r_k}$. According to CEU trio dataset, with the deviations from the Normal distribution at the tails, it triggers us to consider a Laplace distribution for s_k, that is,

$$s_k \sim \text{Laplace}(\mu, \beta),$$

where μ is the location parameter and β is the scale parameter.

More specifically, it has been extensively observed that shorter fragments are subject to greater sizing error. Also, Valouev et al. [18] noticed that sizing error of short fragments ($< 4\,\text{kb}$) does not converge to a normal distribution, so for short fragments, sizing error was modeled as an additive error that is irrelevant to the underlying fragment length. Lin et al. [7] combined a relative error and an additive error to characterize the sizing error. As shown in Fig. 2, we studied

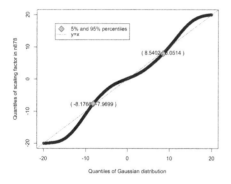

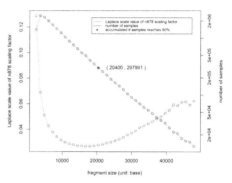

Fig. 1. The Q-Q plot of e_k's of NA-12878 maps ($3600\,\mathrm{bp} \le r_k < 4800\,\mathrm{bp}$) against Normal $(0.1921347, 5.074905)$ (based on Valouev's model).

Fig. 2. The dispersion of the sizing error of NA12878 optical maps against fragment length. (Color figure online)

the dispersion of sizing error against fragment length by looking at the Laplace scale values.[2] The red curve indicates that the sizing error of fragments of length $[1200\,\mathrm{bp}, 2400\,\mathrm{bp})$, $[2400\,\mathrm{bp}, 3600\,\mathrm{bp})$ and $[3600\,\mathrm{bp}, 4800\,\mathrm{bp})$ are less centralized than that of longer fragments of length $(4800\,\mathrm{bp}, \infty)$. Here, $1200\,\mathrm{bp}$ is chosen as the interval unit because it is the imaging resolution in our experimental settings. Consequently, based on different levels of r_k lengths, values of μ and β were estimated separately as below, forming four similar distributions.

$$(\mu, \beta) = \begin{cases} (0.858181, 0.180196), & 1200\,\mathrm{bp} \le r_k < 2400\,\mathrm{bp} \\ (0.980760, 0.071176), & 2400\,\mathrm{bp} \le r_k < 3600\,\mathrm{bp} \\ (1.003354, 0.052800), & 3600\,\mathrm{bp} \le r_k < 4800\,\mathrm{bp} \\ (1.00482, 0.042428), & r_k \ge 4800\,\mathrm{bp} \end{cases}$$

From the perspective of cumulative density and quantile distribution, Fig. 3 exhibits that the scaling factor distribution ($3600\,\mathrm{bp} \le r_k < 4800\,\mathrm{bp}$) estimated from NA12878 fits well with optical maps of NA12891.

Missing Cuts. When a restriction site is incompletely digested by the enzyme, the two flanking fragments will appear concatenated under the microscope. Such restriction sites are called missing cuts. It is natural to regard whether a restriction site is digested or not as a Bernoulli trial, with a probability of p_{digest} to succeed. Upon further investigation, p_{digest} is found to decline when two nearby nicking sites are closer to each other. The red curve in Fig. 4(a) shows that p_{digest} rises as the average distance between this site and its neighbors (denoted as d_{avg})

---

[2] We noticed that the Laplace scale value increases slightly when the fragment length exceeds $20400\,\mathrm{bp}$; however, since the number of samples decreases drastically as the fragment length grows (see the blue curve), it is reasonable to use the scale value of the majority of long fragments in $[4800\,\mathrm{bp}, 20400\,\mathrm{bp})$ to represent the whole.

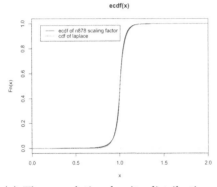
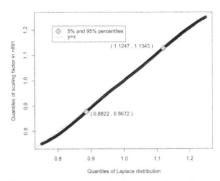

(a) The cumulative density distribution of NA12878 (3600bp $\leq r_k < $ 4800bp) sizing error and the optimal Laplace (1.003354,0.052800).

(b) The Q-Q plot of NA12891 (3600bp $\leq r_k < $ 4800bp) sizing error against Laplace (1.003354, 0.052800) based on NA12878.

Fig. 3. A comparison of NA12878 and NA12891 (3600 bp $\leq r_k < $ 4800 bp) sizing error against the Laplace distribution Laplace (1.003354, 0.052800) estimated from NA12878.

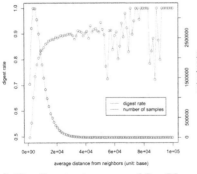
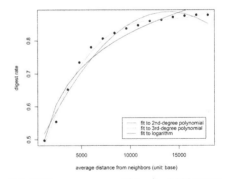

(a) The digest rate p_{digest} of BspQI on NA12878 against d_{avg}.

(b) Fitting $p_{\mathrm{digest}} = f(d_{\mathrm{avg}})$ of NA12891 to NA12878 2nd-degree polynomial, 3rd-degree polynomial and logarithm models.

Fig. 4. The digest rate p_{digest} of enzyme BspQI against the average distance of a restriction site from its adjacent sites d_{avg}. (Color figure online)

increases, and then remains stable at 0.9 when d_{avg} exceeds 18 kbp. $p_{\mathrm{digest}} = f(d_{\mathrm{avg}})$ of NA12878 is fit to 2nd-degree polynomial, 3rd-degree polynomial and logarithm respectively, where the coefficients are predicted by regression. The resulting functions are compared with the digest rate of NA12891 in Fig. 4(b). It is obvious that the best fitting is $p_{\mathrm{digest}} = \alpha_3\, d_{\mathrm{avg}}^3 + \alpha_2\, d_{\mathrm{avg}}^2 + \alpha_1\, d_{\mathrm{avg}} + \alpha_0$, where $\alpha_3 = 3.089 \times 10^{-4}, \alpha_2 = -1.069 \times 10^{-2}, \alpha_1 = 1.253 \times 10^{-1}, \alpha_0 = 3.693 \times 10^{-1}$, and $d_{\mathrm{avg}} = \frac{\text{the average distance to its neighbors}}{1200\,\mathrm{bp}}$.

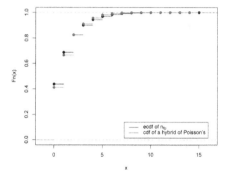

Fig. 5. The cumulative density distribution of NA12878 n_{fp} and the combination of $0.18\,\mathrm{Poisson}(0) + 0.60\,\mathrm{Poisson}(1) + 0.22\,\mathrm{Poisson}(3)$.

False Cuts. False cuts mainly result from random breaks of DNA molecules. A reasonable assumption is that the number of false cuts per unit length (n_{fp}) obeys a Poisson distribution. Taking 200 kb as a unit, n_{fp} is observed to follow Poisson(1.3) with notably more zeros. Figure 5 suggests that a good approximation of such a distribution is $n_{fp} \sim 0.18\,\mathrm{Poisson}(0) + 0.60\,\mathrm{Poisson}(1) + 0.22\,\mathrm{Poisson}(3)$.

Valouev et al. [18] presumed that false cuts occur equally likely at any position, while according to CEU trio data, false cuts are less likely to take place at both ends of optical maps. As Fig. 6(a) shows, the frequency of false cuts drops

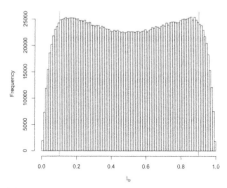

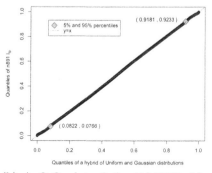

(a) The histogram of the locations of false cuts within NA12878 optical maps. The red vertical lines indicate location = 0.1 and 0.9 respectively.

(b) A Q-Q plot of the NA12891 false cut location and a hybrid of Uniform and Normal distribution derived from NA12878.

Fig. 6. Modeling the locations of NA12878 false cuts as a combination of $U[0.1, 0.9]$, $N(0.100178, 0.044186)$ and $N(0.9, 0.044186)$, where the Union distribution goes for the middle part of an optical map, and the two Normal distributions are fit for the ends. This combination of distributions is tested on NA12891 dataset with a Q-Q plot. (Color figure online)

gradually as it approaches both ends of the molecule. Define the location of a false cut as $l_{fp} = \frac{\text{the distance from this false cut to the 5'-end}}{\text{the total length of the optical map}}$. Figure 6(b) shows that the following distribution perfectly models the scattering of observed false cuts.

$$l_{fp} \sim \begin{cases} U[0.1, 0.9], & 0.1 \leq l_{fp} \leq 0.9, \text{ w.p. } 0.8852 \\ N(0.1, 0.044186), & l_{fp} < 0.1, \text{ w.p. } 0.0574 \\ N(0.9, 0.044186), & l_{fp} > 0.9, \text{ w.p. } 0.0574 \end{cases}$$

Unknown Orientation. The linearization and movement of DNA molecules does not assure its orientation. If we consider 5'-to-3' as the forward direction, a certain proportion (p_{reverse}) of molecules may be reversed (3'-to-5') under the microscope. As reported by CEU trio data, $p_{\text{reverse}} = 0.5$.

2.2   Regions Difficult to Cover

Some regions in a genome are likely to produce optical maps with higher error rate. In this section, we would like to predict these regions. This may affect the quality of downstream genome analysis. In other words, these regions are difficult to be aligned by optical maps produced. This concept will be used in the next section to indirectly validate our error model.

Define $\Delta t_{\text{scaling}}, t_{fn}$ and t_{fp} as high error thresholds for sizing error, missing cuts and false cuts respectively. $\Delta t_{\text{scaling}}$ bounds the difference of scaling factors from 1 and this should not be exceeded by any fragment within a confident alignment. t_{fn} is the maximum missing rate, which means $1 - p_{\text{digest}}$ must be not greater than t_{fn}. t_{fp} is the upper bound of n_{fp}.

Given the site locations of a reference region, according to our error model, p_{e+}, the probability of this region having high error, can be calculated as follows,

$$p_{\text{scaling-}} = Pr(\forall k (1 - \Delta t_{\text{scaling}} < s_k < 1 + \Delta t_{\text{scaling}}))$$

$$= \prod_{k=1}^{n} \left(1 - \frac{1}{2} \exp \frac{\mu - 1 - \Delta t_{\text{scaling}}}{\beta} - \frac{1}{2} \exp \frac{1 - \Delta t_{\text{scaling}} - \mu}{\beta} \right)$$

$$p_{fn-} = Pr(\text{no more than } \lceil n * t_{fn} \rceil \text{ cuts are missing})$$

$$p_{fp-} = Pr(\text{no more than } t_{fp} \text{ false cuts})$$

$$= 0.18 + \sum_{i=0}^{t_{fp}} \left(0.6 \cdot \frac{e^{-1}}{i!} + 0.22 \cdot \frac{3^i e^{-3}}{i!} \right)$$

$$p_{e+} = (1 - p_{\text{scaling-}})(1 - p_{fn-})(1 - p_{fp-})$$

where $p_{\text{scaling-}}, p_{fn-}$ and p_{fp-} are probabilities that reference region $R = (r_1, r_2, \ldots, r_n)$ has small scaling factors, a low missing cut rate and a low false cut rate respectively. Note that p_{fn-} relies on d_{avg}'s to estimate the digest rate p_{digest}^k of each nicking site k ($0 \leq k \leq n$). Given $(p_{\text{digest}}^0, p_{\text{digest}}^1, \ldots, p_{\text{digest}}^n)$, p_{fn-} can be calculated by dynamic programming. Regions with $p_{e+} \geq 0.8$ (i.e. over 80% of molecules originated here will be discarded) can be considered difficult to cover.

3 Experiments

In this section, besides fitting our error model derived from NA128787 into NA12891 (Figs. 3(b), 4(b) and 5(b)), we want to further evaluate our error model using both simulated and real datasets based on alignment. For real datasets, we use the CEU trio dataset, which consists of human optical maps of three individuals, namely NA12878, NA12891 and NA12892, each containing 2164566, 2113707 and 3305770 molecules. For simulated data, we implemented an OM simulator that starts producing an optical map from uniformly picking a starting position of a molecule and setting its length with an exponential random generator. After that, sizing error, missing cuts, and false cuts are successively introduced to this molecule. Finally, a random orientation is applied and an optical map is generated. Both real data and simulated data are digested by Nt.BspQI and share the resolution of 1200 bp. They are aligned to an *in silico* reference map GRCh38 precisely digested by BspQI. The aligner in use is RefAligner from Bio-Nano IrysView kit. Note that RefAligner was not developed based on our error model, thus providing an independent assessment of our model.

Set $p_{e+} \geq 0.8$ as a threshold of regions that are difficult to be covered. Ten simulated datasets of depth $= 10x$ were used to test the difficult regions in GRCh38. The simulation parameters in use were derived from NA12878 (as presented in Sect. 2). For each chromosome, the percentage of difficult regions estimated by error model (called theoretical percentage) and that summarized from ten runs of 10x alignment results (called empirical percentage) are compared in Table 1. We noticed that two other reasons that are unrelated to error model are also causing poorly covered regions. Part of the reference map does not produce any valid molecule because there are too few nicking sites (i.e. nicking sites are too loose) within this range. In addition, an optical map can be

Table 1. The theoretical and empirical percentages of difficult regions ($p_{e+} \geq 0.8$).

Chr id	Theoretical	Empirical	Sensitivity	Chr id	Theoretical	Empirical	Sensitivity
1	0.4169 %	0.4106 %	63.11 %	13	0.1975 %	0.2567 %	62.01 %
2	0.1297 %	0.1428 %	59.56 %	14	0.2476 %	0.2527 %	71.64 %
3	0.3837 %	0.4323 %	72.55 %	15	0	0	NA
4	0.0703 %	0.0732 %	86.44 %	16	0.4462 %	0.5083	100.00 %
5	0.3479 %	0.2135 %	38.30 %	17	0.1402 %	0.1547 %	32.73 %
6	0.0026 %	0.0026 %	100.00 %	18	0	0	NA
7	0.7445 %	0.5682 %	61.59 %	19	0.6399 %	0.7287 %	100.00 %
8	0.3393 %	0.2729 %	58.07 %	20	1.3809 %	1.1334 %	62.37 %
9	0.4445 %	0.2477 %	55.73 %	21	2.7056 %	3.0488 %	95.94 %
10	0.7023 %%	0.6461 %	68.31 %	22	0.6887 %	0.7243 %	100.00 %
11	0.9054 %	0.8158 %	71.50 %	X	0.3472 %	0.3190 %	83.54 %
12	0.4513 %	0.3033 %	64.66 %	Y	0.2583 %	0.2867 %	91.28 %

mistakenly aligned to another region that has highly similar nicking site pattern to the correct one. Both theoretical percentage and empirical percentage are calculated with the exclusion of loose regions and similar regions, that is,

$$\text{theoretical percentage} = \frac{L(\text{regions in chr } c \text{ with } p_{e+} \geq 0.8)}{\text{total length}}$$

$$\text{empirical percentage} = \frac{L(\text{regions in chr } c \text{ covered by less than 3 molecules})}{\text{total length}}$$

$$\text{total length} = L(\text{chr } c) - L(\text{loose or similar regions in chr } c)$$

where $L(\cdot)$ is an abbreviated notation for "the accumulated length of". Based on hundreds of RefAligner experiments, the highest accuracy of 96.88 % can be reached by setting the filtering thresholds of RefAligner to be $\max\left\{\frac{|o_k - r_k|}{\sqrt{r_k}}\right\} = 12.650$, $t_{\text{fn}} = 0.377925$ and $t_{\text{fp}} = 3.9$. This set of thresholds is also adopted in the calculation of theoretical percentage. Since RefAligner measures sizing error as $\frac{|o_k - r_k|}{\sqrt{r_k}}$, $\Delta t_{\text{scaling}}$ is in fact a function of r_k, that is, $\Delta t_{\text{scaling}} = \frac{12.650}{\sqrt{r_k}}$.

Table 1 shows that our error model is quite consistent. Sensitivity is defined as $\frac{\text{the total length of regions that are both theoretically and empirically difficult to cover}}{\text{the total length of regions that are difficult in theory}}$. Evidently, for most chromosomes, our error model identifies 60 % to 100 % of difficult regions correctly. For chromosomes chr5, chr9, and chr17, a relatively smaller fraction of difficult regions than in theory are predicted to be difficult according to the simulated data. Take the most extreme case, i.e. chr17, as an example, among the theoretically difficult regions that do not appear poorly covered on simulated data, 95.74 % has $p_{e+} > 0.77$ in practice. The average p_{e+}(our model) is 0.7792, which does not deviate much from 0.8.

In our experiment on real data, we justify the accuracy of p_{e+} calculation focusing on a list of manually-verified non-SV regions [8]. We did not present the percentage of difficult regions because there are too few verified non-SV regions with $p_{e+} \geq 0.8$. The total length of these non-SV regions accumulates to be 6760268, 7420612 and 7884938 bases for NA12878, NA12891 and NA12892 respectively. Instead, we compared the accuracy of p_{e+} estimated by our model with the accuracy of p_{e+} based on Valouev's model. Please note that to calculate the theoretical p_{e+} for each individual, the parameters are estimated solely from the other two.

Table 2 summarizes the average difference of p_{e+}(our model) $- p_{e+}$(trio) and p_{e+}(valouev's model) $- p_{e+}$(trio) for non-SV regions on each chromosome. p_{e+} (our model) is calculated as formulated in Sect. 2.2. p_{e+}(valouev's model) follows a similar computation except that $p_{scaling-}$ is calculated in the light of Valouev's Normal distribution. p_{e+}(trio) $= \frac{\text{the number of molecules mapped to this region}}{\text{optical mapping depth}}$, where the depth is 100x. Evidently, the p_{e+} based on our error model is more accurate than that of Valouev's error model. On average, Valouev's model has four times greater probability difference against real data than ours. Moreover, p_{e+}(our model) $- p_{e+}$(trio) is much stabler than p_{e+}(valouev's model) $- p_{e+}$(trio)), with a standard deviation of 0.0764 over Valouev's 0.1512.

Table 2. The average of p_{e+}(our model) $- p_{e+}$(trio) and p_{e+}(valouev's model) $- p_{e+}$(trio) for verified non-SV regions on each chromosome.

Chr id	NA12878		NA12891		NA12892	
	Our model	Valouev's	Our model	Valouev's	Our model	Valouev's
1	−0.0301	−0.1954	−0.0410	−0.2760	0.0287	0.0936
2	−0.0529	−0.2195	−0.0501	−0.1070	0.0138	−0.1107
3	−0.0644	−0.1003	−0.0646	−0.1200	−0.0078	−0.1294
4	−0.0973	−0.1374	−0.0989	0.1579	−0.0406	−0.1635
5	−0.0724	−0.1085	−0.0630	−0.1182	−0.0103	−0.1314
6	0.0664	−0.1148	−0.0469	−0.1246	−0.0083	−0.1298
7	−0.0532	−0.2184	−0.0466	−0.1032	0.0134	−0.1101
8	−0.0591	−0.2225	−0.0662	−0.1207	−0.0039	−0.1247
9	−0.0437	0.0908	0.0343	0.0881	0.0211	0.1008
10	−0.0319	−0.1985	−0.0263	0.0823	0.0259	0.0976
11	−0.0448	−0.0908	−0.0462	−0.1031	0.0291	0.0946
12	−0.0485	−0.0960	−0.0470	0.1210	0.0159	−0.1048
13	−0.0412	−0.0939	−0.0284	−0.1428	0.0301	−0.1550
14	−0.0495	−0.0996	−0.0414	−0.0981	0.0152	−0.1090
15	0.0033	−0.1649	−0.0067	0.0591	0.0427	−0.0512
16	−0.0001	−0.1685	0.0163	0.0363	0.0574	−0.0490
17	0.0082	−0.1587	0.0195	−0.0339	0.0896	0.0347
18	−0.0383	0.0883	−0.0349	−0.0917	0.0043	−0.1194
19	−0.0332	−0.1400	−0.0048	−0.1023	0.0175	−0.0725
20	0.0304	0.0031	0.0145	−0.0084	0.1038	0.0194
21	−0.776	0.0395	−0.0064	−0.1237	−0.0084	−0.1347
22	0.0036	−0.1859	0.0156	−0.0342	0.0736	−0.0755

4 Discussion

In this paper, we present a probabilistic error model based on the alignment results of BioNano RefAligner on CEU trio maps. We use both simulated and real datasets to verify our model. Based on our model, we compute a probability for each non-SV region showing how likely an optical map produced from this region has high error. By comparing these probabilities with the percentages of molecules covering these regions in real data, we show that our model is more accurate than the popular error model being used in the community. On the other hand, we admit that we should conduct a more comprehensive evaluation once more real optical map data are available in order to get a more accurate and robust error model. We also want to remark that the validation method used in the paper may not be the only method. For example, we may compare our error model with the error rates indicated by some accurate assemblers, such as Gentig [2] and the assembler from IrysView kit. To conclude, although we are still not able to capture all possible errors in our model, it should provide a better model for subsequent development of downstream analysis tools to make full use of BioNano Irys optical maps. Also, with our error model, it is possible to compute a lower bound on the required depth for each region in the genome which may guide practitioners how much optical map data we should produce. This research was supported by NSFC/RGC joint research scheme (N_HKU 729/13).

References

1. Anantharaman, T.S., Mishra, B., Schwartz, D.C.: Genomics via optical mapping II: ordered restriction maps. J. Comput. Biol. **4**(2), 91–118 (1997)
2. Anantharaman, T.S., Mishra, B., Schwartz, D.C.: Genomics via optical mapping III: contiging genomic dna and variations. In: The Seventh International Conference on Intelligent Systems for Molecular Biology, vol. 7, pp. 18–27. Citeseer (1999)
3. Das, S.K., Austin, M.D., Akana, M.C., Deshpande, P., Cao, H., Xiao, M.: Single molecule linear analysis of DNA in nano-channel labeled with sequence specific fluorescent probes. Nucleic Acids Res. **38**(18), e177 (2010)
4. Ganapathy, G., Howard, J.T., Ward, J.M., Li, J., Li, B., Li, Y., Xiong, Y., Zhang, Y., Zhou, S., Schwartz, D.C., et al.: High-coverage sequencing and annotated assemblies of the budgerigar genome. GigaScience **3**(1), 1–9 (2014)
5. Hastie, A.R., Dong, L., Smith, A., Finklestein, J., Lam, E.T., Huo, N., Cao, H., Kwok, P.Y., Deal, K.R., Dvorak, J., et al.: Rapid genome mapping in nanochannel arrays for highly complete and accurate de novo sequence assembly of the complex aegilops tauschii genome. PloS one **8**(2), e55864 (2013)
6. Kim, Y., Kim, K.S., Kounovsky, K.L., Chang, R., Jung, G.Y., Jo, K., Schwartz, D.C., et al.: Nanochannel confinement: DNA stretch approaching full contour length. Lab Chip **11**(10), 1721–1729 (2011)
7. Lin, H.C., Goldstein, S., Mendelowitz, L., Zhou, S., Wetzel, J., Schwartz, D.C., Pop, M.: Agora: assembly guided by optical restriction alignment. BMC Bioinf. **13**(1), 189 (2012)
8. Mak, A.C., Lai, Y.Y., Lam, E.T., Kwok, T.P., Leung, A.K., Poon, A., Mostovoy, Y., Hastie, A.R., Stedman, W., Anantharaman, T., et al.: Genome-wide structural variation detection by genome mapping on nanochannel arrays. Genetics **202**(1), 351–362 (2016)
9. Meng, X., Benson, K., Chada, K., Huff, E.J., Schwartz, D.C.: Optical mapping of lambda bacteriophage clones using restriction endonucleases. Nat. Genet. **9**(4), 432–438 (1995)
10. Muggli, M.D., Puglisi, S.J., Boucher, C.: Efficient indexed alignment of contigs to optical maps. In: Brown, D., Morgenstern, B. (eds.) WABI 2014. LNCS, vol. 8701, pp. 68–81. Springer, Heidelberg (2014)
11. Myers, E.W., Huang, X.: An $\mathcal{O}(n^2 \log n)$ restriction map comparison and search algorithm. Bull. Math. Biol. **54**(4), 599–618 (1992)
12. Nagarajan, N., Read, T.D., Pop, M.: Scaffolding and validation of bacterial genome assemblies using optical restriction maps. Bioinformatics **24**(10), 1229–1235 (2008)

13. Ramirez, M.S., Adams, M.D., Bonomo, R.A., Centrón, D., Tolmasky, M.E.: Genomic analysis of acinetobacter baumannii A118 by comparison of optical maps: identification of structures related to its susceptibility phenotype. Antimicrob. Agents Chemother. **55**(4), 1520–1526 (2011)
14. Ray, M., Goldstein, S., Zhou, S., Potamousis, K., Sarkar, D., Newton, M.A., Esterberg, E., Kendziorski, C., Bogler, O., Schwartz, D.C.: Discovery of structural alterations in solid tumor oligodendroglioma by single molecule analysis. BMC Genomics **14**(1), 505 (2013)
15. Sarkar, D.: On the analysis of optical mapping data. Ph.D. thesis, University of Wisconsin-Madison (2006)
16. Sarkar, D., Goldstein, S., Schwartz, D.C., Newton, M.A.: Statistical significance of optical map alignments. J. Comput. Biol. **19**(5), 478–492 (2012)
17. Teague, B., Waterman, M.S., Goldstein, S., Potamousis, K., Zhou, S., Reslewic, S., Sarkar, D., Valouev, A., Churas, C., Kidd, J.M., et al.: High-resolution human genome structure by single-molecule analysis. Proc. Natl. Acad. Sci. **107**(24), 10848–10853 (2010)
18. Valouev, A., Li, L., Liu, Y.C., Schwartz, D.C., Yang, Y., Zhang, Y., Waterman, M.S.: Alignment of optical maps. J. Comput. Biol. **13**(2), 442–462 (2006)
19. Valouev, A., Schwartz, D.C., Zhou, S., Waterman, M.S.: An algorithm for assembly of ordered restriction maps from single DNA molecules. Proc. Natl. Acad. Sci. **103**(43), 15770–15775 (2006)
20. Waterman, M.S., Smith, T.F., Katcher, H.L.: Algorithms for restriction map comparisons. Nucleic Acids Res. **12**(1 Part 1), 237–242 (1984)
21. Zhou, S., Bechner, M.C., Place, M., Churas, C.P., Pape, L., Leong, S.A., Runnheim, R., Forrest, D.K., Goldstein, S., Livny, M., et al.: Validation of rice genome sequence by optical mapping. BMC Genomics **8**(1), 278 (2007)
22. Zhou, S., Deng, W., Anantharaman, T.S., Lim, A., Dimalanta, E.T., Wang, J., Wu, T., Chunhong, T., Creighton, R., Kile, A., et al.: A whole-genome shotgun optical map of yersinia pestis strain kim. Appl. Environ. Microbiol. **68**(12), 6321–6331 (2002)

HapIso: An Accurate Method for the Haplotype-Specific Isoforms Reconstruction from Long Single-Molecule Reads

Serghei Mangul[1]([✉]), Harry (Taegyun) Yang[1]([✉]), Farhad Hormozdiari[1],
Elizabeth Tseng[2], Alex Zelikovsky[3], and Eleazar Eskin[1]([✉])

[1] Department of Computer Science, University of California,
Los Angeles, CA, USA
`smangul@ucla.edu, harrydgnt@g.ucla.edu, eeskin@cs.ucla.edu`
[2] Pacific Biosciences, Menlo Park, CA, USA
[3] Department of Computer Science, Georgia State University,
Atlanta, GA, USA

Abstract. Sequencing of RNA provides the possibility to study an individual's transcriptome landscape and determine allelic expression ratios. Single-molecule protocols generate multi-kilobase reads longer than most transcripts allowing sequencing of complete haplotype isoforms. This allows partitioning the reads into two parental haplotypes. While the read length of the single-molecule protocols is long, the relatively high error rate limits the ability to accurately detect the genetic variants and assemble them into the haplotype-specific isoforms. In this paper, we present HapIso (**H**aplotype-specific **I**soform Reconstruction), a method able to tolerate the relatively high error-rate of the single-molecule platform and partition the isoform reads into the parental alleles. Phasing the reads according to the allele of origin allows our method to efficiently distinguish between the read errors and the true biological mutations. HapIso uses a k-means clustering algorithm aiming to group the reads into two meaningful clusters maximizing the similarity of the reads within cluster and minimizing the similarity of the reads from different clusters. Each cluster corresponds to a parental haplotype. We use family pedigree information to evaluate our approach. Experimental validation suggests that HapIso is able to tolerate the relatively high error-rate and accurately partition the reads into the parental alleles of the isoform transcripts. Furthermore, our method is the first method able to reconstruct the haplotype-specific isoforms from long single-molecule reads.

The open source Python implementation of HapIso is freely available for download at https://github.com/smangul1/HapIso/.

1 Introduction

Advances in the RNA sequencing technologies and the ability to generate deep coverage data in the form of millions of reads provide an exceptional

---

Serghei Mangul and Harry Yang contributed equally to this work.

© Springer International Publishing Switzerland 2016
A. Bourgeois et al. (Eds.): ISBRA 2016, LNBI 9683, pp. 80–92, 2016.
DOI: 10.1007/978-3-319-38782-6_7

opportunity to study the functional implications of the genetic variability [4,16,17]. RNA-Seq has become a technology of choice for gene expression studies, rapidly replacing microarray approaches [20]. RNA-Seq provides sequence information, which aids in the discovery of genetic variants and alternatively spliced isoforms within the transcripts. RNA-Seq has the potential to quantify the relative expression of two alleles in the same individual and determine the genes subject to differential expression between the two alleles. Comparison of the relative expression of two alleles in the same individual as a phenotype influenced by the cis-acting genetic variants helps determine the cis-acting nature of the individual polymorphism [23].

There are three major difficulties in current approaches to identify allele specific expression using RNA-Seq data. First, short read protocols [15] cut genetic material into small fragments and destroy the linkage between genetic variants. Short reads obtained from the fragments are well suited to access the allele-specific expression on the single variant level. However, complexity of the higher eukaryotic genomes makes it hard to phase the individual variants into the full-length parental haplotypes (haplotype-specific isoforms). A common technique to assess the allele-specific expression (ASE) is to count the number of reads with the reference allele and the number of reads with alternate allele. However, this approach works on individual variant level and is not well suited to determine the allele-specific expression on the isoform level. Second, mapping the short reads onto the reference genome introduces a significant bias toward higher mapping rates of the reference allele at the heterozygous loci. Masking known loci in the genome does not completely remove the inherent bias [6]. Aside from the allele-specific expression, mapping biases may affect the QTL mapping and the discovery of new sequence variants. Third, the high sequence similarity between alternatively spliced variants of the same gene results in a significant number of short reads to align in multiple places of the reference transcriptome [9].

Multi-kilobase reads generated by single-molecule protocols [8] are within the size distribution of most transcripts and allow the sequencing of full-length haplotype isoforms in a single pass [14]. The reads cover multiple genomic variants across the gene, eliminating the necessity to phase the individual variants into the isoform haplotypes. Additionally, the extended length of the reads makes it simple to map the reads uniquely and eliminate the read-mapping biases. However, the relatively high error rates of the single-molecule protocols limit the application of the long single-molecule protocol to studies of the allele specific variants. There are currently no methods able to accurately detect the genetic variants from the long single-molecule RNA-Seq data and connect them to haplotype-specific isoforms.

In this paper, we present HapIso (**H**aplotype-specific **I**soform Reconstruction), a comprehensive method for the accurate reconstruction of the haplotype-specific isoforms of a diploid cell that uses the splice mapping of the long single-molecule reads and partitions the reads into parental haplotypes. The single molecule reads entirely span the RNA transcripts and bridge the single nucleotide variation (SNV) loci across a single gene. Our method starts with mapping the reads onto the reference genome. Aligned reads are partitioned into the genes as a clusters of overlapping reads. To overcome gapped coverage and splicing structures of the gene, the haplotype reconstruction procedure is applied independently for regions of contiguous coverage defined as *transcribed segments*. Restricted reads from the transcribed regions are partitioned into two local clusters using the 2-mean clustering. Using the linkage provided by the long single-molecule reads, we connect the local clusters into two global clusters. An error-correction protocol is applied for the reads from the same cluster. To our knowledge, our method is the first method able to reconstruct the haplotype-specific isoforms from long single-molecule reads. We applied HapIso to publicly available single-molecule RNA-Seq data from the GM12878 cell line [18]. Circular-consensus (CCS) single-molecule reads were generated by Pacific Biosciences platform [8]. Parental information (GM12891 and GM12892 cell lines) is used to validate the accuracy of the isoform haplotype reconstruction (i.e. assignment of RNA molecules to the allele of origin). We use short read RNA-Seq data for the GM12878 sample to validate the detected SNVs using a different sequencing platform (short reads). Our method discovered novel SNVs in the regions that were previously unreachable by the short read protocols.

Discriminating the long reads into the parental haplotypes allows to accurately calculate allele-specific gene expression and determine imprinted genes [7,19]. Additionally it has a potential to improve detection of the effect of cis- and trans-regulatory changes on the gene expression regulation [5,21]. Long reads allow to access the genetic variation in the regions previously unreachable by the short read protocols providing new insights into the disease sustainability.

2   Methods

2.1   Overview

Very similarly to the genome-wide haplotype assembly problem, the problem of haplotype-specific isoform assembly aims to infer two parental haplotypes given the collection of the reads [1,12]. While those problems are related, the allele expression ratio between RNA haplotypes is a priori unknown and may be significantly different from 1:1. An additional difference is due to the RNA-Seq gapped alignment profile and alternative splice structures of the gene. Overall, the problem of reconstruction of the haplotype-specific isoforms of a diploid transcriptome represents a separate problem requiring novel computational approaches.

We apply a single-molecule read protocol to study the allele-specific differences in the haploid transcriptome (Fig. 1). The single molecule protocol skips the amplification step and directly sequences the poly (A) selected RNA molecules. The reads generated by the protocol entirely span the RNA transcripts bridging the single nucleotide variation (SNV) loci across a single gene.

We introduce a method able to reconstruct the haploid transcriptome of a diploid organism from long single-molecule reads (Fig. 2). This method is able to tolerate the relatively high error-rate of the single-molecule sequencing and to partition the reads into the parental alleles of the isoform transcript. The errors in the long single-molecule reads typically are predominantly one-base

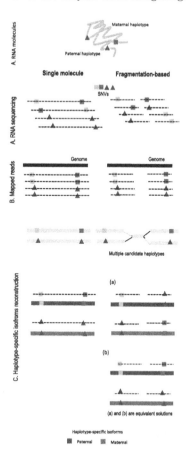

Fig. 1. Overview of long single-molecule protocol. (A) Unamplified, polyA-selected RNA molecules are sequenced by the single-molecule protocol able to entirely span the RNA transcripts to produce long single-molecule reads. The fragmentation-based protocols shred the amplified and poly (A) selected RNA into short fragments appropriately sized for sequencing. Short reads destroys the linkage between the SNVs. (B) Reads are mapped onto the reference genome. (C) SNVs are assembled into the two parental haplotype isoforms.

deletions and insertions [3]. Both insertions and deletions are corrected through the alignment with the reference genome. The remaining mismatch errors are further passed to the downstream analysis.

Our method starts with mapping the reads onto the reference genome (Fig. 2A). Long reads allow us to identify the unique placement of the read (99.9 % of the reads from GM12878 sample are mapped to a single location in the genome). The reads are partitioned into the genes as clusters of overlapping reads. The haplotype reconstruction procedure is applied independently for every gene. First, we identify the *transcribed segments* corresponding to contiguous regions of equivalently covered positions. Two positions are equivalently covered if any read covering one position also covers the other one.

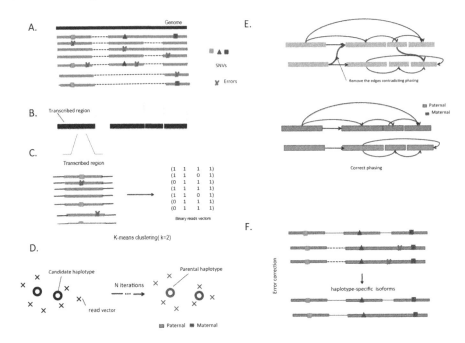

Fig. 2. Overview of HapIso. (A) The algorithm takes long single-molecule reads that have been mapped to the reference genome as an input. (B) The transcribed segments are identified as contiguous regions of equivalently covered positions. (C) Aligned nucleotides of the transcribed segment are condensed into the binary matrix whose width equals the number of variable positions. The entry "1" corresponds to the position with the observed mismatch, the entry is encoded as "0" if it matches the reference allele. (D) Reads restricted to the transcribed segment (rows of the binary matrix) are partitioned into two clusters, using the 2-means clustering algorithm. Each cluster corresponds to a local haplotype. (E) The segment graph is constructed to incorporate the linkage between the alleles. The edges of the graph connect the local haplotypes. The minimum number of corrections to the graph is applied to partition the graph into two independent components corresponding to full-length parental gene haplotypes. (F) An error-correction protocol is applied for the reads from the same cluster. The protocol corrects the sequencing errors and produce corrected haplotype-specific isoforms.

To account for gapped coverage and splicing structures of the gene, we cluster the reads into two parental haplotypes for every transcribed segments independently (Fig. 2C). The clustering procedure first condenses the aligned nucleotides of the transcribed segment into a binary matrix with a width equal to the number of variable positions. The entry "1" corresponds to the position which mismatches the reference allele, while the entry is encoded as "0" if it matches the reference allele. We partition the rows (reads restricted to the transcribed segment) into two clusters, using the 2-means clustering algorithm (Fig. 2D). The result from the 2-means clustering partitions the restricted reads into local parental haplotypes. Using the linkage provided by the long single-molecule reads, we reconstruct the full-length gene haplotypes. We build the segment graph encoding the linkage between the alleles in form of edges (Fig. 2E). The minimum number of corrections to the graph is applied to partition the graph into two independent components corresponding to two parental haplotypes. The transcript reads are then grouped according to the allele of origin (Fig. 2F). An error-correction protocol is applied for the reads from the same cluster. The protocol corrects the sequencing errors and produces corrected haplotype-specific isoforms.

2.2 Single-Molecule RNA-Seq

We use publicly available single-molecule RNA-Seq data generated from the peripheral blood lymphocyte receptors for B-lymphoblastoid cell lines (GM12878 cell line) [18]. Additionally, we use parental long read RNA-Seq data from GM12891 and GM12892 cell lines to validate the accuracy of the proposed approach. Libraries were sequenced using the Pacific Bioscience platform [8] able to produce long single-molecule reads for all three samples in the trio. Unamplified, polyA-selected RNA was sequenced by the circular molecules. Circular-consensus (CCS) single-molecule read represent a multi-pass consensus sequence, where each base pair is covered on both strands at least once and the multiple low-quality base calls can be used to derive a high-quality calls.

2.3 Read Mapping

The first step of the analysis is to place sequencing reads onto the reference genome. Long read length provided by the single molecule protocol provides enough confidence to find unique position in the genome where the reads were generated from without using the existing gene structure annotation. The 715,902 CCS reads were aligned to the human reference genome (hg19) using the GMAP aligner, which maps cDNA sequences to a genome [22]. GMAP was originally designed to map both messenger RNAs (mRNAs) and expressed sequence tags (ESTs) onto to genome. The tool is able to tolerate high number of sequence errors in long cDNA sequences which makes it perfect fit for Pac Bio single-molecule platform.

GMAP is able to identify up to two placements in the genome for 99.6 % reads. Only a small portion of those are mapped to two locations of the genome (1.6 %). However in many case two placement in the genome have a evident differences thus making it easy to select the most preferable placement. In this way, vast majority of the CCS reads have single high-confidence mapping covering the entire exon-intron structure of an isoform transcript.

2.4 Haplotype-Specific Isoform Reconstruction

Having the reads spanning the full-length isoform transcripts, the problem of the haplotype-specific isoform reconstruction aims to discriminate the transcript reads into two parental haplotypes. The problem is equivalent to the read error-correction problem. If all the errors are corrected, the long reads provide the answer to the reconstruction problem, i.e. each non redundant read is the haplotype-specific isoform. Since the long reads are error prone, it is required to fix the errors or equivalently call single nucleotide variants (SNVs). Rather than phasing each isoform separately, it is preferable to agglomerate all the isoforms from a single gene and cluster the reads by haplotype of origin. All the reads from a single haplotype contain the same alleles of shared SNVs, thus all the differences between reads in shared transcribed segments should be corrected. We propose the following optimization formulation minimizing the number of errors.

Phasing Problem. Given a set of the long reads R corresponding to the transcripts from the same gene g. Partition the reads into two haplotype clusters such that the number of sequencing errors in the reads is minimized.

Typically, the errors in the long single-molecule reads are dominated by one-base insertions and deletions. Both are corrected through the alignment to the reference genome – insertions are deleted and deletions are imputed from the reference. The remaining mismatch errors are further passed to the downstream analysis.

Since long reads are uniquely aligned, the aligned reads are uniquely partitioned into clusters corresponding to the genes. Further, the haplotype are independently reconstructed for every gene. First we split the genes into *transcribed segments* corresponding to contiguous regions of equivalently covered positions. Two positions are equivalently covered if any read covering one position also covers the other one. To overcome gapped coverage and splicing structure of the gene, we cluster the reads into two parental haplotype for every transcribed segments independently. The clustering procedure first condenses the aligned nucleotides of the transcribed segment into the binary matrix whose width equals the number of polymorphic positions. The entry "1" corresponds to each position whose allele mismatches the reference and the entry is encoded as "0" if it matches the reference allele. We partition the rows (reads restricted to the transcribed segment) into two clusters, using the 2-means clustering algorithm. The clustering algorithm returns a set of centroids, one for each of the two clusters. An observation vector is classified with the cluster number or centroid index of the centroid closest to it. A vector r belongs to cluster i if it is closer to centroid i than any other centroids. If r belongs to i, centroid i is refereed as dominating centroid of r.

Given a set of observations $(r_1, r_2, , r_n)$, where each observation is a binary read vector, k-means clustering aims to partition the n reads into two sets $S = S_1, S_2$ so as to minimize the distortion defined as

$$D = \sum_{i=0}^{2} \sum_{x \in S_i} \|x - \mu_i\|^2 \text{ where } \mu_i \text{ is the dominating centroid in } S_i. \quad (1)$$

Each step of the k-means algorithm refines the choices of centroids to minimize distortion. The clustering algorithm uses change in distortion as stopping condition.

Haplotypes for each transcribed segment obtained by local read clustering are further linked using the long single-molecule reads as follows.

First, we build a graph G in which each vertex corresponds to a transcribed segment and two vertices are adjacent if they belong to the same read. Two transcribed segments A and B with pairs of haplotypes $(A1, A2)$ and $(B1, B2)$, respectively, can be either linked in parallel $A1B1$ and $A2B2$ or in cross $A1B2$ and $A2B1$. If among reads containing A and B there are more reads consistent with the parallel linkage than the reads consisting with the cross linkage, then the parallel linkage is more likely and vice versa. The larger skew between number of parallel and cross reads gives the higher confidence in the corresponding linkage. Therefore the weight of an edge between A and B in the graph G is set to

$$w(A, B) = \left| \log \left(\frac{\#\text{parallel } (A - B) - \text{reads}}{\#\text{cross } (A - B) - \text{reads}} \right) \right| \tag{2}$$

Then we find the maximum-weight spanning tree T of the graph G consisting of links with the highest confidence in the chosen (cross or parallel) linkage [2]. The tree T is split into two trees T1 and T2 uniquely identifying two global haplotypes for the corresponding gene as follows. Each vertex A of T is split into two vertices $A1$ and $A2$ corresponding to two haplotypes of the transcribed segment A. Each edge (A, B) of T with the positive weight $w(A, B)$ is split into two edges $(A1, B1)$ and $(A2, B2)$ and each edge (A, B) with the negative weight is split into $(A1, B2)$ and $(A2, B1)$. Starting with $A1$ we traverse the tree $T1$ concatenating all haplotypes corresponding to its vertices into a single global haplotype. Similarly, starting with $A2$, we traverse the complementary tree $T2$ concatenating its haplotypes into the complimentary global haplotype.

Long reads are grouped according to the haplotype of origin. An error-correction protocol is applied for the reads from the same cluster. The protocol corrects the sequencing errors and produce corrected haplotype-specific isoforms.

Finally, the resulting two haplotypes are different in heterozygous loci allowing our method to determine the SNVs. Long reads provide one to one mapping between the reconstructed haplotypes and the isoform haplotypes. Reads counts are used to determine allelic expression of each haplotype copy of the isoform.

The absence of the systematic errors allows us to successfully correct the randomly distributed errors and accurately reconstruct the isoform haplotypes. Comparing to other approaches requiring trio family data, we are able to correct the error and reconstruct the parental haplotypes from sequencing data.

3   Results

3.1   HapIso is able to Accurately Reconstruct Haplotype-Specific Isoforms

We used trio family long single-molecule RNA-seq data to validate the reconstructed haplotype-specific isoforms. Single molecule RNA-Seq data was generated from the peripheral blood lymphocyte receptors for B-lymphoblastoid cell

A. Mendelian consistency   B. Mendelian inconsistency   C. Ambiguity

Fig. 3. Allele inheritance model. (A) Mendelian consistency; Each parent provides one copy of the allele to a child; child inherits red allele from mother and a green allele from father. (B) Mendelian inconsistency; Child does not inherit green allele from any of the parents. (C) Ambiguity: Green allele is not expressed or covered in the right parent due to lack of sequencing coverage. Mendelian (in)consistency can not be verified. (Color figure online)

lines from a complete family trio composed of a father, a mother and a daughter. We reconstructed the haplotype-specific isoforms of each individual independently using HapIso method. Family pedigree information makes it possible for us to detect Mendelian inconsistencies in the data. We use reconstructed haplotypes to infer the heterozygous SNVs determined as position with non-identical alleles with at least 10x coverage.

According to Mendelian inheritance, one allele in the child should be inherited from one parent and another allele from the other parent (Fig. 3A). Mendelian inconsistencies correspond to loci from the child with at least one allele not confirmed by parents (Fig. 3B). We also separately account for the missing alleles from the parental haplotypes due to insufficient expression or coverage of the alternative allele (Fig. 3C). Such loci are ambiguous since Mendelian consistency or inconsistency cannot be verified.

HapIso was able to detect 921 genes with both haplotypes expressed among 9,000 expressed genes. We observed 4,140 heterozygous loci corresponding to position with non-identical alleles among inferred haplotypes. 53 % of detected SNVs follow Mendelian inheritance. The number of variants with Mendelian inconsistencies accounts for 10 % of the heterozygous SNVs. The remaining SNVs are ambiguous and the Mendelian consistency cannot be verified.

Additionally we check the number of recombinations in the inferred haplotypes. Our approach can theoretically identify recombinations in the transmitted haplotypes. Crossovers between the parental haplotypes result in recombination events in the child's haplotypes (Fig. 4B). Since recombination events are rare, most of the time they manifest switching errors in phasing. Single-molecule reads are long enough to avoid switching errors, which are confirmed by lack of recombination events, observed in the reconstructed haplotypes.

3.2   SNV Discovery and Cross Platform Validation

The single-molecule RNA-Seq was complemented by 101-bp paired-end RNA-Seq data of the child. Short RNA-seq reads are used for cross-platform validation of the detected SNVs. The haplotypes assembled by the HapIso were scanned with the 10x coverage threshold to detect the heterozygous loci passed for the validation. The short RNA-Seq reads were mapped onto the hg19 reference genome

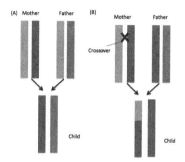

Fig. 4. Gene inheritance model. (A) Each parent provides one copy of the gene to a child; child inherits blue haplotype from mother and a green haplotype from father. (B) Haplotypes of mother pair up with each other and exchange the segments of their genetic material to form recombinant haplotype (recombination of orange and blue haplotypes). The child inherits the recombined haplotype. Haplotype from father is inherited with no recombination. (Color figure online)

complemented by the gene annotations (tophat 2.0.13, GRCh37 ENSEMBL). GATK [11] variant caller was used to call the SNVs from the short RNA-seq reads following the publicly available best practices. Additionally, the public catalogue of variant sites (dbSNP), which contains approximately 11 million SNVs, was used to validate genomic position identified as SNVs by single-molecule and short read protocols.

We compared genomic positions identified as SNVs from long single-molecule reads and short reads (Fig. 5). First, we compared the positions identified as SNVs by both platforms. 279 genomic positions were reported as SNV by both platforms. Those positions were also confirmed by dbSNP. Of those SNVs, 94 % are concordant between the platform i.e. contain identical alleles. Among the detected SNVs by the single-molecule protocols, 23 positions are identified as SNVs only by the single-molecule protocol. We investigated the coverage of those SNVs by the short reads. Those SNVs are covered by the short reads with the alternative allele expression under the SNV calling threshold, while the remaining SNVs are not covered by short reads.

We compared haplotypes assembled from long and short RNA-Seq reads (child sample, GM12878). We use HapCUT [1] to assemble the haplotypes from the short RNA-Seq reads. HapCUT is a max-cut based algorithm for haplotype assembly from the two chromosomes of an individual. GATK is used to generate vcf file with genomic variants required by HapCUT. HapCUT produces multiple contigs per gene shorter than the transcript isoforms, thus limiting the possibility to access haplotype-specific isoforms.

Unfortunately we could not compare our method with HapCUT for the long single-molecule reads. HapCUT is originally designed for the short reads. We are not able to generate the genomic variants (vcf format) required by HapCUT. The GATK tool doesn't have the best practice pipeline for the Pac Bio RNA-Seq reads.

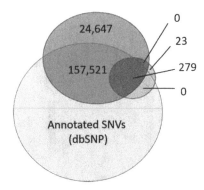

Fig. 5. Venn diagram showing the numbers of genomic position identified as SNVs from long single-molecule reads (green) and short reads (orange). SNVs calls from both platforms were match against the dbSNP catalogue of variant sites (blue). (Color figure online)

4   Discussion

RNA molecules represent an essential piece of the cell identity, playing an important role as a messenger and regulatory molecule [13]. Long single-molecule protocols provide an unprecedented allele-specific view of the haploid transcriptome. Partitioning the long reads into the parental haplotypes allows us to accurately calculate allele-specific transcript and gene expression and determine imprinted genes [7,19]. Additionally, it has the potential to improve detection of the effect of cis- and trans-regulatory changes on the gene expression regulation [5,21]. Long reads allow us to access the genetic variation in the regions previously unreachable by the short read protocols providing new insights into the disease sustainability. Availability of full-length haplotype-specific isoforms opens a wide avenue for the accurate assessment of allelic imbalance to study molecular mechanisms regulating genetic or epigenetic causative variants, and associate expression polymorphisms with the disease susceptibility.

We have presented HapIso, an accurate method for the reconstruction of the haplotype-specific isoforms of a diploid cell. Our method uses the splice mapping and partitions the reads into parental haplotypes. The proposed method is able to tolerate the relatively high error-rate of the single-molecule sequencing and discriminate the reads into the paternal alleles of the isoform transcript. Phasing the reads according to the allele of origin allows efficiently distinguish between the read errors and the true biological mutations. HapIso uses the 2-means clustering algorithm aiming to group the reads into two meaningful clusters maximizing the similarity of the reads within cluster, and minimizing the similarity of the reads from different clusters. Clustering is applied locally for the transcribed regions, which are further reconnected in the segment graph. Each cluster corresponds to the parental haplotype. An error-correction protocol is applied for the reads from the same cluster allowing to correct the errors and reconstruct haplotype-specific isoforms.

Traditional haplotype assembly [1,12] approaches are designed for whole-genome sequencing and are not well suited for gapped alignment profile offed by the RNA sequencing. Genome-wide haplotype assembly aims to assemble two haplotypes for a chromosome given the collection of sequencing fragments. In contrast, RNA haplotype reconstruction requires to assemble multiple haplotypes of a gene, which each isoform having two parental haplotype copies. ASE detection methods [10] are well suited to determine the allele-specific expression on the on individual variant level further aggregated into gene-level estimates. However, those methods are originally designed for SNV-level read counts and are not applicable to reconstruct full-length haplotype-specific isoforms of a gene.

To our knowledge, our method is the first method able to reconstruct the haplotype-specific isoforms from long single-molecule RNA-seq data. Other approaches [18] quantify the allele-specific expression of the genes using trio family data, while only being able to provide the ratio between allele expression of the genes. Such approaches are not suited to reconstruct the haplotype-specific isoforms and correct sequencing errors. Experimental validation based on the trio family data and orthogonal short read protocol suggests that HapIso is able to tolerate the relatively high error-rate and accurately reconstruct the haplotype-specific isoforms for genes with at least 10x coverage. Deeper sequencing is required to assemble haplotype-specific isoforms of genes with low expression level.

References

1. Bansal, V., Bafna, V.: HapCUT: an efficient and accurate algorithm for the haplotype assembly problem. Bioinformatics **24**(16), i153–i159 (2008)
2. Brinza, D., Zelikovsky, A.: 2SNP: scalable phasing based on 2-SNP haplotypes. Bioinformatics **22**(3), 371–373 (2006)
3. Chaisson, M.J.P., Huddleston, J., Dennis, M.Y., Sudmant, P.H., Malig, M., Hormozdiari, F., Antonacci, F., Surti, U., Sandstrom, R., Boitano, M., Landolin, J.M., Stamatoyannopoulos, J.A., Hunkapiller, M.W., Korlach, J., Eichler, E.E.: Resolving the complexity of the human genome using single-molecule sequencing. Nature **517**, 608–611 (2014)
4. Cloonan, N., Forrest, A.R., Kolle, G., Gardiner, B.B., Faulkner, G.J., Brown, M.K., Taylor, D.F., Steptoe, A.L., Wani, S., Bethel, G., et al.: Stem cell transcriptome profiling via massive-scale mRNA sequencing. Nat. Methods **5**(7), 613–619 (2008)
5. Cowles, C.R., Hirschhorn, J.N., Altshuler, D., Lander, E.S.: Detection of regulatory variation in mouse genes. Nat. Genet. **32**(3), 432–437 (2002)
6. Degner, J.F., Marioni, J.C., Pai, A.A., Pickrell, J.K., Nkadori, E., Gilad, Y., Pritchard, J.K.: Effect of read-mapping biases on detecting allele-specific expression from RNA-sequencing data. Bioinformatics **25**(24), 3207–3212 (2009)
7. DeVeale, B., Van Der Kooy, D., Babak, T.: Critical evaluation of imprinted gene expression by RNA-Seq: a new perspective. PLoS Genet. **8**(3), e1002600 (2012)
8. Eid, J., Fehr, A., Gray, J., Luong, K., Lyle, J., Otto, G., Peluso, P., Rank, D., Baybayan, P., Bettman, B., et al.: Real-time DNA sequencing from single polymerase molecules. Science **323**(5910), 133–138 (2009)

9. Li, B., Ruotti, V., Stewart, R.M., Thomson, J.A., Dewey, C.N.: RNA-seq gene expression estimation with read mapping uncertainty. Bioinformatics **26**(4), 493–500 (2010)

10. Mayba, O., Gilbert, H.N., Liu, J., Haverty, P.M., Jhunjhunwala, S., Jiang, Z., Watanabe, C., Zhang, Z.: Mbased: allele-specific expression detection in cancer tissues and cell lines. Genome Biol. **15**(8), 405 (2014)

11. McKenna, A., Hanna, M., Banks, E., Sivachenko, A., Cibulskis, K., Kernytsky, A., Garimella, K., Altshuler, D., Gabriel, S., Daly, M., et al.: The genome analysis toolkit: a mapreduce framework for analyzing next-generation DNA sequencing data. Genome Res. **20**(9), 1297–1303 (2010)

12. Patterson, M., Marschall, T., Pisanti, N., van Iersel, L., Stougie, L., Klau, G.W., Schönhuth, A.: WHATSHAP: haplotype assembly for future-generation sequencing reads. In: Sharan, R. (ed.) RECOMB 2014. LNCS, vol. 8394, pp. 237–249. Springer, Heidelberg (2014)

13. Saliba, A.E., Westermann, A.J., Gorski, S.A., Vogel, J.: Single-cell RNA-seq: advances and future challenges. Nucleic Acids Res. **42**, 8845–8860 (2014)

14. Sharon, D., Tilgner, H., Grubert, F., Snyder, M.: A single-molecule long-read survey of the human transcriptome. Nat. Biotechnol. **31**(11), 1009–1014 (2013)

15. Steijger, T., Abril, J.F., Engström, P.G., Kokocinski, F., Hubbard, T.J., Guigó, R., Harrow, J., Bertone, P., Consortium, R., et al.: Assessment of transcript reconstruction methods for RNA-seq. Nat. Methods **10**, 1177–1184 (2013)

16. Sultan, M., Schulz, M.H., Richard, H., Magen, A., Klingenhoff, A., Scherf, M., Seifert, M., Borodina, T., Soldatov, A., Parkhomchuk, D., et al.: A global view of gene activity and alternative splicing by deep sequencing of the human transcriptome. Science **321**(5891), 956–960 (2008)

17. Tang, F., Barbacioru, C., Wang, Y., Nordman, E., Lee, C., Xu, N., Wang, X., Bodeau, J., Tuch, B.B., Siddiqui, A., et al.: mRNA-Seq whole-transcriptome analysis of a single cell. Nat. Methods **6**(5), 377–382 (2009)

18. Tilgner, H., Grubert, F., Sharon, D., Snyder, M.P.: Defining a personal, allele-specific, and single-molecule long-read transcriptome. Proc. Nat. Acad. Sci. **111**(27), 9869–9874 (2014)

19. Wang, X., Miller, D.C., Harman, R., Antczak, D.F., Clark, A.G.: Paternally expressed genes predominate in the placenta. Proc. Nat. Acad. Sci. **110**(26), 10705–10710 (2013)

20. Wang, Z., Gerstein, M., Snyder, M.: RNA-Seq: a revolutionary tool for transcriptomics. Nat. Rev. Genet. **10**(1), 57–63 (2009)

21. Wittkopp, P.J., Haerum, B.K., Clark, A.G.: Evolutionary changes in cis and trans gene regulation. Nature **430**(6995), 85–88 (2004)

22. Wu, T.D., Watanabe, C.K.: GMAP: a genomic mapping and alignment program for mRNA and EST sequences. Bioinformatics **21**(9), 1859–1875 (2005)

23. Xu, X., Wang, H., Zhu, M., Sun, Y., Tao, Y., He, Q., Wang, J., Chen, L., Saffen, D.: Next-generation DNA sequencing-based assay for measuring allelic expression imbalance (AEI) of candidate neuropsychiatric disorder genes in human brain. BMC Genomics **12**(1), 518 (2011)

Protein-Protein Interactions
and Networks

Genome-Wide Structural Modeling of Protein-Protein Interactions

Ivan Anishchenko[1,2], Varsha Badal[1], Taras Dauzhenka[1], Madhurima Das[1], Alexander V. Tuzikov[2], Petras J. Kundrotas[1], and Ilya A. Vakser[1(✉)]

[1] Department of Molecular Biosciences, Center for Computational Biology, University of Kansas, Lawrence, KS 66047, USA
vakser@ku.edu
[2] United Institute of Informatics Problems, National Academy of Sciences, 220012 Minsk, Belarus

Abstract. Structural characterization of protein-protein interactions is essential for fundamental understanding of biomolecular processes and applications in biology and medicine. The number of protein interactions in a genome is significantly larger than the number of individual proteins. Most protein structures have to be models of limited accuracy. The structure-based methods for building the network of protein interactions have to be fast and insensitive to the inaccuracies of the modeled structures. This paper describes our latest development of the docking methodology, including global docking search, scoring and refinement of the predictions, its systematic benchmarking on comprehensive sets of protein structures of different accuracy, and application to the genome-wide networks of protein interactions.

Keywords: Protein recognition · Modeling of protein complexes · Protein docking · Structure prediction

1 Introduction

Many cellular processes involve protein-protein interactions (PPI). Structural characterization of PPI is essential for fundamental understanding of these processes and for applications in biology and medicine. Because of the inherent limitations of experimental techniques and rapid development of computational power and methodology, computational modeling is increasingly a tool of choice in many biological studies [1]. Protein docking is defined as structural modeling of protein-protein complexes from the structure of the individual proteins [2]. Protein docking techniques are largely based on structural and physicochemical complementarity of the proteins. Rigid-body approximation, neglecting the internal (conformational) degrees of freedom, is a common approach, allowing exhaustive exploration of the docking search space (three translations and three rotations in Cartesian coordinates), often using pattern recognition techniques like correlation by Fast Fourier Transform [3]. The free docking is increasingly complemented by comparative (template-based) docking, where the docking pose of the

© Springer International Publishing Switzerland 2016
A. Bourgeois et al. (Eds.): ISBRA 2016, LNBI 9683, pp. 95–105, 2016.
DOI: 10.1007/978-3-319-38782-6_8

target proteins is inferred by experimentally determined complexes of proteins that are similar to the targets, either in sequence or in structure [4].

Current docking approaches are targeted towards protein structures determined outside of the complex (unbound structures). Thus the rigid-body approximation requires a degree of tolerance to conformational changes in proteins upon binding. Increasingly, the object of docking is the modeled rather than experimentally determined structures. Generally, such structures have lower accuracy than the ones determined by experiment (primarily, X-ray crystallography and nuclear magnetic resonance) [1]. Thus conformational search becomes a necessary part of the docking routine, especially in light of the rapid development of computing power, such as accessible and inexpensive GPU computing.

Structural modeling of PPI also addresses the problem of reconstruction and characterization of the network of connections between proteins in a genome [5]. The number of protein interactions in a genome is significantly larger than the number of individual proteins. Moreover, most protein structures have to be models of limited accuracy. Thus, structure-based methods for building this network have to be fast and insensitive to significant inaccuracies of the modeled structures. The precision of these methods may be correlated with the precision of the protein structures – lower for less accurate models and higher for more exact models.

This paper describes our latest development of the docking methodology, including global docking search, scoring and refinement of the predictions, its systematic benchmarking on comprehensive sets of protein structures of different accuracy, and application to the genome-wide networks of protein interactions.

2   Docking

2.1   Comparative and Free Docking

Our free docking method is based on the systematic grid search by correlation techniques using Fast Fourier Transformation (FFT) [3, 6]. The algorithm and its subsequent development are implemented in software GRAMM (http://vakser.compbio.ku.edu), which is freely available and widely used in the biomedical community. New comparative docking techniques are being developed following the rapidly increasing availability of the structural templates, and statistics on residue-residue propensities, along with the coarse-grained potentials accounting for structural flexibility.

The comparative modeling of protein complexes relies on target/template relationships based on sequence or structure similarity, with the latter showing a great promise in terms of availability of the templates [7]. Docking assumes knowledge of the structures of the interacting proteins. Thus, the templates for protein-protein complex may be found by structure alignment of the target proteins to full structures of proteins in the co-crystallized complexes. Dissimilar proteins may have similar binding modes. Thus, docking can also be performed by the structure alignment of the target proteins with interfaces of the co-crystallized proteins [8].

Important element in structure alignment is the diversity, non-redundancy and completeness of the template libraries. While selecting all pairwise protein-protein complexes from Protein Data Bank (PDB) [9] would produce the complete set, such "brute force" set will have many identical or highly similar complexes, and overrepresentation of some types of complexes. The set would also have low-quality, erroneous, and biologically irrelevant structures. Thus, to retain only the relevant interactions, it is important to generate a template library by filtering PDB.

We generated a carefully curated, nonredundant library of templates containing 4,950 full structures of protein-protein complexes and 5,936 protein-protein interfaces extracted at 12 Å distance cut-off [10]. Redundancy was removed by clustering proteins based on structural similarity. The clustering threshold was determined from the analysis of the clusters and the docking performance. An automated procedure and manual curation yielded high structural quality of the interfaces in the template and validation sets. The library is part of the DOCKGROUND resource for molecular recognition studies (http://dockground.compbio.ku.edu).

2.2 Constraints

An important problem in protein-protein docking is identification of a near-native match among very large number of putative matches produced by a global docking search. To detect the near-native matches at the docking post-processing stage, a scoring procedure is performed by re-ranking the matches, often by functions that are too computationally expensive or impossible to include in the global scan. Such scoring may be based on structural, physicochemical, or evolutionary considerations [11]. Often information on the docking mode (e.g. one or more residues at the protein-protein interfaces) is available prior to the docking. If this information is reliable, the global scan may not be necessary, and the search can be performed in the sub-space that satisfies the constraints. However, if the probability of such information is less than certain, it may rather be included in the post-processing scoring. Given the inherent uncertainties of the global-search predictions, such information on the binding modes is very valuable. For docking server predictions, which can be used by the biological community, an automated search for such data can be of great value.

We developed the first approach to generating text-mining (TM) [12] constraints for protein-protein docking [13]. Our methodology, by design, is a combination and extension of two well-developed TM fields: (1) prediction of interactors in PPI networks, and (2) detection of protein functional sites for small ligands. The first one was used as the source of expertise on TM of PPI (existing approaches predict the fact, not the mode of interaction), and the second one as the source of expertise on TM for prediction of the binding sites on protein structures (existing approaches are developed for small non-protein ligands).

The procedure retrieves published abstracts on specific PPI and extracts information relevant to docking. The procedure was assessed on protein complexes from the DOCKGROUND resource. The results show that correct information on binding residues can be extracted for about half of the complexes. The amount of irrelevant information was reduced by conceptual analysis of a subset of the retrieved abstracts. Support Vector Machine models were trained and validated on the subset. The extracted constraints were incorporated in the docking protocol and tested on the DOCKGROUND unbound benchmark set, significantly increasing the docking success rate (Fig. 1).

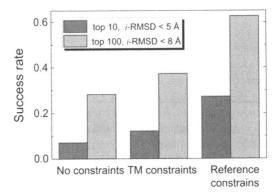

Fig. 1. Docking with TM constraints. The results of benchmarking on the unbound X-ray set from DOCKGROUND. A complex was predicted successfully if at least one in top ten matches had ligand C^α interface RMSD < 5 Å (light gray), and one in top hundred had RMSD < 8 Å (dark gray). The success rate is the percentage of successfully predicted complexes in the set. The low-resolution geometric scan output (20,000 matches) from GRAMM docking, with no post-processing, except removal of redundant matches, was scored by the TM results. The reference bars show scoring by the actual interface residues.

2.3 Refinement

Comparative docking lacks explicit penalties for inter-molecular penetration as opposed to the free docking where such penalty is inherently based on the shape complementarity paradigm. Thus, the template-based docking models are commonly perceived as requiring special treatment to obtain usable PPI structural models without significant interatomic clashes. In this study, we compared the clashes in the template-based models with the same targets produced by the free docking. The resulting clashes in the two types of docking, in fact, were similar according to all considered parameters, due to the higher quality of the comparative docking predictions. This indicates that the refinement techniques developed for the free docking, can be successfully applied to the refinement of the comparative models.

A new method for protein-protein docking refinement was developed and implemented. This method is a compromise between a complete exhaustive search and polynomial-time approximation schemes, in a sense that it avoids combinatorial explosion, but does not provide polynomial-time separation of solution and its approximation. A grid-based method was implemented in C++/CUDA programming languages. A rotamers library [14] was employed as a set of primitives for combinatorial geometric optimization of the side-chains. The method was validated on a subset of protein complexes from the DOCKGROUND unbound benchmark 3.0. The results (Fig. 2) suggest that the method effectively removes steric clashes from the docked unbound proteins. The GPU-based implementation was 38 faster than the single-thread C++ implementation of the same algorithm (GPU Quadro K4000, CPU Intel Xeon 2.7 GHz).

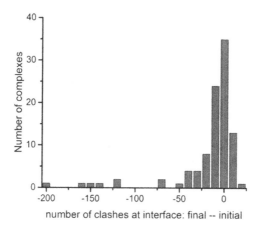

Fig. 2. Clashes removal by an exhaustive search in rotameric space of the interface residues for 99 complexes of the DOCKGROUND unbound benchmark 3.0. A residue was considered at the interface if any of its atoms was within 6 Å from any atom of the other protein. Steric clash was defined as any two atoms of different proteins within 2 Å, including atoms of the backbone. For each complex a conformation with the minimal number of clashes was selected.

3   Benchmarking

Because of limitations of the experimental techniques the vast majority of protein structures in a genome would be models. However, sensitivity of the docking methods to the inherent inaccuracies of protein models, as opposed to the experimentally determined high-resolution structures, has remained largely untested, primarily due to the absence of appropriate benchmark sets. Protein models in such sets should have pre-defined accuracy levels and, at the same time, be similar to actual protein models in terms of structural motifs and packing. The sets should also be large enough to ensure statistical reliability of the benchmarking. The traditional protein-protein benchmark sets contain only the X-ray structures. An earlier study on low-resolution free docking of protein models utilized simulated (not actual) protein models – artificially distorted structures with limited similarity to actual models.

We developed a set of protein models based on 63 binary protein-protein complexes from DOCKGROUND, which have experimentally resolved unbound structures for both proteins. This allowed comparison to the traditional docking of unbound X-ray structures. However, only one third of proteins in the dataset were true homology models and the rest was generated by the Nudged Elastic Band method [15]. In the new release [16], we report a new, 2.5 times larger set of protein models with six levels of accuracy. All structures were built by the I-TASSER modeling procedure [16] without any additional generation of intermediate structures. Thus, the new set contains a larger number of complexes, all of them *bona fide* models, providing an objective, statistically significant benchmark for systematic testing protein-protein docking approaches on modeled structures.

In this paper, we address the problem of models' utility in protein docking using our benchmark set of protein models. The quality of free and template-based docking predictions built from these models was thoroughly assessed to reveal the tolerance limits of docking to structural inaccuracies of the protein models. The predictive power of the currently available rigid-body and flexible docking approaches is similar [11]. Thus in this study we used basic rigid-body approaches that would clearly reveal the general similarities and differences in free and template-based docking performance depending on the accuracy of the interacting protein models.

The results (Fig. 3) show that the existing docking methodologies can be successfully applied to protein models with a broad range of structural accuracy; the template-based docking is much less sensitive to inaccuracies of protein models than the free docking; and docking can be successfully applied to entire proteomes where most proteins are models of different accuracy.

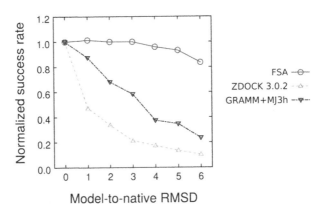

Fig. 3. Normalized success rates for the template-based and free docking. The free docking at low resolution was performed by GRAMM [6] with Miyazawa-Jernigan potentials [17], and at high resolution by ZDOCK 3.0.2 [18]. Template-based docking was performed by the full structure alignment (FSA) [8] using TM-align [19]. The complex was predicted successfully if one out of top 10 predictions was correct (acceptable, medium or high quality, according to CAPRI criteria). All success rates are normalized by the ones for the co-crystallized X-ray structures. The numbers above the data points show the absolute number of successful docking outcomes (out of 165 complexes in Benchmark 2).

4   Genome-Wide Database of Protein Complexes

4.1   GWIDD Design and Content

The progress in 3D modeling of PPI is reflected in the Genome-Wide Docking Database (GWIDD), which provides annotated collection of experimental and modeled PPI structures from the entire universe of life from viruses to humans. The resource has user-friendly search interface, providing preview and download options for experimental and modeled PPI structures. Since its introduction in 2006 and major overhaul in 2010 [20], GWIDD database targets large-scale genome-wide modeling of PPI. Here, we report a major update of GWIDD (version 2.0), which includes an addition of 47,896 proteins and 673,468 PPI from new external PPI sources as well as new functionalities in the Web user interface and data processing pipeline. The database spans 1,652 organisms in all kingdoms of life, from viruses to human. The available structures can be accessed, visualized and downloaded through the user-friendly web interface at http://gwidd.compbio.ku.edu. GWIDD 2.0 has been populated with the PPI imported from BIOGRID [21] and INTACT [22] repositories of non-structural PPI data. Out of 117,697 imported proteins, 92,389 non-redundant proteins, involved in 800,365 binary protein-protein complexes, with sequence length 30-1000 residues were selected for further modeling. Additional screening was performed to remove proteins containing unknown amino acid residues (filtering out 1117 proteins). A summary of the GWIDD content is in Table 1.

Table 1. Overview of GWIDD content. Total numbers in corresponding category are in bold.

Kingdoms of life	Species	Proteins*	Interactions**
Archaea	**59**	**733**	**831**
Bacteria	**609**	**15663**	**92853**
α-proteobacteria	68	518	1143
β-proteobacteria	34	99	137
γ-proteobacteria	95	818	1634
Eukaryota	**504**	**73001**	**601330**
Animals	239	51493	263962
Plants	126	9510	29185
Fungi	92	10944	306725
Viruses	**448**	**1736**	**10274**
Unclassified***	**32**	**139**	**289**

* Proteins with 30-1000 amino acids
** Both monomers are from the same organism
*** Taxonomy IDs correspond to vector sequences and do not have an organism specified by NCBI

The user interface is shown in Fig. 4 and has structure-homology models incorporated in the search procedure. The interface offers search by keywords, sequences or structures for one or both interacting proteins. Alternatively, advanced search can be used for a more specific query, based on different menus and boxes. The additional search options of choosing the model type, an upgrade in the current release, are (i) *X-ray structures*: X-ray/NMR structures as deposited in RCSB PDB; (ii) *Sequence homology models*: PPI models built by homology docking utilizing known structure of a homologous complex found at the sequence level; (iii) *Structure homology models*: Interacting monomers are independently modeled by the sequence homology modeling and subsequently docked by full structural alignment; and (iv) *no model structure (yet)*: model PPI structures to be generated. The Web interface has been tested on all platforms in Windows, Linux and Mac and runs best in Safari, Chrome and Firefox.

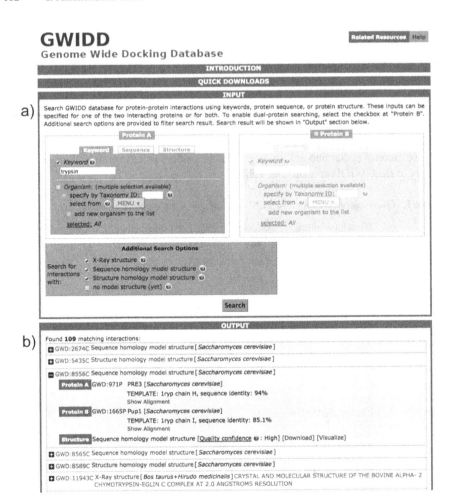

Fig. 4. GWIDD database browser. (a) Search panel with a keyword "trypsin." (b) PPI structure results for the query. The "Download" button provides the atomic coordinates. The "Visualize" button opens a new pop-up window that displays the model structure along with the modeling parameters (Fig. 5).

4.2 Visualizing PPI

The visualization screen of the database (Fig. 5) has advanced and mobile-friendly JavaScript implementation of Jmol, JSmol, to view the 3D structures of PPIs. The default view is a cartoon of the PPI model colored by the chain with interface residues between the monomers in ball-and-stick representation colored by the atom type. The interface residues can be displayed as sticks-only, ball-and-sticks, or spacefill representations. The header contains the name of proteins and the number of residues along with the modeling parameters. The protein name is also mapped either to RCSB (X-ray structures) or UNIPROT (sequence/structure homology models). The interactive list of interface residues, provided below the main JSmol screen, enables highlighting individual interface residues. There is also an option of displaying the van der Waals dot surface or the solvent accessible surface area of the interface residues.

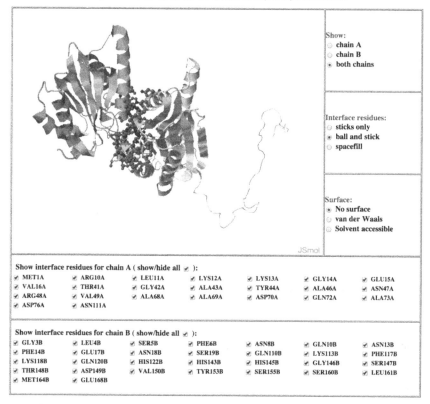

Sequence-homology model of complex
Chain A (215 residues): PRE3 [UNIPROT ID: P38624]
Chain B (261 residues): Pup1 [UNIPROT ID: P25043.1]

MODELING PARAMETERS

Template 1RYP (H,I) Seq. Identity with chain H:94%, Seq. Identity with chain I:85%

Fig. 5. Visualization screen for a sequence homology model of protein complex with trypsin as one of the monomers.

4.3 Future GWIDD Development

The future GWIDD development will primarily focus on improvement of the modeling pipeline. Full sequences of individual proteins will be split into domains according to the UNIPROT data and the structures of interacting domains will be modeled. Models of individual proteins will be taken from the external databases of protein models (e.g.

MODBASE). Proteins that do not have externally precompiled models will be first clustered and a representative protein of each cluster will be modeled by state-of-the-art standalone package I-TASSER. The rest of the proteins in the cluster will be generated from the representative models by simple residue replacements. Other docking techniques will be incorporated, including partial structural alignment and free docking. The complexes will be further annotated by their functionality, place in biological pathways, etc., and search engine will be modified accordingly to enable the new functionality.

Acknowledgments. This study was supported by NIH grant R01GM074255 and NSF grant DBI1262621. Calculations were conducted in part on ITTC computer cluster at The University of Kansas.

References

1. Vakser, I.A.: Low-resolution structural modeling of protein interactome. Curr. Opin. Struct. Biol. **23**, 198–205 (2013)
2. Vakser, I.A.: Protein-protein docking: From interaction to interactome. Biophys. J. **107**, 1785–1793 (2014)
3. Katchalski-Katzir, E., Shariv, I., Eisenstein, M., Friesem, A.A., Aflalo, C., Vakser, I.A.: Molecular surface recognition: Determination of geometric fit between proteins and their ligands by correlation techniques. Proc. Natl. Acad. Sci. U.S.A. **89**, 2195–2199 (1992)
4. Szilagyi, A., Zhang, Y.: Template-based structure modeling of protein–protein interactions. Curr. Opin. Struct. Biol. **24**, 10–23 (2014)
5. Zhang, Q.C., Petrey, D., Deng, L., Qiang, L., Shi, Y., Thu, C.A., et al.: Structure-based prediction of protein-protein interactions on a genome-wide scale. Nature **490**, 556–560 (2012)
6. Vakser, I.A.: Protein docking for low-resolution structures. Protein Eng. **8**, 371–377 (1995)
7. Kundrotas, P.J., Zhu, Z., Janin, J., Vakser, I.A.: Templates are available to model nearly all complexes of structurally characterized proteins. Proc. Natl. Acad. Sci. U.S.A. **109**, 9438–9441 (2012)
8. Kundrotas, P.J., Vakser, I.A.: Global and local structural similarity in protein-protein complexes: Implications for template-based docking. Proteins **81**, 2137–2142 (2013)
9. Berman, H.M., Westbrook, J., Feng, Z., Gilliland, G., Bhat, T.N., Weissig, H., et al.: The Protein Data Bank. Nucleic Acids Res. **28**, 235–242 (2000)
10. Anishchenko, I., Kundrotas, P.J., Tuzikov, A.V., Vakser, I.A.: Structural templates for comparative protein docking. Proteins. **83**, 1563–1570 (2015)
11. Lensink, M.F., Wodak, S.J.: Docking, scoring, and affinity prediction in CAPRI. Proteins. **81**, 2082–2095 (2013)
12. Krallinger, M., Valencia, A.: Text-mining and information-retrieval services for molecular biology. Genome Biol. **6**, 224 (2005)
13. Badal, V.D., Kundrotas, P.J., Vakser, I.A.: Text mining for protein docking. PLoS Comp Biol. **11**, e1004630 (2015)
14. Kirys, T., Ruvinsky, A., Tuzikov, A.V., Vakser, I.A.: Rotamer libraries and probabilities of transition between rotamers for the side chains in protein-protein binding. Proteins. **80**, 2089–2098 (2012)
15. Elber, R., Karplus, M.: A method for determining reaction paths in large molecules - application to myoglobin. Chem. Phys. Lett. **139**, 375–380 (1987)

16. Anishchenko, I., Kundrotas, P.J., Tuzikov, A.V., Vakser, I.A.: Protein models docking benchmark 2. Proteins. **83**, 891–897 (2015)

17. Miyazawa, S., Jernigan, R.L.: Self-consistent estimation of inter-residue protein contact energies based on an equilibrium mixture approximation of residues. Proteins **34**, 49–68 (1999)

18. Pierce, B.G., Hourai, Y., Weng, Z.: Accelerating protein docking in ZDOCK using an advanced 3D convolution library. PLoS ONE **6**, e24657 (2011)

19. Zhang, Y., Skolnick, J.: TM-align: A protein structure alignment algorithm based on the TM-score. Nucl. Acid Res. **33**, 2302–2309 (2005)

20. Kundrotas, P.J., Zhu, Z., Vakser, I.A.: GWIDD: Genome-Wide Protein Docking Database. Nucl Acid Res. **38**, D513–D517 (2010)

21. Stark, C., Breitkreutz, B.J., Chatr-Aryamontri, A., Boucher, L., Oughtred, R., Livstone, M.S., et al.: The BioGRID Interaction Database: 2011 update. Nucl Acid Res. **39**, D698–D704 (2011)

22. Orchard, S., et al.: The MIntAct project-IntAct as a common curation platform for 11 molecular interaction databases. Nucl Acid Res. **42**, D358–D363 (2014)

Identifying Essential Proteins by Purifying Protein Interaction Networks

Min Li[1(✉)], Xiaopei Chen[1], Peng Ni[1], Jianxin Wang[1], and Yi Pan[1,2]

[1] School of Information Science and Engineering, Central South University,
Changsha 410083, China
{limin,chenxiaopei,nipeng,jxwang}@csu.edu.cn,
pan@cs.gsu.edu
[2] Department of Computer Science, Georgia State University,
Atlanta, GA 30302-4110, USA

Abstract. Identification of essential proteins based on protein interaction network (PIN) is a very important and hot topic in the post genome era. In this paper, we propose a new method to identify essential proteins based on the purified PIN by using gene expression profiles and subcellular location information. The basic idea behind the proposed purifying method is that two proteins can physically interact with each other only if they appear together at the same subcellular location and are active together at least at a time point in the cell cycle. The original static PIN is marked as S-PIN and the final PIN purified by our method is marked as TS-PIN. To evaluate whether the constructed TS-PIN is more suitable to being used in the identification of essential proteins, six network-based essential protein discovery methods (DC, EC, SC, BC, CC, and IC) are applied on it to identify essential proteins. It is the same way with S-PIN and NF-APIN. NF-APIN is a dynamic PIN constructed by using gene expression data and S-PIN. The experimental results on the protein interaction network of S.cerevisiae shows that all the six network-based methods achieve better results when being applied on TS-PIN than that being applied on S-PIN and NF-APIN.

1 Introduction

With the developments of high-throughput technologies, such as yeast-two-hybrid, tandem affinity purification, and mass spectrometry, a large number of protein-protein interactions have been accumulated. The protein-protein interactions are generally constructed as an undirected network. It has become a hot topic to identify essential proteins from protein interaction networks by using various topological characters. Generally, we said a protein is essential for an organism if its knock-out results in lethality or infertility, i.e., the organism cannot survive without it [1, 2]. As the biological

Y. Pan—This work was supported in part by the National Natural Science Foundation of China under Grants (No.61370024, No.61232001, and No.61428209) and the Program for New Century Excellent Talents in University(NCET-12-0547).

A. Bourgeois et al. (Eds.): ISBRA 2016, LNBI 9683, pp. 106–116, 2016.
DOI: 10.1007/978-3-319-38782-6_9

experiment-based methods, such as gene knockouts [3], RNA interference [4] and conditional knockouts [5], are relatively expensive, time consuming and laborious, computational methods for identifying essential proteins offer the essential candidates in an easier way and furnish the experimental evidence for further researching.

Up to now, a number of network-based essential protein discovery methods have been proposed. The simplest one of all the network-based methods is the degree centrality (DC), known as a centrality-lethality rule, which was proposed by Jeong et al. [6]. It has been observed in several species, such as S. cerevisiae, C.elegans, and D. melanogaster, that the highly connected proteins tend to be essential [7–10]. Some researchers also investigated the reason why the highly connected proteins tend to be essential. Though there are still some disputes, the centrality-lethality rule has been used widely in the identification of essential proteins. Besides DC, several other popular centrality measures used in complex networks, such as betweenness centrality (BC) [11], closeness centrality (CC) [12], subgraph centrality (SC) [13], eigenvector centrality (EC) [14], information centrality (IC) [15], were also been used for the identification of essential proteins. BC is a global metric which calculates the fraction of shortest paths going through a given node. CC is also a global metric, which evaluates the closeness of a given node with all the rest proteins in a given protein interaction network. SC accounts for the participation of a node in all subgraphs of the network. EC simulates a mechanism in which each node affects all of its neighbors in the network and IC describes how information might flow through many different paths. In recent years, Yu et al. [16] studied the importance of bottlenecks in protein interaction networks and investigated its correlation with gene essentiality by constructing a tree of shortest paths starting from each node in the network. Lin et al. proposed two neighborhood-based methods [17] (maximum neighborhood component (MNC) and density of maximum neighborhood component (DMNC)) to identify essential proteins from PIN. In our previous studies, we also proposed two neighborhood-based methods: LAC [18] and NC [19]. LAC predicts essential proteins by using a local average connectivity and NC identifies essential proteins considering how many neighbors a protein has but also the edge clustering coefficient of the interaction which connects the protein and its neighbor.

Though great progresses have been made in the network-based essential protein discovery methods, it is still a challenge to improve the predicted precision as most of these methods are sensitive to the reliability of the constructed PIN. It is well known that the protein-protein interactions generated by high-throughput technologies include high false positives [20, 21]. von Mering et al. [20] investigated the quality of protein-protein interactions and found that there are about 50 % false positives under circumstances and in the nearly 80,000 interactions they studied there are merely 3 % of protein-protein interactions can be detected by more than two experimental methods. In order to overcome the effects of high false positives, some researchers began to construct weighted PINs [10], dynamic PINs [22–24] or propose new methods by integrating the network with different biological information, such as gene expression profiles [25–27], gene ontology annotations [28], domain types [29], orthology [30], protein complexes [31, 32].

In this paper, we propose a new method to purify the PIN with high false positives by using gene expression profiles and subcellular location information to identify essential proteins more accurately. The basic idea behind the proposed purifying method is that two proteins can physically interact with each other only if they are active together at least at a time point in the cell cycle and appear together at the same subcellular location. The original static protein interaction network is marked as S-PIN, and the final protein interaction network purified by our method is denoted by TS-PIN. To test the effectiveness of the purified network TS-PIN, we applied six typical network-based essential protein discovery methods (DC [6], BC [11], CC [12], SC [13], EC [14], and IC [15]) on it and compared the results with those on the original network S-PIN and a dynamic network NF-APIN. The experimental results on the protein interaction network of S.cerevisiae shows that all the six network-based methods achieve better results when being applied on TS-PIN than on S-PIN and NF-APIN in terms of the prediction precision, sensitivity, specificity, positive predictive value, negative predictive value, F-measure, accuracy rate, and a jackknifing methodology. It has been proved that the purified method contributes to filtering false positives in the protein interaction network (PIN) and can help to identify essential proteins more accurately.

2   Methods

In this section, we first introduce how to construct a high-quality network by purifying the protein-protein interactions based on integration of gene expression profiles and subcellular location information.

In this study, S-PIN(Static Protein Interaction Network) referred to is the original network which includes all the protein-protein interactions occur at different time points and locations. The S-PIN can be described as an undirected graph $G(V,E)$, which $V = \{v_1,...,v_n\}$ is the set of proteins, and $E \subseteq V \times V$ is the set of protein-protein interactions.

2.1   Purification by Using Gene Expression Data

It is well known that protein-protein interactions in a cell are changing over time, environments and different stages of cell cycle [33]. Considering that the expression profiles under different time points and conditions provide the information of a protein's dynamic, some researchers have proposed different methods to construct dynamic protein interaction (PIN) network by integrating gene expression data with PIN. It is only possible that two proteins physically interact with each other if their corresponding genes are both expressed at the same time point. For a time point, how to determine whether a gene is expressed? Generally, a potential threshold is used as a cutoff to determine whether a gene is expressed at a time point [22]. Considering the fact that some proteins with low expression values will be filtered even if it is active at some time points, in our previous study [23] we proposed a three-sigma-based method to determine an active threshold for each protein based on the characteristics of its expression curve. Given a gene v, its corresponding gene's expression value at time point i is denoted by $EV(v,i)$, the algorithmic mean of its expression values over times 1 to m is denoted by $\mu(v)$ and the standard deviation of its expression values is denoted by $\sigma(v)$. Here, we adopted the same strategy used in [24] to calculate the active threshold for each gene. For a gene v, its active threshold is computed by using the following formula:

$$Active\_th(v) = \mu(v) + k\sigma(v) \times (1 - F(v)) \tag{1}$$

$$\mu(v) = \frac{\sum_{i=1}^{m} EV(v, i)}{m} \tag{2}$$

$$\sigma(v) = \sqrt{\frac{\sum_{i=1}^{m} (EV(v, i) - \mu(v))^2}{m - 1}} \tag{3}$$

$$F(v) = \frac{1}{1 + \sigma(v)^2} \tag{4}$$

where $k = 2.5$ is used in this paper according to the analysis in [24].

A protein is regarded as active at a time point i if and only if the expression level of its corresponding gene $EV(v, i)$ is larger than its active threshold (i.e., $EV(v, i) > Active\_th(v)$). If there are m time points in the gene expression data, we can use a m-dimensional vector $T(v) = \{t_i(v), i = 1 \text{ to } m\}$ to describe a protein's active time points, where $t_i(v) = 1|$ if $EV(v, i) > Active\_th(v)$ and 0 otherwise.

To reduce effects of noise in the gene expression data, we used the same method described in [24] to filter noisy genes based on time-dependent model and time-independent model [34]. After the processing, S-PIN is purified by using the filtered gene expression data and the three-sigma principle. For an *edge $(u,v) \in E$* in S-PIN, if there exists a time point i that both proteins u and v are active (i.e., $t_i(u) = t_i(v) = 1$), we say that they may interact with each other. If not, the *edge (u,v)* will be removed from the edge set E.

2.2 Purification by Using Subcellular Location Information

It is well known that proteins must be localized at their appropriate subcellular compartment to perform their desired function. The basic idea that we use subcellular location information to purify PIN is that two proteins should be at the same subcellular location if they interact with each other. The subcellular localization database of COMPARTMENT [35] developed by Binder et al., provides the subcellular location information of several species, including yeast, human, mouse, rat, fly, worm and arabidopsis. We download the subcellular location information of yeast. There are 11 different subcellular locations for yeast proteins: cytoskeleton, golgi apparatus, cytosol, endosome, mitochondrion, plasma membrane, nucleus, extracellular space, vacuole, endoplasmic, reticulum, peroxisome. For a protein u, its subcellular location information can be viewed as an *r-dimensional vector* $(r = 11)$, denoted as $L(u) = (l_1,...,l_i, ...,l_r)$. If a protein u is localized at the ith subcellular compartment, $l_i(u) = 1$. For an edge *$(u,v) \in E$*, only there exists a subcellular localization i where $l_i(u) = l_i(v) = 1$, we say that they may interact with each other at subcellular localization i. If not, the *edge* (u,v) will be removed from the edge set E.

2.3   Network-Based Essential Protein Discovery Methods

In the past decades, a number of network-based essential protein discovery methods have been proposed. In this paper, we collect six typical network-based methods and test whether their performances on prediction of essential proteins are improved when being applied on the purified (TS-PIN). As there is no differences for the network-based methods when being applied on S-PIN or TS-PIN, only undirected graph $G(V,E)$ is used in the following definitions. Actually, the edge set of TS-PIN is a subset of E.

Given a protein interaction network $G(V,E)$ and a protein $u \in V$, let A be the adjacency matrix of the network and N_u be the set of its neighbors. The six network-based methods (DC [6], BC [11], CC [12], SC [13], EC [14], and IC [15]) are defined as following:

$$DC(u) = |Nu| \tag{5}$$

$$BC(u) = \sum_{s \neq u \neq t} \frac{\rho(s,u,t)}{\rho(s,t)} \tag{6}$$

$$CC(u) = \frac{|N_u| - 1}{\sum_{v \in V} dist(u,v)} \tag{7}$$

$$SC(u) = \sum_{l=0}^{\infty} \frac{\mu_l(u)}{l!} \tag{8}$$

$$EC(u) = \alpha_{max}(u) \tag{9}$$

$$IC(u) = \left[\frac{1}{|V|} \sum_{v \in V} \frac{1}{I_{u,v}} \right]^{-1} \tag{10}$$

where $\rho(s,t)$ is the total number of shortest paths from node s to node t and $\rho(s,u,t)$ denotes the number of those shortest paths that pass through u, $dist(u,v)$ represents the distance of the shortest path from node u to node v, $\mu_l(u)$ is the number of closed walks of length l which starts and ends at node u, α_{max} is the eigenvector corresponding to the largest eigenvalue of the adjacency matrix A, $I_{u,v} = (c_{u,u} + c_{v,v} - 2 * c_{u,v})^{-1}$ and $C = (c_{u,v}) = (D - A + J)^{-1}$, D is the diagonal matrix of all nodes' degree, J is a matrix whose all elements are 1.

3   Results and Discussion

In order to evaluate whether the proposed TS-PIN is effective for predicting essential proteins, we applied six typical network-based essential protein discovery methods (DC [6], BC [11], CC [12], SC [13], EC [14], and IC [15]) on it and compare the results with those of the same methods applied on the original S-PIN, and NF-APIN. Moreover, the sensitivity (SN), specificity (SP), F-measure (F), positive predictive value (PPV), negative predictive value (NPV), and accuracy (ACC) are calculated for each method on all the three networks. Finally, different proteins identified by six essential proteins discovery methods from TS-PIN and that from S-PIN, and NF-APIN are studied.

3.1 Experimental Data

All the experiments in this study are based on the protein-protein interaction data of S. cerevisiae, which is now the most complete data in all species and has widely been used in the validation of essential protein discovery. The protein-protein interaction data of S.cerevisiae is downloaded from DIP database [36] of 20101010, which contains 5093 proteins and 24743 interactions after the repeated interaction and the self-interactions are removed. S-PIN in this study is the network constructed by the 5093 proteins and 24743 interactions. Other biological information used in this study is as following:

Essential Proteins: The essential proteins of S.cerevisiae are obtained from the following databases: MIPS(Mammalian Protein-Protein Interaction Database) [37], SGD (Saccharomyces Genome Database) [38], DEG(Database of Essential Genes) [39] and SGDP(Saccharomyces Genome Deletion Project) [40]. Out of all the 5093 proteins in S-PIN, 1167 proteins are essential.

Subcellular Localization Data: The Subcellular localization information of S.cerevisiae is obtained from COMPARTMENT [35] database. After removing the repeated information, we finally obtain 11 subcellular locations.

Gene Expression Data: GSE3431 from gene Expression Omnibus (GEO) [41] is used in this paper. GSE3431 is a gene expression profiling of S.cerevisiae over three successive metabolic cycles. For each cycle there are 12 time time points, and the time interval between two time points is 25 min. The 6,777 gene products in the gene expressing profile cover 95 % of the proteins in S-PIN. That is to say, 4846 gene expression profiles are used in our experiment.

3.2 Identification of Essential Proteins from TS-PIN, S-PIN, and NF-APIN

To compare the constructed TS-PIN with S-PIN and NF-APIN, we applied six popular network-based essential protein discovery methods (DC, IC, EC, SC, BC, and CC) on them. Similar to most validation methods for identifying essential proteins, we ranked all the proteins by using each essential protein discovery method and selected a certain top number of proteins as essential candidates. Then, the number of true essential proteins was counted. A comparison of the number of true essential proteins identified from TS-PIN and that identified from S-PIN and NF-APIN by using DC, EC, SC, BC, CC, and IC was shown in Fig. 1.

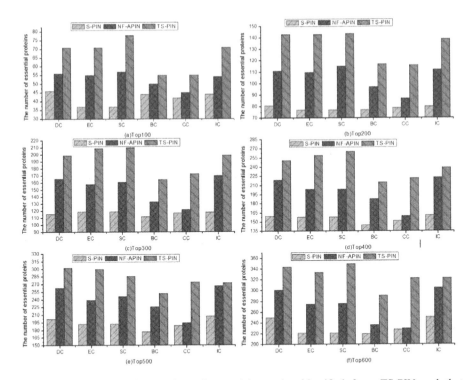

Fig. 1. Comparison of the number of essential proteins identified from TS-PIN and that identified from S-PIN and NF-APIN by using six network-based essential protein discovery methods: DC, EC, SC, BC, CC, and IC.

From Fig. 1 we can see that the performance of all the six network-based methods is significantly improved when being applied on TS-PIN compared to being applied on S-PIN and NF-APIN in terms of a number of true essential proteins identified. Taking the typical centrality measure DC as an example, the improvements of its application on TS-PIN are about 54.35 % and 26.79 %. Respectively, compared to the case applied on S-PIN and NF-APIN when predicting the top 100 essential candidates. For EC and SC, the improvements of them are both more than 90 % when being applied on TS-PIN than on S-PIN for predicting the top 100 essential candidates. When compared to NF-APIN, there are still 29.09 % and 36.84 % improvements for EC and SC when being applied on TS-PIN. The experimental results show that the network-based essential protein discovery methods are sensitive to false positives in PIN and the purification of PIN can help to identify essential proteins more accurately.

3.3   Validated by Accuracy

In recent years, the sensitivity (Sn), specificity (Sp), F-measure (F), positive predictive value (PPV), negative predictive value (NPV), and accuracy (ACC) have also been used for the validation of essential protein discovery [19]. Let P be the number of predicted essential protein candidates and N be the number of the rest proteins. TP represents the number of true essential proteins and FP is the number of non-essential proteins in the prediction, respectively. FN denotes the number of the false negatives, i.e., that true essential proteins are ignored by a method and TN is the number of true negatives that non-essential proteins are correctly predicted to be nonessential. Then, the sensitivity (Sn), specificity (Sp), F-measure (F), positive predictive value (PPV), negative predictive value (NPV), and accuracy (ACC) are defined as following:

$$Sn = \frac{TP}{TP + FN} \tag{11}$$

$$Sp = \frac{TN}{TN + FP} \tag{12}$$

$$F = \frac{2Sn \times Sp}{Sn + Sp} \tag{13}$$

$$PPV = \frac{TP}{TP + FP} \tag{14}$$

$$NPV = \frac{TN}{TN + FN} \tag{15}$$

$$ACC = \frac{TP + TN}{P + N} \tag{16}$$

The sensitivity (Sn), specificity (Sp), F-measure (F), positive predictive value (PPV), negative predictive value (NPV), and accuracy (ACC) of the six methods (DC, EC, SC, BC, CC, IC, LAC, NC, BN and DMNC) applied on TS-PIN, S-PIN and NF-APIN were calculated and the results were shown in Table 1.

Table 1. The sensitivity(Sn), specificity (Sp), F-measure (F), positive predictive value (PPV), negative predictive value (NPV), and accuracy (ACC) of the six methods (DC, EC, SC, BC, CC, and IC) applied on TS-PIN, and that on NF-APIN, and S-PIN

Centrality	Network	Sn	Sp	F	PPV	NPV	ACC
DC	**TS-PIN**	**0.4513**	**0.8146**	**0.5808**	**0.451**	**0.8148**	**0.7229**
	NF-PIN	0.4148	0.8023	0.5469	0.4145	0.8025	0.7045
	S-PIN	0.393	0.7949	0.526	0.3927	0.7951	0.6935
EC	**TS-PIN**	**0.4358**	**0.8093**	**0.5665**	**0.4355**	**0.8096**	**0.7151**
	NF-PIN	0.3837	0.7918	0.5169	0.3834	0.792	0.6888
	S-PIN	0.3665	0.786	0.4999	0.3663	0.7862	0.6801
IC	**TS-PIN**	**0.4389**	**0.8104**	**0.5694**	**0.4386**	**0.8106**	**0.7167**
	NF-PIN	0.4101	0.8007	0.5424	0.4098	0.8009	0.7021
	S-PIN	0.3938	0.7952	0.5267	0.3935	0.7954	0.6939

(Continued)

Table 1. (*Continued*)

Centrality	Network	Sn	Sp	F	PPV	NPV	ACC
SC	**TS-PIN**	**0.4537**	**0.8154**	**0.583**	**0.4533**	**0.8156**	**0.7241**
	NF-PIN	0.3883	0.7933	0.5214	0.388	0.7935	0.6911
	S-PIN	0.3665	0.786	0.4999	0.3663	0.7862	0.6801
BC	**TS-PIN**	**0.4358**	**0.8093**	**0.5665**	**0.4355**	**0.8096**	**0.7151**
	NF-PIN	0.3712	0.7876	0.5046	0.3709	0.7878	0.6825
	S-PIN	0.3385	0.7765	0.4715	0.3383	0.7767	0.666
CC	**TS-PIN**	**0.4389**	**0.8104**	**0.5694**	**0.4386**	**0.8106**	**0.7167**
	NF-PIN	0.3588	0.7834	0.4922	0.3585	0.7836	0.6762
	S-PIN	0.351	0.7807	0.4843	0.3507	0.7809	0.6723

From Table 1, we can see that the sensitivity (Sn), specificity (Sp),F-measure (F), positive predictive value (PPV), negative predictive value (NPV), and accuracy (ACC) of the six methods applied on TS-PIN are consistently higher than that they S-PIN and NF-APIN.

4   Conclusion

In the postgenome era, it has become a hot spot in the systems biology and bioinformatics to identify essential proteins from PINs with the accumulation of protein-protein interactions. However, it is still a challenge to improve the predicted precision of the network-based methods for the available protein-protein interaction data are inevitable to contain many false positives, and the network-based methods are very sensitive to false positives. In this paper, we therefore proposed a new purifying method to filtering the false positives based on the assumption that two proteins can physically interact with each other only if they are at the same subcellular location and active together at least at a time point in the cell cycle. Correspondingly, we constructed TS-PIN and try to improve the accuracy of identifying essential proteins based on the new network. To test the effectiveness of the purified network TS-PIN, we applied six network-based essential protein discovery methods (DC [6], BC [11], CC [12], SC [13], EC [14], and IC [15]) on it and compared the results with that on S-PIN and on a dynamic network NF-APIN. The experimental results of the six network-based methods from TS-PIN are consistently better than from S-PIN and NF-APIN. The experimental results also demonstrates that the quality of TS-PIN is much better than that of S-PIN and NF-APIN, the network-based methods will achieve better results on a more precision PIN. The proposed purifying method can also be used in the PIN of other species if its gene expression profiles and subcellular location information are available. Moreover, some other network-based methods by integration of other biological information can also be applied and tested the constructed TS-PIN. As future work, it would be interesting to apply the purifying method and TS-PIN to other studies, such as identification of protein complexes and functional modules.

References

1. Winzeler, E.A., Shoemaker, D.D., Astromo, A., Liang, H., Anderson, K., et al.: Functional characterization of the S. cerevisiae genome by gene deletion and parallel analysis. Science **285**, 901–906 (1999)
2. Kamath, R.S., Fraser, A.G., Dong, Y., Poulin, G., Durbin, R., et al.: Systematic functional analysis of the Caenorhabditis elegans genome using RNAi. Nature **421**, 231–237 (2003)
3. Giaever, G., Chu, A.M., Ni, L., et al.: Functional profiling of the Saccharomyces cerevisiae genome. Nature **418**(6896), 387–391 (2002)
4. Cullen, L.M., Arndt, G.M.: Genome-wide screening for gene function using RNAi in mammalian cells. Immunol. Cell Biology **83**(3), 217–223 (2005)
5. Roemer, T., Jiang, B., Davison, J., et al.: Large-scale essential gene identification in Candida albicans and applications to antifungal drug discovery. Mol. Microbiol. **50**(1), 167–181 (2003)
6. Jeong, H., Mason, S.P., Barabási, A.L., et al.: Lethality and centrality in protein networks. Nature **411**(6833), 41–44 (2001)
7. Lin, C.C., Juan, H.F., Hsiang, J.T., et al.: Essential Core of Protein − Protein Interaction Network in Escherichia coli. J. Proteome Res. **8**(4), 1925–1931 (2009)
8. Liang, H., Li, W.H.: Gene essentiality, gene duplicability and protein connectivity in human and mouse. TRENDS in Genet. **23**(8), 375–378 (2007)
9. Zhao, B., Wang, J., Li, M., et al.: Prediction of essential proteins based on overlapping essential modules. IEEE Trans. Nanobiosci. **13**(4), 415–424 (2014)
10. Li, M., Wang, J.X., Wang, H., et al.: Identification of essential proteins from weighted protein–protein interaction networks. J. Bioinform. Comput. Biol. **11**(03), 1341002 (2013)
11. Joy, M.P., Brock, A., Ingber, D.E., et al.: High-betweenness proteins in the yeast protein interaction network. BioMed. Res. Int. **2005**(2), 96–103 (2005)
12. Wuchty, S., Stadler, P.F.: Centers of complex networks. J. Theor. Biol. **223**(1), 45–53 (2003)
13. Estrada, E., Rodríguez-Velázquez, J.A.: Subgraph centrality in complex networks. Phy. Rev. E **71**(5), 056103 (2005)
14. Bonacich, P.: Power and centrality: a family of measures. Am. J. Sociol. **92**(5), 1170–1182 (1987)
15. Stevenson, K., Zelen, M.: Rethinking centrality: methods and examples. Soc. Netw. **11**(1), 1–37 (1989)
16. Yu, H., Kim, P.M., Sprecher, E., et al.: The importance of bottlenecks in protein networks: correlation with gene essentiality and expression dynamics. PLoS Comput. Biol. **3**(4), e59 (2007)
17. Lin, C.Y., Chin, C.H., Wu, H.H., et al.: Hubba: hub objects analyzer—a framework of interactome hubs identification for network biology. Nucleic Acids Res. **36**(suppl 2), W438–W443 (2008)
18. Li, M., Wang, J., Chen, X., et al.: A local average connectivity-based method for identifying essential proteins from the network level. Comput. Biol. Chem. **35**(3), 143–150 (2011)
19. Wang, J., Li, M., Wang, H., et al.: Identification of essential proteins based on edge clustering coefficient. IEEE/ACM Trans. Comput. Biol. Bioinform. **9**(4), 1070–1080 (2012)
20. Von Mering, C., Krause, R., Snel, B., et al.: Comparative assessment of large-scale data sets of protein–protein interactions. Nature **417**(6887), 399–403 (2002)
21. Brohee, S., Van Helden, J.: Evaluation of clustering algorithms for protein-protein interaction networks. BMC Bioinform. **7**(1), 488 (2006)

22. Tang, X., Wang, J., Liu, B., et al.: A comparison of the functional modules identified from time course and static PPI network data. BMC Bioinform. **12**(1), 339 (2011)

23. Wang, J., Peng, X., Li, M., et al.: Construction and application of dynamic protein interaction network based on time course gene expression data. Proteomics **13**(2), 301–312 (2013)

24. Xiao, Q., Wang, J., Peng, X., et al.: Detecting protein complexes from active protein interaction networks constructed with dynamic gene expression profiles. Proteome Sci. **11** (suppl 1), S20 (2013)

25. Li, M., Zhang, H., Wang, J., et al.: A new essential protein discovery method based on the integration of protein-protein interaction and gene expression data. BMC Syst. Biol. **6**(1), 15 (2012)

26. Li, M., Zheng, R., Zhang, H., et al.: Effective identification of essential proteins based on priori knowledge, network topology and gene expressions. Methods **67**(3), 325–333 (2014)

27. Tang, X., Wang, J., Zhong, J., Pan, Y.: Predicting essential proteins based on weighted degree centrality. IEEE/ACM Trans. Comput. Biol. Bioinform. **11**(2), 407–418 (2014)

28. Kim, W.: Prediction of essential proteins using topological properties in GO-pruned PPI network based on machine learning methods. Tsinghua Sci. Technol. **17**(6), 645–658 (2012)

29. Peng, W., Wang, J., Cheng, Y., et al.: UDoNC: an algorithm for identifying essential proteins based on protein domains and protein-protein interaction networks. IEEE/ACM Trans. Comput. Biol. Bioinform. **12**(2), 276–288 (2015)

30. Peng, W., Wang, J., Wang, W., et al.: Iteration method for predicting essential proteins based on orthology and protein-protein interaction networks. BMC Syst. Biol. **6**(1), 87 (2012)

31. Li, M., Lu, Y., Niu, Z., et al.: United complex centrality for identification of essential proteins from PPI networks (2015). doi:10.1109/TCBB.2015.2394487

32. Ren, J., Wang, J., Li, M., et al.: Discovering essential proteins based on PPI network and protein complex. Int. J. Data Mining Bioinform. **12**(1), 24–43 (2015)

33. Przytycka, T.M., Singh, M., Slonim, D.K.: Toward the dynamic interaction: it's about time. Brief Bioinform. **11**, 15–29 (2010)

34. Wu, F.X., Xia, Z.H., Mu, L.: Finding significantly expresses genes from timecourse expression profiles. Int. J. Bioinform. Res. Appl. **5**(1), 50–63 (2009)

35. Binder, J.X., Pletscher-Frankild, S., Tsafou, K., et al.: COMPARTMENTS: unification and visualization of protein subcellular localization evidence. Database, 2014: bau012 (2014)

36. Xenarios, I., Rice, D.W., Salwinski, L., et al.: DIP: the database of interacting proteins. Nucleic Acids Res. **28**(1), 289–291 (2000)

37. Mewes, H.W., Frishman, D., Mayer, K.F.X., et al.: MIPS: analysis and annotation of proteins from whole genomes in 2005. Nucleic Acids Res. **34**(suppl 1), D169–D172 (2006)

38. Cherry, J.M., Adler, C., Ball, C., et al.: SGD: Saccharomyces genome database. Nucleic Acids Res. **26**(1), 73–79 (1998)

39. Zhang, R., Ou, H.Y., Zhang, C.T.: DEG: a database of essential genes. Nucleic Acids Res. **32**(suppl 1), D271–D272 (2004)

40. Saccharom yces Genome Deletion Project. http://www.sequence.stanford.edu/group/yeast_deletion_project

41. Tu, B.P., Kudlicki, A., Rowicka, M., et al.: Logic of the yeast metabolic cycle: temporal compartmentalization of cellular processes. Science **310**(5751), 1152–1158 (2005)

42. Holman, A.G., Davis, P.J., Foster, J.M., et al.: Computational prediction of essential genes in an unculturable endosymbiotic bacterium, Wolbachia of Brugia malayi. BMC Microbiol. **9**(1), 243 (2009)

Differential Functional Analysis and Change Motifs in Gene Networks to Explore the Role of Anti-sense Transcription

Marc Legeay[1,2], Béatrice Duval[1(✉)], and Jean-Pierre Renou[2]

[1] LERIA - Université d'Angers - UNAM, 2 bd Lavoisier, 49045 Angers, France
beatrice.duval@univ-angers.fr
[2] Institut de Recherche en Horticulture et Semences (IRHS),
UMR1345 INRA-Université d'Angers-AgroCampus Ouest, Centre Angers-Nantes,
42 rue Georges Morel - BP 60057, 49071 Beaucouzé, France
marc.legeay@univ-angers.fr

Abstract. Several transcriptomic studies have shown the widespread existence of anti-sense transcription in cell. Anti-sense RNAs may be important actors in transcriptional control, especially in stress response processes. The aim of our work is to study gene networks, with the particularity to integrate in the process anti-sense transcripts. In this paper, we first present a method that highlights the importance of taking into account anti-sense data into functional enrichment analysis. Secondly, we propose the differential analysis of gene networks built with and without anti-sense actors in order to discover interesting change motifs that involve the anti-sense transcripts. For more reliability, our network comparison only studies the conservative causal part of a network, inferred by the C3NET method. Our work is realized on transcriptomic data from apple fruit.

1 Introduction

Understanding the regulation mechanisms in a cell is a key issue in bioinformatics. As large-scale expression datasets are now available, gene network inference (GNI) is a useful approach to study gene interactions [1], and a lot of methods have been proposed in the literature for this reverse engineering task [2–4]. Going a step further, the field of differential network analysis [5,6] proposes to decipher the cellular response to different situations. In medicine the comparison of interaction maps observed in cancerous tissues and healthy tissues may reveal network rewiring induced by the disease [7]. In these approaches, the comparative analysis is performed on networks that involve the same set of actors, namely the genes or proteins of the studied organism.

The aim of our work is to study gene networks, with the particularity to integrate in the process anti-sense transcripts. Anti-sense RNAs are endogenous RNA molecules whose partial or entire sequences exhibit complementarity to other transcripts. Their different functional roles are not completely known

© Springer International Publishing Switzerland 2016
A. Bourgeois et al. (Eds.): ISBRA 2016, LNBI 9683, pp. 117–126, 2016.
DOI: 10.1007/978-3-319-38782-6_10

but several studies suggest that they play an important role in stress response mechanisms [8]. A recent study with a full genome microarray for the apple has detected significant anti-sense transcription for 65 % of expressed genes [9], which suggests that a large majority of protein coding genes are actually concerned by this process.

The work described in this paper proposes a large-scale analysis of apple transcriptomic data, with measures of anti-sense transcripts. To highlight the impact of anti-sense transcription, we propose to compare context-specific gene networks that involve different kinds of actors, on one hand the sense transcripts that are usually used in gene networks and on the other hand the sense and anti-sense transcripts. GNI methods generally find many false positive interactions, and some authors have proposed to study the core part of a gene network [10], by only computing for each gene the most significant interaction with another gene. We follow this line in order to discover which interactions of the core network are modified when we integrate in our GNI method the anti-sense transcripts. To characterize these modifications, we define the notions of change motifs for the comparison graph. Our preliminary results on the apple datasets show that relevant information is provided by this approach.

In Sect. 2, we present the motivations of this work and the apple datasets that are used in our study. In Sect. 3, we present a differential functional analysis that reveals the interest of taking into account anti-sense data. In Sect. 4, we present our method to compare two core gene networks and to detect motifs that underline the role of anti-sense transcripts.

2   Motivations and Biological Material

Several studies have revealed the widespread existence of anti-sense RNAs in many organisms. Anti-sense transcripts can have different roles in the cell [8]. A significant effect is the post-transcriptional gene silencing: the self-regulatory circuit where the anti-sense transcript hybridizes with the sense transcript to form a double strand RNA (dsRNA) that is degraded in small interfering RNAs (siRNA). Previous studies on *Arabidopsis Thaliana* showed that sense and anti-sense transcripts for a defense gene (RPP5) form dsRNA and generate siRNA which presumably contributes to the sense transcript degradation in the absence of pathogen infection [11].

In [9], the authors have combined microarray analysis with a dedicated chip and high-throughput sequencing of small RNAs to study anti-sense transcription in eight different organs (seed, flower, fruit, ...) of apple (*Malus* × *domestica*). Their atlas of expression shows several interesting points. Firstly, the percentage of anti-sense expression is higher than that reported in other studies, since they identify anti-sense transcription for 65 % of the sense transcripts expressed in at least one organ, while it is about 30 % in previous *Arabidopsis Thaliana* studies. Secondly, the anti-sense transcript expression is correlated with the presence of short interfering RNAs. Thirdly, anti-sense expression levels vary depending on both organs and Gene Ontology (GO) categories. It is higher for genes belonging to the "defense" GO category and on fruits and seeds.

In order to study the impact of anti-sense transcripts, we use data of apple fruit during fruit ripening. The fruit ripening is a stress-related condition involving "defense" genes. We analyse RNA extracted from the fruit of apple thanks to the chip AryANE v1.0 containing 63011 predicted sense genes and 63011 complementary anti-sense sequences. This chip allows us to study the role of anti-sense transcripts at the genome-wide level by supplying transcriptional expression on both sense and anti-sense transcripts. We study the fruit ripening process described by two conditions: harvest (H) and 60 days after harvest (60DAH), and for each condition, 22 samples of apple fruit have been analysed. We first identify transcripts displaying significant differences between the two conditions (p-val<1%). With a further threshold of 1 log change between the two conditions, we found 931 sense (S) and 694 anti-sense transcripts (AS) differentially expressed, with among them, 200 transcripts ($S \cap AS$) for which both sense and anti-sense fulfil the condition. In the following, these 1625 transcripts will be called *transcripts of interest* for our study of apple ripening.

3   Differential Functional Analysis

A lot of tools are available to identify which GO categories are statistically over-represented in a set of genes. The Cytoscape plugin BiNGO [12] performs this task in a flexible and interactive way and moreover, the output of BiNGO is a graph where nodes represent GO categories and arcs represent the hierarchy between categories. In this visualisation, the size of a node is proportional to the number of genes in the test set annotated by this category, and the color of a node codes the over-representation: dark orange categories are most significantly over-represented, whereas white nodes are not significant but are included to show the hierarchy linking the dark categories.

In our experiment about apple ripening, the analysis of probes that are differentially expressed shows an important proportion of anti-sense actors: 694 AS probes for 931 S probes. Therefore it is relevant to question the role of these anti-sense actors in the ripening process. To look at this point, we propose a differential functional analysis where we compare the functional categories over-represented in the set of S probes and the functional categories over-represented in the set of probes $S \cup AS$. Anti-sense probes are not associated with a GO category. We decided to associate an anti-sense probe with the category of its corresponding sense probe[1]. This decision is based on the fact that due to its sequence complementarity, an anti-sense transcript may interact with the corresponding sense transcript, or at least with a very close member of the gene family.

The differential functional analysis is performed as follows. We apply BiNGO on the set S containing 931 genes, and we apply BiNGO on the set SAS ($S \cup AS$) containing 494 supplementary genes. These 494 genes are the genes associated with the AS probes of interest for which the corresponding sense probe is not

---

[1] Because there is no GO category for apple transcripts, we use *Arabidopsis Thaliana* orthologs in order to associate apple transcripts with GO categories.

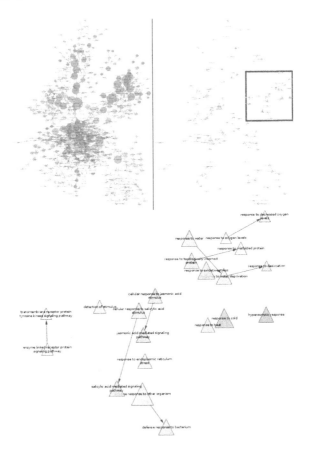

Fig. 1. BiNGO outputs for SAS probes (left), the revealed-by-AS terms (right) with a zoom (bottom) of the red square of the revealed-by-AS terms. Node size denotes the number of genes in the GO category. Node color denotes the over-representation of the GO category (yellow : low, orange : high). Arcs denote the hierarchy between nodes. (Color figure online)

differentially expressed. We thus obtain two sets of GO categories represented as two sub-graphs of the GO. Our proposal is then to compute the difference of these two sets; this provides a set of functional terms that are over-represented only when we include the AS probes in the functional analysis and we call these terms the *revealed-by-AS* terms.

Figure 1 shows the SAS ontology and the difference with the S ontology that gives the revealed-by-AS terms. We can see on this representation that the nodes of the difference occur on many branches of the SAS ontology, meaning that the revealed-by-AS terms are not specific to a GO category.

This differential analysis gives us 125 revealed-by-AS terms, associated with their p-values. We present the top 10 terms in Table 1, where we report how many genes are associated with the terms. As pointed out by the authors of

Table 1. Top 10 of revealed-by-AS terms, sorted by p-values. For each term, we indicate the number of genes associated to transcripts of interest, and if the term is a most specific revealed-by-AS term.

GO category	p-value	# genes	Most specific
hyperosmotic response	4.4644e-05	30	yes
response to cold	5.4256e-05	56	yes
multicellular organismal process	1.1794e-04	225	no
response to high light intensity	5.2641e-04	25	yes
growth	1.6007e-03	56	no
cellular biosynthetic process	1.6007e-03	329	no
cell growth	1.8010e-03	51	no
regulation of response to stimulus	1.8376e-03	58	no
salicylic acid mediated signaling pathway	2.0524e-03	30	yes
jasmonic acid mediated signaling pathway	2.2286e-03	27	yes

BiNGO, due to the interdependency between GO categories in the hierarchy, the most relevant terms of the output are the terms located farthest down the hierarchy, that correspond to more specific functions. Therefore the interpretation will focus on the most specific terms according to the ontology hierarchy, as indicated in Table 1. These revealed-by-AS terms highlight biological functions that are over-represented in our probes of interest only when we include AS informations. Our experiment concerns the complex process of apple ripening. In this experiment, between harvest (H) and 60 days after harvest (60DAH), fruits are stored in cold rooms and have to react to cold stress. We notice that the `response to cold` term is a revealed-by-AS term. Therefore, if we do not consider the anti-sense data, we loose important information for a functional analysis. Moreover, the `response to cold` term is represented by 56 sense or anti-sense actors in our probes of interest; if we examine the differential expression of these transcripts, we notice that 24 of them are anti-sense probes with a diminution of their expression between H and 60DAH, while the corresponding sense probes have no differential expression. The differential functional analysis that we propose is thus a way to focus on interesting anti-sense transcripts that deserve a further biological study.

4   Network Comparison

4.1   Inference of the Core Part of a Gene Network

Many models have been proposed to infer gene networks from transcriptomic data. Reviews of the reverse engineering methods can be found in [2,3,13]. A family of inference methods reconstruct pairwise gene interaction networks by measuring with a statistical criterion whether two genes are co-expressed or co-regulated. This statistical measure can be the Spearman or Pearson correlation [14], or the mutual information [10,15]. Mutual information enables the detection of non-linear relationships. These methods need a step of thresholding to decide which values of the statistical measure are significant. One major drawback of these methods is that many of the predicted interactions are false positives. We can differentiate two types of false positive interactions: an interaction that does not biologically exist, and an indirect interaction. If two genes g_2 and g_3 are regulated by g_1, then mutal information (as well as correlation) between g_2 and g_3 is high and an indirect interaction is put in the inferred network. Indirect interactions lead in large gene networks difficult to interpret by biologists and they must be pruned from the output networks [15,16]. To avoid this pitfall, the method C3NET proposes to compute the conservative causal core of a gene network, by selecting for each gene a unique interaction. Figure 2 decomposes the C3NET algorithm, that we use in this work. From the mutual information matrix, for each line corresponding to a gene g, the algorithm identifies the maximal mutual information which defines the best neighbor that will be connected to g in the network. Experiments have shown the good ability of this conservative method to capture the causal structure of a regulatory network.

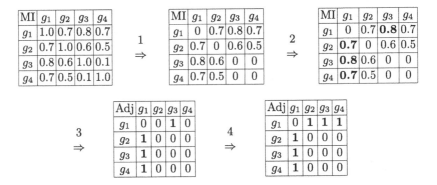

Fig. 2. C3NET procedure from the mutual information matrix to the network adjacency matrix. The mutual information matrix is computed from transcriptomic data. Step 1: non-significant and diagonal values are suppressed. Step 2: the maximal mutual information is identified for each row. Step 3: the matrix is transformed into a boolean matrix. Step 4: the resulting adjacency matrix is made symmetric because mutual information does not provide directional information. This is the adjacency of the computed network.

4.2   Comparison of Core Networks : Change Motifs

To study the role of anti-sense transcripts in gene regulation networks, we propose to compare two networks obtained by C3NET, the N_S network using only sense actors, and the N_{SAS} network using sense and anti-sense actors. Our goal is to identify which direct interactions are modified in the core network when

MI	s_1	s_2	s_3	s_4
s_1	0	0.7	0.8	0.7
s_2	0.7	0	0.6	0.5
s_3	0.8	0.6	0	0
s_4	0.7	0.5	0	0

MI	s_1	s_2	s_3	s_4	as_1	as_2	as_3	as_4
s_1	0	0.7	0.8	0.7	0.6	0.5	0.7	0.6
s_2	0.7	0	0.6	0.5	0.4	0.6	0	0.5
s_3	0.8	0.6	0	0	0.8	0.9	0.3	0
s_4	0.7	0.5	0	0	0.6	0	0.5	0.6
as_1	0.6	0.4	0.8	0.6	0	0.6	0.4	0.7
as_2	0.5	0.6	0.9	0	0.6	0	0.6	0.4
as_3	0.7	0	0.3	0.5	0.4	0.6	0	0.5
as_4	0.6	0.5	0	0.6	0.7	0.4	0.5	0

Fig. 3. Mutual information matrices from S (left) and SAS (right). With anti-sense data, the maximal mutual information changes (green values). s_3 is connected with s_1 in N_S (red value) and with as_2 in N_{SAS}. (Color figure online)

we consider anti-sense transcripts. To achieve this, we need the first three steps of C3NET (we do not need the symmetric matrix). Figure 3 illustrates modifications of interactions in the matrix. In N_S for s_3 the direct interaction occurs with s_1 but in N_{SAS}, the direct interaction occurs with as_2. This is this type of modification that we want to identify.

When we integrate anti-sense actors in the core network computation, we focus on sense nodes which become connected with an anti-sense node. It means that an arc from N_S between two sense nodes is now an arc from a sense to an anti-sense in N_{SAS}.

In order to highlight those modifications, we construct a comparison graph G by adding N_S arcs to N_{SAS}. We visualize this graph with Cytoscape [17], where we color arcs of G depending on the network they belong to: an arc is green if it only exists in N_{SAS}, red if it only exists in N_S and grey if it exists in both networks. With this color code, an interaction from N_S replaced in N_{SAS} is represented by a sense node with a green and a red arc (Fig. 4a).

Around this elementary motif, named M_0, we observe richer configurations represented in Fig. 4b. M_1 motif denotes a strong link between a sense and an anti-sense. M_2 motif reveals that the interaction between $S1$ and $S2$ observed in N_S is in fact an indirect one that involves the anti-sense $AS3$.

We have constructed comparison graphs from H and 60DAH experiments. The 60DAH comparison graph can be visualized in Fig. 4c. This visualization allows biologists to explore what gene links are impacted by the anti-sense transcripts. In a more quantitative way, we give in Table 2 the count of the motifs existing in each of the graphs. The most important information is the number of M_0 motifs that indicates how many sense actors are connected to an anti-sense. We notice that there are about 380 M_0 motifs, which means that 40 % of the 931 S transcripts are involved in a M_0 motif, for 50 % of the 694 AS transcripts. Among these M_0 motifs, about 30 % are M_1 motifs, where sense and anti-sense are strongly connected. Richer motifs M_2 are less represented. As the core network tries to capture the most important gene interactions, the fact that 40 % of S network is impacted by anti-sense actors shows that anti-sense transcripts play a role in fruit ripening.

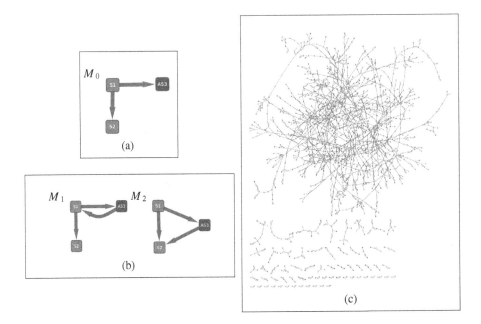

Fig. 4. Motifs and comparison graph of a Sense network with a Sense and Anti-sense network. Blue nodes denote sense nodes and purple nodes denote anti-sense nodes. **(4a)** Elementary motif. **(4b)** Richer motifs observed. **(4c)** Comparison graph between N_S and N_{SAS} networks from 60DAH experiment. (Color figure online)

Table 2. Number of motifs, number of motifs containing at least one revealed-by-AS transcript and number of these transcripts being in motifs. The number of revealed-by-AS transcripts associated with the term is noted in parentheses.

Experiment		H			60DAH		
Motif		M_0	M_1	M_2	M_0	M_1	M_2
Global	# motifs	371	107	18	384	116	19
hyperosmotic response	# motifs	18	3	1	21	8	0
(37 transcripts)	# transcripts	17	3	1	18	8	0
response to cold	# motifs	25	4	2	32	12	1
(63 transcripts)	# transcripts	24	4	2	29	13	1
response to high light	# motifs	14	5	1	13	7	0
intensity (31 transcripts)	# transcripts	12	5	2	12	7	0
salicylic acid mediated	# motifs	17	3	0	14	3	1
signaling pathway (36 transcripts)	# transcripts	15	3	0	13	3	1
jasmonic acid mediated	# motifs	14	2	0	13	3	0
signaling pathway (31 transcripts)	# transcripts	13	2	0	12	3	0

4.3 Change Motifs and Functional Analysis

In Sect. 2 we have proposed a differential functional analysis and defined the revealed-by-AS terms. We now combine the information provided by the functional analysis with the information provided by motifs. To illustrate this, we select in Table 1 the most specific GO categories from the top 10 revealed-by-AS terms and study what kind of motifs are related to these terms. Table 2 counts, for each term, the number of motifs which contain at least one transcript associated to the term, and the number of transcripts associated to the term present in the motifs. We notice that in both experiments, for each term, around 40 % of the transcripts are involved in a M_0 motif. This observation encourages us to study the gene regulatory networks related to the revealed-by-AS terms, which will be the next step of this work.

5 Conclusion

The aim of our work is to study gene networks, with the particularity to integrate in the process anti-sense transcripts. Firstly we propose a method that highlights biological functions impacted by anti-sense transcription. Biological functions are identified by computing the difference between two ontologies. Secondly we propose a differential gene network analysis allowing to identify which direct interactions are modified. We combine these two methods to submit limited sets of anti-sense transcripts to the biological interpretation.

The field of differential network analysis is certainly a promising approach to study context-specific regulation networks. For example, the method proposed in [18] compares two networks computed by C3NET corresponding to different cell conditions (disease versus normal). The aim is to identify the disease network, that is the interactions that only appear in disease-related cells. This method can not be applied in our case. In fact, we rely on the same algorithm to compute a network of direct interactions. But in our case, we compare two networks that involve different actors, the sense transcripts in one hand and the sense and anti-sense transcripts in the other hand. These two networks concern the same experimental condition and the change motifs that we compute aim to highlight potential anti-sense actions.

References

1. Brazhnik, P., de la Fuente, A., Mendes, P.: Gene networks: how to put the function in genomics. Trends Biotechnol. **20**(11), 467–472 (2002)
2. Bansal, M., Belcastro, V., Ambesi-Impiombato, A., di Bernardo, D.: How to infer gene networks from expression profiles. Mol. Syst. Biol. **3**(1), 78 (2007)
3. Marbach, D., Costello, J.C., Küffner, R., Vega, N.M., Prill, R.J., Camacho, D.M., Allison, K.R., The DREAM5 Consortium, Kellis M., Collins J.J., Stolovitzky G.: Wisdom of crowds for robust gene network inference. Nat. Methods **9**(8), 796–804 (2012)

4. Emmert-Streib, F., Glazko, G., Altay, G., de Matos, S.R.: Statistical inference and reverse engineering of gene regulatory networks from observational expression data. Bioinform. Comput. Biol. **3**, 8 (2012)
5. Sharan, R., Ideker, T.: Modeling cellular machinery through biological network comparison. Nat. Biotechnol. **24**(4), 427–433 (2006)
6. Ideker, T., Krogan, N.J.: Differential network biology. Mol. Syst. Biol. **8**(1), 565 (2012)
7. Barabási, A.-L., Gulbahce, N., Loscalzo, J.: Network medicine: a network-based approach to human disease. Nat. Rev. Genet. **12**(1), 56–68 (2011)
8. Pelechano, V., Steinmetz, L.M.: Gene regulation by antisense transcription. Nat. Rev. Genet. **14**(12), 880–893 (2013)
9. Celton, J.-M., Gaillard, S., Bruneau, M., Pelletier, S., Aubourg, S., Martin-Magniette, M.-L., Navarro, L., Laurens, F., Renou, J.-P.: Widespread anti-sense transcription in apple is correlated with siRNA production and indicates a large potential for transcriptional and/or post-transcriptional control. New Phytologist, 287–299 (2014)
10. Altay, G., Emmert-Streib, F.: Inferring the conservative causal core of gene regulatory networks. BMC Syst. Biol. **4**(1), 132 (2010)
11. Yi, H., Richards, E.J.: A cluster of disease resistance genes in Arabidopsis is coordinately regulated by transcriptional activation and RNA silencing. Plant Cell **19**, 2929–2939 (2007)
12. Maere, S., Heymans, K., Kuiper, M.: BiNGO: a Cytoscape plugin to assess over-representation of gene ontology categories in biological networks. Bioinformatics **21**(16), 3448–3449 (2005)
13. Friedel, S., Usadel, B., von Wiren, N., Sreenivasulu, N.: Reverse Engineering: A Key Component of Systems Biology to Unravel Global Abiotic Stress Cross-Talk. Front. Plant Sci. **3**, 294 (2012)
14. Langfelder, P., Horvath, S.: WGCNA: an R package for weighted correlation network analysis. BMC Bioinform. **9**(1), 559 (2008)
15. Margolin, A.A., Nemenman, I., Basso, K., Wiggins, C., Stolovitzky, G., Favera, R.D., Califano, A.: ARACNE: an algorithm for the reconstruction of gene regulatory networks in a mammalian cellular context. BMC Bioinform. **7**(Suppl 1), S7 (2006)
16. Zhang, X., Liu, K., Liu, Z.-P., Duval, B., Richer, J.-M., Zao, X.-M., Hao, J.-K., Chen, L.: NARROMI: a noise and redundancy reduction technique improves accuracy of gene regulatory network inference. Bioinformatics **29**(1), 106–113 (2012)
17. Shannon, P., Markiel, A., Ozier, O., Baliga, N.S., Wang, J.T., Ramage, D., Amin, N., Schwikowski, B., Ideker, T.: Cytoscape: a software environment for integrated models of biomolecular interaction networks. Genome Res. **13**(11), 2498–2504 (2003)
18. Altay, G., Asim, M., Markowetz, F., Neal, D.E.: Differential C3NET reveals disease networks of direct physical interactions. BMC Bioinform. **12**(1), 296 (2011)

Predicting MicroRNA-Disease Associations by Random Walking on Multiple Networks

Wei Peng[1,2(✉)], Wei Lan[3], Zeng Yu[2], Jianxin Wang[3], and Yi Pan[2]

[1] Computer Center, Kunming University of Science and Technology,
Kunming 650050, China
weipeng1980@gmail.com
[2] Department of Computer Science, Georgia State University,
Atlanta, GA 30302-4110, USA
[3] School of Information Science and Engineering, Central South University,
Changsha 410083, China

Abstract. MicroRNA refers to a set of small non-coding RNA which plays important roles in regulating specific mRNA targets and suppressing their expression. Previous researches have verified that the deregulations of microRNA are closely associated with human disease. However it is still a big challenge to design an effective computational method which can integrate multiple biological information to predict microRNA-disease associations. Based on the observation that microRNAs with similar functions tend to associate with common diseases, the diseases sharing similar phenotypes are likely caused by common microRNAs and similar environment factors also affect microRNAs with similar functions and diseases with similar phenotypes. In this work, we design a computational method which can combine microRNA, disease and environmental factors to predict microRNA-disease associations. The method namely ThrRWMDE, takes several steps of random walking on three different biological networks, microRNA-microRNA functional similarity network(MFN), disease-disease similarity network(DSN) and environmental factor similarity network(ESN) respectively so as to get microRNA-disease association information from the neighbors in corresponding networks. In the course of walking, the microRNA-disease association information will also be transferred from one network to another according to the interactions between the nodes in different networks. Our method is not only a framework which can effectively integrate different types of biological methods but also can easily treat these information differently with respect to the topological and structural difference of the three networks. The results of experiment show that our method achieves better prediction performance than other state-of-the-art methods.

1 Introduction

MicroRNAs are a set of non-coding RNA which are approximately 22 nucleotides in length. MicroRNAs regulate the gene expressions at the post-transcriptional level. They suppress protein synthesis or initiate mRNA degradation by binding

© Springer International Publishing Switzerland 2016
A. Bourgeois et al. (Eds.): ISBRA 2016, LNBI 9683, pp. 127–135, 2016.
DOI: 10.1007/978-3-319-38782-6_11

to the 3'UTR of targeted mRNA. MicroRNAs are involved in a wide range of biological functions, such as cell proliferation, cell death, stem cell differentiation, hematopoiesis. Recently some researches have pointed out some microRNAs have close associations with human diseases including cancers, heart disease and neurological disease. With the development of high throughput techniques, plenty of microRNAs have been detected. However the functions of most of these microRNAs still remain to be discovered. Lots of works are needed to identify which microRNAs are associated to diseases. Experimental methods to detect disease related microRNA are expensive and time-consuming. Therefore, some computational methods have been proposed.

Most of computational methods are based on the assumption that microRNAs with similar functions tend to association with common disease and the diseases shared similar phenotypes are likely caused by common microRNA [1]. Jiang et al. [1] have proposed first microRNA-disease association prediction method, which has constructed a functionally related microRNA network and a human phenome-microRNA network. After that, they calculated a score for each microRNA by the cumulative hypergeometric distribution. Xuan et al. [2] have proposed a new method, HDMP, which calculates the similarity between microRNAs according to the similarity of disease their associated. Then they selected the weighted k most similar neighbors to predict disease-related microRNAS. Considering that previous method only use local network information to predict microRNA-disease association, Chen et al. [3] have adopted global network information and applied a Random walk with Restart method (RWRMDA) on microRNA-microRNA functional similarity network to predict microRNA-disease association. Recently, Chen's group [4] have proposed a new method RLSMDA, which is a semi-supervised and global method to get microRNA-disease association information from microRNA functional similarity network and disease semantic similarity network simultaneously. This method can work without known microRNA-disease associations.

Since microRNAs regulate diseases through their target genes, there is a high probability that the microRNA will affect the disease if an microRNA target gene is associated with a disease related genes. Based on this assumption, Jiang et al. [5] have used a Naive Bayes model to integrate multiple types of data source to calculate the functional similarity between genes. There are associations between microRNA and genes, between diseases and genes. They used the functional similarity between the microRNA target genes and disease related genes to priority the disease related microRNAs. Shi et al. [6] have mapped microRNA targeted genes and disease related genes to protein-protein interaction (PPI)network respectively. They obtained two ranked list of genes by random walk with restart algorithm with different seeds. Then they used the p-value to measure the significant that a microRNA is associated with a disease. This type of methods highly depend on the correct associations between genes and microRANs or diseases. However it still is a big challenge to identify microRAN target genes and disease related genes. Chen et al. [7] focused on computing similarity of two microRNAs from two separated perspectives, microRNA-based similarity inference and phenotype-based inference. They calculated the

Pearson correlation scores between the two types of similarities to infer the associations between microRNA and disease. The methods mentioned above have used the microRNA-microRNA similarity and disease-disease similarity to make prediction. They have employed different methods to calculate the similarities, i.e. microRNA functional similarity, microRNA sequence similarity, disease phenotype similarity, disease function similarity, disease semantic similarity. Lan et al. [8] have proposed a kernelized Bayesian matrix factorization(KBMFMDI) method to integrate multiple similarities to get better prediction performance.

More introduction and discussion about the computational methods of predicting microRNA-disease associations are in [9]. Although previous works have done many efforts to improve the accuracy of disease related microRNA prediction on the base of combining microRNA similarity and disease similarity. In order to further improve the prediction performance, more biological knowledge should be utilized [9]. Recent studies show that microRNA expression can be altered by environment factors (EF) [10], such as diet,stress,drug,alcohol etc. Diseases also have relationships with environment factors [9]. The database miREnvironment [11] collects manually curated and experimentally supported associations among microRNAs, environment factors and disease phenotypes. Qiu et al. [12] have revealed the microRNA-EF interaction patterns and proposed a new computational method to predict new EF-disease associations. Chen et al. [13] have made use of environmental factor data to predict EF-microRNA associations. Li et al. [14] have incorporated environmental factors to predict microRNA networks. However, few works have combines microRNAs, diseases and Environment factors to predict microRNA-disease associations. Consequently, in this work, we propose a new method named by ThrRWMDE, which can both use the inter- and intra- relationships of the three types of information to predict disease related microRNAs.

Our method is based on following observations: (1) two functional similar microRNAs tend to affect a common disease and vice versa, (2) two functional similar microRNAs tend to interact with similar environmental factors and vice versa [7]. (3) two similar diseases are highly caused by a common environmental factors and vice versa [12]. With respect to similarity of each biological property, three types of biological networks can be constructed, microRNA-microRNA functional similarity network (MFN), disease-disease similarity network (DSN) and environmental factor similarity network (ESN). There are intricate associations within and between these networks. Our method ThrRWMDE implements several random walking steps on the three biological networks respectively so as to get microRNA-disease association information from the neighbors in corresponding networks. In the course of walking, the microRNA-disease association information will be transferred from one network to another according to the interactions between the nodes in different networks. Our method is not only a framework which can effectively integrate different types of biological method but also can treat these information differently with respect to the topological and structural difference of the three networks. The results of experiment show that our method achieves better prediction performance than other two state-of-the-art methods KBMFMDI [8] and RLSMDA [4] in terms of area under the curves(AUC).

2   Methods

2.1   Experimental Data

We obtain known human microRNA-disease association data from [15], which is also downloaded from HMDD database [16]. The dataset includes 271 microR-NAs, 137 diseases and 1395 miRNA-disease interactions.

The microRNA functional similarity data is downloaded from http://www.cuilab.cn/misim.zip [15]. The microRNA functional similarity network is constructed based on the microRNA functional similarity data.

Disease similarity network is constructed based on disease function similarity which is calculated by SemFunSim [17]. This method assumed that similar diseases tend to be related to genes with similar functions.

The environmental factor similarity network is constructed based on the chemical structure similarity of environmental factor. The chemical structure similarities are downloaded from supplemental material of [13], which are calculated by SIMCOMP [18]. There are 138 EFs in EF similarity network. The association between the EF and microRNA, EF and diseases are downloaded from miREnironment database. There are 1019 associations between the 138 EFs and the 271 microRNAs. There are 978 associations between the 138 EFs and the 137 diseases.

Since miREnironment database collects the associations between EF and disease phenotypes. We map the phenotypes in miREnironment to the disease in HMDD database by using the information downloaded from Disease Ontology http://aber-owl.net/aber-owl/diseasephenotypes/data/ and Medical Subject Headings(MeSH, http://www.nlm.nil.gov)

2.2   Three Random Walk Algorithm on Three Biological Networks

ThrRWMDE method mainly takes two steps to predict the associations between microRNAs and diseases. Firstly, construct three different biological networks. They are microRNA functional similarity network(MSN), disease similarity network(DSN) and environmental factor similarity network(ESN). Secondly several random walk steps are taken in ESN, DSN and MSN iteratively so as to obtain the information of level-k neighbors in corresponding network. Moreover, our method can easily walks different steps on the three networks with respect to their difference topologies and structures. In the course of iteration, some potential associations between microRNAs and diseases can not only be explored from the inter and intra association between microRNAs and diseases but also be inferred according to the associations between microRNAs and EFs and the associations between EFs and diseases. To formally define our method, some variables are introduced.

Let $M(m*m)$, $D(d*d)$ and $E(e*e)$ be the adjacency matrix of MSN, DSN and ESN respectively. Let matrix $Y1(m*d)$, $Y2(e*m)$ and $Y3(e*d)$ store known microRNA-disease associations, known EF-microRNA associations and EF-disease associations respectively. The values of elements in these matrixes are 1, if there exist associations between corresponding nodes, 0 otherwise. Matrix $Rmd(m*d)$ and $Red(e*d)$ denote the predicted microRNA-disease associations and predicted EF-disease associations respectively.

Our work aims to get matrix Rmd according to matrix M, D, E, $Y1$, $Y2$ and $Y3$. The values in matrix Rmd can be updated through three ways. Firstly, several random walk steps (l_1) are taken in MSN network to get disease related information from level- l_1 neighbors of microRNA (see Formula 1). Secondly, several random walk steps (r_1) are taken in DSN to get microRNA related information from level- r_1 neighbors of disease (see Formula 2). Thirdly, some potential microRNA-disease associations can be inferred by passing the disease-EF associations through the known microRNA-EF associations (see Formula 3). Finally, the three types of predicted microRNA-disease associations can be weighed synthesized to get the final predicted microRNA-disease associations of each iteration (see Formula 4).

Similarly, the predicted EF-disease associations stored in matrix Red are also updated in three ways. Some potential EF-disease associations can be explored by extending EF path and disease path in ESN and DSN respectively (see Formulas 5 and 6). The predicted EF-disease associations can also be updated by transferring the microRNA-disease associations to EF through the associations between microRNAs and EFs (see Formula 7). Finally, the final predicted EF-disease associations of each iteration are calculated by weighted sum of the three types of predicted EF-disease associations(see Formula 8).

In order to infer information from different levels of neighbors in the three networks, random walking is iteratively taken on corresponding networks. Parameters l_1, r_1, l_2 and r_2 are used to control the walking steps in the three networks in the course of iteration. Their values can be easily set differently with respect to the difference among the three networks. In summary, Algorithm 1 outlines the algorithm of ThrRWMDE.

3   Results

In order to assess the effectiveness of ThrRWMDE, we compare it with other two methods KBMFMDI [8] and RLSMDA [4]. The parameters α, l_1, r_1, l_2 and r_2 of ThrRWMDE are set to 0.9, 1,1,1 and 1 respectively. The parameter w in RLSMDA is set to 0.9 which is selected from those recommended by the authors.

3.1   Five-Fold Cross Validation of Performance

To evaluate the performance of our method, five-fold cross validation is adopted, which divides the known microRNA-disease associations into five folds. One of the five fold is put into test set and the rest associations are selected as train set. According to known microRNA-disease associations, two metrics, true positive rate (TPR)and false positive rate(FPR) are utilized to evaluate the accuracy of the prediction. For each disease, microRNAs ranked in top t are considered as disease related. With different values of t selected, the ROC (TPR-FPR)

Algorithm 1. ThrRWMDE

1: **Input:** Matrix M, D, E, $Y1$,$Y2$, $Y3$, parameter α, iteration steps l_1, r_1, l_2 and r_2, test set S;

2: **Output:** predicted association matrix Rmd, Red ;

3: Clear the values i,j of matrix $Y1$, if i,j in S

4: $Rmd^0 = Y1 = \frac{Y1}{sum(Y1)}$

5: $Red^0 = Y2 = \frac{Y2}{sum(Y2)}$

6: **for** $(t = 1$ to $\max(l_1 , r_1, l_2 , r_2))$ **do**

7:     $\lambda_{m1} = \lambda_{d1} = \lambda_{e1} = \lambda_{m2} = \lambda_{d2} = \lambda_{e2} = 0$;

8:     **if** $(t <= l_2)$ **then**

9:         $Red_e^t = \alpha * E * Red^{t-1} + (1 - \alpha) * Y2$ (Formula 5)

10:         $Red_m^t = Y3 * Rmd^{t-1}$ (Formula 7)

11:         $\lambda_{e2} = 1$

12:         $\lambda_{m2} = 1$

13:     **end if**

14:     **if** $(t<=r_2)$ **then**

15:         $Red_d^t = \alpha * Red^{t-1} * D + (1 - \alpha) * Y2$ (Formula 6)

16:         $\lambda_{d2} = 1$

17:     **end if**

18:     $Red^t = ((\lambda_{e2} * Red_e^t + \lambda_{d2} * Red_d^t + \lambda_{m2} * Red_m^t)/(\lambda_{e2} + \lambda_{d2} + \lambda_{m2}))$ (Formula 8)

19:     **if** $(t<=l_1)$ **then**

20:         $Rmd_m^t = \alpha * M * Rmd^{t-1} + (1 - \alpha) * Y1$ (Formula 1)

21:         $Rmd_e^t = Y2' * Red^{t-1}$ (Formula 3)

22:         $\lambda_{m1} = 1$

23:         $\lambda_{e1} = 1$

24:     **end if**

25:     **if** $(t<=r_1)$ **then**

26:         $Rmd_d^t = \alpha * Rmd^{t-1} * D + (1 - \alpha) * Y1$ (Formula 2)

27:         $\lambda_{d1} = 1$

28:     **end if**

29:     $Rmd^t = ((\lambda_{m1} * Rmd_m^t + \lambda_{d1} * Rmd_d^t + \lambda_{e1} * Rmd_e^t)/(\lambda_{m1} + \lambda_{d1} + \lambda_{e1}))$ (Formula 4)

30: **end for**

31: **return** (Rmd, Red)

---

curve of each method is plotted and corresponding AUC score is calculated. The process is repeated 500 times. For each method, their average AUC value over all diseases is calculated. As Fig. 1 shown, compared with RLSMD(AUC is 0.7862)and KBMF(AUC is 0.7877), ThrRWMDE has the highest AUC value, which is 0.8461. The outperforms of ThrRWMDE suggests its success in integrating three different biological network resources.

3.2   Effect of Parameter on Performance of ThrRWMDE

In ThrRWMDE, parameter α controls the weight of the regulation of known associations in the course of iteration. When α is set to 1, ThrRWMDE explores potential MicroRNA-disease associations without considering known ones. In order to test the effect of parameter α on performance of ThrRWMDE,we set α to

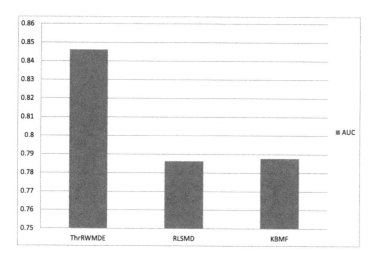

Fig. 1. Comparison of each method performance in terms of AUC value of ROC curve.

Table 1. Comparison of AUC scores over all diseases with respect to different α values

α	0.1	0.2	0.3	0.4	0.5	0.6	0.7	0.8	0.9
AUC	0.8461	0.8473	0.8461	0.8467	0.8463	0.8468	0.8475	0.8469	0.8467

different values ranging from 0.1 to 0.9. The average AUC values of ThrRWMDE with respect to different α are listed in Table 1. The results in Table 1 show that the change of α value has little influence on the predict performance. Consequently, in this work, we set the value of parameter α to 0.9 as same as RLSMDA.

4 Conclusion

In this work, we propose a new method named ThrRWMDE to predict microRNA-disease associations based on three biological networks, MSN, DSN and ESN. This method walks several steps in the three networks respectively. In the course of walking, the microRNA-disease association or EF-disease information can be inferred from the neighbors in corresponding network but also the information of microRNA-disease association is transferred from one network to another through the associations between the nodes in them. Compared to previous methods, our method incorporates environmental factors to predict microRNA-disease associations. Moreover, our method makes good use of the interactions within and between the microRNAs, diseases and environmental factors simultaneously. Additionally, our method is flexible to deal with the structure and topological difference of the three networks. Compared with two state-of-the-art methods KBMFMDI [8] and RLSMDA [4], our method achieves better performance, which verifies the effectiveness of our method on integrating multiple biological information to predict microRNA-disease association.

Acknowledgement. This work is supported in part by the National Natural Science Foundation of China under grant no. 61502214, 31560317, 61472133, 61502166, 81460007 and 81560221.

References

1. Jiang, Q., Hao, Y., Wang, G., Juan, L., Zhang, T., Teng, M., Liu, Y., Wang, Y.: Prioritization of disease micrornas through a human phenome-micrornaome network. BMC Syst. Biol. **4**(Suppl. 1), S2 (2010)
2. Xuan, P., Han, K., Guo, M., Guo, Y., Li, J., Ding, J., Liu, Y., Dai, Q., Li, J., Teng, Z., et al.: Prediction of micrornas associated with human diseases based on weighted k most similar neighbors. PloS One **8**(8), e70204 (2013)
3. Chen, X., Liu, M.X., Yan, G.Y.: RWRMDA: predicting novel human microrna-disease associations. Mol. BioSyst. **8**(10), 2792–2798 (2012)
4. Chen, X., Yan, G.Y.: Semi-supervised learning for potential human microrna-disease associations inference. Sci. Rep. **4**, Article No.5501 (2014)
5. Jiang, Q., Wang, G., Wang, Y.: An approach for prioritizing disease-related micrornas based on genomic data integration. In: 2010 3rd International Conference on Biomedical Engineering and Informatics (BMEI), vol. 6, pp. 2270–2274. IEEE (2010)
6. Shi, H., Xu, J., Zhang, G., Xu, L., Li, C., Wang, L., Zhao, Z., Jiang, W., Guo, Z., Li, X.: Walking the interactome to identify human miRNA-disease associations through the functional link between miRNA targets and disease genes. BMC Syst. Biol. **7**(1), 101 (2013)
7. Chen, H., Zhang, Z.: Similarity-based methods for potential human microRNA-disease association prediction. BMC Med. Genom. **6**(1), 12 (2013)
8. Lan, W., Wang, J., Li, M., Liu, J., Pan, Y.: Predicting microRNA-disease associations by integrating multiple biological information. In: 2015 IEEE International Conference on Bioinformatics and Biomedicine (BIBM), pp. 183–188. IEEE (2015)
9. Zeng, X., Zhang, X., Zou, Q.: Integrative approaches for predicting microRNA function and prioritizing disease-related microRNA using biological interaction networks. Brief. Bioinform. **17**, 193–203 (2015)
10. Das, U.N.: Obesity: genes, brain, gut, and environment. Nutrition **26**(5), 459–473 (2010)
11. Yang, Q., Qiu, C., Yang, J., Wu, Q., Cui, Q.: miREnvironment database: providing a bridge for microRNAs, environmental factors and phenotypes. Bioinformatics **27**(23), 3329–3330 (2011)
12. Qiu, C., Chen, G., Cui, Q.: Towards the understanding of microRNA and environmental factor interactions and their relationships to human diseases. Sci. Rep. **2**, Article No.318 (2012)
13. Chen, X., Liu, M.X., Cui, Q.H., Yan, G.Y.: Prediction of disease-related interactions between microRNAs and environmental factors based on a semi-supervised classifier. PloS One **7**(8), e43425 (2012)
14. Li, J., Wu, Z., Cheng, F., Li, W., Liu, G., Tang, Y.: Computational prediction of microRNA networks incorporating environmental toxicity and disease etiology. Sci. Rep. **4**, Article No.5576 (2014)
15. Wang, D., Wang, J., Lu, M., Song, F., Cui, Q.: Inferring the human microRNA functional similarity and functional network based on microRNA-associated diseases. Bioinformatics **26**(13), 1644–1650 (2010)

16. Li, Y., Qiu, C., Tu, J., Geng, B., Yang, J., Jiang, T., Cui, Q.: Hmdd v2. 0: a database for experimentally supported human microRNA and disease associations. Nucleic Acids Res. **gkt1023**, 1–5 (2013)

17. Cheng, L., Li, J., Ju, P., Peng, J., Wang, Y.: SemFunSim: a new method for measuring disease similarity by integrating semantic and gene functional association. PloS One **9**(6), e99415 (2014)

18. Hattori, M., Okuno, Y., Goto, S., Kanehisa, M.: Development of a chemical structure comparison method for integrated analysis of chemical and genomic information in the metabolic pathways. J. Am. Chem. Soc. **125**(39), 11853–11865 (2003)

Progression Reconstruction from Unsynchronized Biological Data using Cluster Spanning Trees

Ryan Eshleman and Rahul Singh[✉]

Department of Computer Science, San Francisco State University,
San Francisco, CA 94132, USA
rahul@sfsu.edu

Abstract. Identifying the progression-order of an unsynchronized set of biological samples is crucial for comprehending the dynamics of the underlying molecular interactions. It is also valuable in many applied problems such as data denoising and synchronization, tumor classification and cell lineage identification. Current methods that attempt solving this problem are ultimately based either on polynomial and piece-wise approximation of the unknown generating function or its reconstruction through the use of spanning trees. Such approaches face difficulty when it is necessary to factor-in complex relationships within the data such as partial ordering or bifurcating or multifurcating progressions. We propose the notion of Cluster Spanning Trees (CST) that can model both linear as well as the aforementioned complex progression relationships in data. Through a number of experiments on synthetic data sets as well as datasets from the cell cycle, cellular differentiation, and phenotypic screening, we show that the proposed CST approach outperforms the previous approaches in reconstructing the temporal progression of the data.

1 Introduction

Biochemical processes are dynamic processes expressed over time (and space). In terms of characterizing their temporal progression, a small set of generating functions can characterize such processes. For example, linear or polynomial functions (cell growth [11]), cyclical functions (cell cycle [12]), and branching (bifurcating or multifurcating) functions (cancer progression [13]). If the system under study can be sufficiently synchronized, as with cell synchrony methods [22], characterizing the underlying progression is relatively straightforward. Often however, this is not possible and the temporal order has to be reconstructed from a sampling of the process. We focus on this latter case and note that it is complicated due to epistemic and intrinsic factors such as the unknown nature of the molecular mechanisms of action, their (putative) non-linearity, phase shifts, and rate heterogeneity, as well as extrinsic factors such as under-sampling, and noise.

Formally, if we think of a biological process as a series of states evolving with respect to time, the problem of constructing the temporal ordering for a set of samples requires specifying the function $f(t) = [x_1(t), x_2(t), \ldots, x_d(t)]$, where $x_i(t)$ is the value of dimension i at time t, so the output $f(t)$ is a point in d dimensions representing the state of the process at time t. This function has to be reconstructed from the samples

© Springer International Publishing Switzerland 2016
A. Bourgeois et al. (Eds.): ISBRA 2016, LNBI 9683, pp. 136–147, 2016.
DOI: 10.1007/978-3-319-38782-6_12

$S = \{s_1, \ldots s_n\}$, where $s_i = f(i) + \varepsilon$ with ε denoting the noise. Noise modeling is often simplified by using well characterized distributions, such as a Gaussian. Graph-theoretic representation of the biological data provides a powerful formalism, especially for representing non-linear progressions. In such a representation, the complete data set is represented by a graph $G_c = (V, E)$ with each data point corresponding to a vertex in V and the edges in E connecting the vertices based on some criterion. Within this framework, Minimum Spanning Trees (MST) constitute a powerful representation for progression reconstruction [9, 10, 13]. MST-based methods assume that the tree with the minimum total edge weight best represents the underlying process. This does not account for relationships present in the data, such as groupings corresponding to subprocesses. Furthermore, the connectivity of a tree can be sensitive to how edges are selected and a poor choice may misrepresent relationships in the data. To illustrate this point, we use three different methods to reconstruct the progression of gene expression during the cell cycle. In this example 20 proteins associated with different phases of the cell cycle are chosen from the cell cycle cDNA expression micro array dataset [12]. Figure 1 shows the temporal ordering reconstructed by the MST-based method [9], the Sample Progression Discovery (SPD) method [10] and the proposed Cluster Spanning Tree (CST) approach. All three methods accurately group proteins from the G1/S, S, and G2/M phases, however only CST correctly groups the G2 phase proteins. Moreover, the CST is the only method that arranges the proteins in the proper order that reflects the stages of the cell cycle: G1/S, S, G2, G2/M. A more complete evaluation on this dataset is presented in the results section.

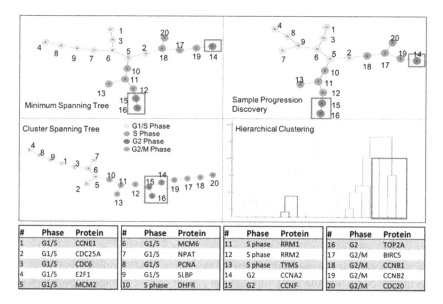

#	Phase	Protein	#	Phase	Protein	#	Phase	Protein	#	Phase	Protein
1	G1/S	CCNE1	6	G1/S	MCM6	11	S phase	RRM1	16	G2	TOP2A
2	G1/S	CDC25A	7	G1/S	NPAT	12	S phase	RRM2	17	G2/M	BIRC5
3	G1/S	CDC6	8	G1/S	PCNA	13	S phase	TYMS	18	G2/M	CCNB1
4	G1/S	E2F1	9	G1/S	SLBP	14	G2	CCNA2	19	G2/M	CCNB2
5	G1/S	MCM2	10	S phase	DHFR	15	G2	CCNF	20	G2/M	CDC20

Fig. 1. Progressions reconstructions from applying three reconstruction methods (MST, SPD, and CST) to a subset of the cell cycle micro array dataset in [12].

2   Background

Given a sampling S of size n of f, one way of reconstructing the underlying generating function is through polygonal approximation. Polygonal reconstruction [1] builds a connected graph $G = (V, E)$, where the vertices V are points from S and edges E connect the vertices such that each vertex has degree of 1 or 2 and for each set of adjacent vertices $[v_i\ v_j]$ corresponding to points $[f(i)\ f(j)]$, there does not exist a v_k: $f(i) \leq f(k) \leq f(j)$. This can be achieved by determining a traveling salesman path. The notion of principal curves can also be used to order data points when the manifold on which they lie has a curvature. Principal curves were introduced in [2] and constitute a non-linear generalization of principal components. For the set S, a principal curve is defined as a smooth function f_c that passes through the center of mass of the sample set S and is self-consistent, as defined by Eq. (1):

$$f_c(t) = E[S|t_f(S) = t] \tag{1}$$

In Eq. (1), $t_f(S)$ denotes points in S that are projected to point t. That is, each point on the principal curve coincides with the expectation of the data points that are mapped to it. As noted in [9], principle curves may require sampling at a denser rate than is provided in many biological contexts.

Neither polygonal reconstruction nor principal curves can be used to model branching processes. In such cases the system at time t has more than one possible state at time $t+1$. To address such issues, piece-wise representations, such as a spanning trees, have been employed that create tessellated representations of the data and reconstruct temporal ordering in each tessellate. A spanning tree of a complete graph $G_c = (V, E)$ is the connected graph $G_s = (V, E')$ where $E' \subseteq E$ and $\exists u \in V : (u, v) \in E' \lor (v, u) \in E' \forall v \in V$. Plainly, the subset E' contains edges that span all vertices in V. Because of the limited number of edges, a spanning tree enforces a unique path between vertices. Per Cayley's formula [3] there are n^{n-2} spanning trees on any complete graph. Therefore we must add constraints to find those trees which are biologically meaningful. An MST on G_c is a spanning tree with the additional constraint that $\sum_{e \in E'} e$ is the minimum across all spanning trees on G_c. MSTs can be constructed with one of many greedy algorithms, such as Kruskal's [4] or Boruvka's [5] that iteratively collect edges with the least weight to build the tree. The methods described in [9, 10] employ variations on the MST approach. In [9], the diameter path through the MST (or multiple candidate diameters with a PQ tree in the presence of noise) are used to determine the progression. In [10], an automated feature selection step is incorporated where MSTs are constructed on subspaces of the original feature space. The subspaces that generate the most similar MST topologies are merged to form the final putative MST progression.

As discussed earlier, the MST formulation cannot represent interrelationships such as natural sub-processes or groupings in the data. However, hierarchical clustering methods (like UPGMA [6]) may be used to identify data clusters which should be maintained in the resulting temporal reconstruction. Indeed, a method like UPGMA may be used directly for reconstructing temporal progression as in phylogenetics. A generic application of phylogenetic methods to this problem is however precluded, since such methods always impose a bifurcating structure on the data.

3 Methods

We propose the idea of cluster spanning trees (CST) that can maintain temporal and hierarchical clustering structure of the data and investigate three algorithmic variations for CST construction. At the fundamental level, this method is a process of traversing a hierarchical tree which represents the relations in the data and iteratively adding edges between nodes or groupings thereof. A binary hierarchical tree $G_b = (V_b, E_b)$ in our formulation contains $2n\text{-}1$ vertices, n is the number of data points being clustered. The n leaf vertices represent the data points. Each of the $n\text{-}1$ internal vertices represent the union of its descendants. Accordingly, the root is a set of size n. Each internal vertex v_i has two children, c_{i1} and c_{i2} each containing disjoint sets where $v_i = \{c_{i1} \cup c_{i2}\}$. Figure 2 shows an example. The CST is constructed as follows: beginning with G_b and a graph of disconnected vertices $G_{CST} = (V_{CST}, E_{CST})$ where V_{CST} is the set of original n data points, e.g. data points in the root node of G_b and E_{CST} is the empty set. For each non leaf vertex v_i in V_b an edge is added to E_{CST} from the child vertices of v_i, c_{i1} and c_{i2}, that connects a point in c_{i1} to a point in c_{i2} and minimizes a distance function $d(c_{i1}, c_{i2})$. While the order in which the vertices are traversed is arbitrary and does not affect the resulting CST, if an in-order traversal is performed, this algorithm can be understood as the iterative merging of a set of trees into a single tree.

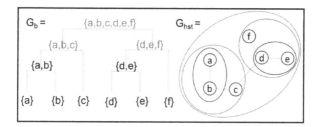

Fig. 2. **Left**, dendrogram and subsets assigned to the binary tree. Each internal node contains the union of the two child nodes. **Right**, Cluster Spanning Tree constructed from the hierarchical tree. For every internal vertex of G_b there is a connected subtree of G_{CST}.

When this operation has been performed over all internal nodes, we are guaranteed that for every internal node v_i in V_b there exists a connected sub-tree of G_{CST}, $G'_{CST} = (V'_{CST}, E'_{CST})$ where $V'_{CST} \subseteq V_{CST}, E'_{CST} \subseteq E_{CST}$ and $V'_{CST} = v_i$. Accordingly, the hierarchical clusters identified at the clustering stage are represented as sub-trees of the CST. As a general framework for the downward projection of a binary hierarchical tree of $2n\text{-}1$ vertices into a tree of $n\text{-}1$ vertices, there are two major algorithmic components to consider, namely, hierarchical data clustering and cluster merging.

3.1   Hierarchical Data Clustering

There are a number of established hierarchical clustering techniques that can be utilized to perform the initial data clustering. Methods we have investigated include the Unweighted Average (UPGMA) [6], Weighted Average (WPGMA) [6], Complete Linkage [6], Centroid [7], Median [7] and Incremental Sum of Squares (Ward) [8] methods. Details on these methods can be found in the references. All of these methods induce a hierarchical structure on the data that can be used to obtain a hierarchical clustering of the data. Non-hierarchical clustering techniques can also be employed for this problem. To limit the scope of this paper, they are not discussed.

3.2   Cluster Merging

The second algorithmic component is the strategy used to draw edges between points in the subsets at each bifurcation of the hierarchical binary tree. This consists primarily of choosing a distance function to minimize. The first vertex merging strategy is the nearest neighbor approach. An edge is drawn from the point in c_{i1} to the point in c_{i2} that are nearest in terms of some distance measure, for example Euclidean (used in the next three examples). Formally,

$$\text{argmin}_{a \in c_{i1}, b \in c_{i2}} d(a, b) = \sqrt{\sum_{j=1}^{t} (a_j - b_j)^2} \tag{2}$$

This method is similar in principle to the traditional MST approach, except edges are constructed between the hierarchically derived subsets. This approach can be sensitive to outliers, for example if two outlying points in adjacent clusters happen to present the minimum distance. To minimize the influence of outliers, we employ the second merging method, called weighted centroids (defined in Eq. (3)) where we incorporate into the objective function, the distance from the centroid of the corresponding cluster point. This gives us the convex combination described in Eq. (3).

$$\text{argmin}_{a \in c_{i1}, b \in c_{i2}} d(a, b) = (1 - \lambda)\sqrt{\sum_{j=1}^{t} (a_j - b_j)^2} +$$
$$\lambda \left(\sqrt{\sum_{j=1}^{t} (a_j - \overline{c_{i1}})^2} + \sqrt{\sum_{j=1}^{t} (b_j - \overline{c_{i2}})^2} \right) \tag{3}$$

Here, $\overline{c_{i1}}$ is the mean value of points in c_{i1} equivalent to the centroid of points in the set and λ is a mixing value between 0 and 1. At $\lambda = 0$ this becomes the same as the nearest neighbor strategy. Our third method, centroid points, explicitly encourages the best alignment to cluster centroids by choosing a point in c_{i1} closest to the centroid of c_{i2}.

$$\text{argmin}_{a \in c_{i1}, b \in c_{i2}} d(a, b) = \sqrt{\sum_{j=1}^{t} (a_j - \overline{c_{i2}})^2} + \sqrt{\sum_{j=1}^{t} (b_j - \overline{c_{i1}})^2} \tag{4}$$

While the above methods do not guarantee the construction of a *minimum* spanning tree they do guarantee that higher groupings within the dataset are maintained.

4 Results

We evaluated our methods on two synthetic datasets including simulated state transitions and data generated through a noisy polynomial generating function. We also evaluated on biological datasets from cellular differentiation, the cell cycle, and phenotypic screening. That the method can be successfully employed on widely differing data sets, underscores its generic nature and broad applicability.

4.1 Synthetic Datasets

To show how CST captures the larger internal structures of a dataset, we generated synthetic data by sampling six discrete states that have an implicit ordering along the abscissa. Gaussian noise was introduced at varying intensities, as shown in Fig. 3. In this example, we see that the diameter of the CST correctly passes through each of the six states in order because it encourages the path to pass through local centers of mass. The MST takes a simpler path and does not pass through all states. The dendrogram, number 2 in Fig. 3, shows the hierarchical structure found by the UPGMA algorithm that was used to guide the tree construction.

The previous dataset allowed us to observe the reconstruction of state transitions. For a more rigorous evaluation we constructed a synthetic dataset by sampling the polynomial $y = x^3 + 3x^2 - 6x - 8$ with Gaussian noise. This allows us to measure the reconstruction error of our methods and quantify the effect of increasing noise on deviation from the ground truth polynomial as shown in Fig. 4. The CST method consistently outperforms the MST based approaches proposed in [9, 10]. Interestingly, all trees, including MSTs, are rather robust to noise except for a significant initial spike. This phenomenon occurs because when the noise level is low enough, the diameter path will pass through every point. The reconstruction error will increase with noise as long as the diameter path passes through every point, however when noise increases and outlier points are no longer on the diameter path, the outlier error no longer contributes to the reconstruction error, and reconstruction error stabilizes.

4.2 Reconstruction of Embryonic Stem Cell Differentiation Data

The two previous examples showed the method's ability to reconstruct processes that are non-branching by representing the progression as the diameter path in the tree. However, many biological progressions are characterized by branching processes. For example, the pluripotent embryonic stem cell (ESC) differentiation data set from [10] contains 44 samples of mouse stem cells at different stages of differentiation. Interventions were performed on these samples to induce differentiation into trophoblasts, neural cells, endoderm lineages, and embryonic carcinoma. Each sample contains 25,164 gene expression measurements. After application of CST, all differentiation lineages are

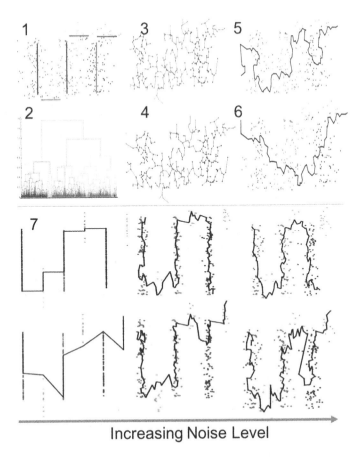

Increasing Noise Level

Fig. 3. Minimum spanning tree and UPGMA spanning tree path reconstruction for a noisy (additive Gaussian noise) synthetic data set composed of six states with an implicit horizontal ordering. **1.** The dataset showing mean values of the 6 states. **2.** The UPGMA dendrogram that shows the hierarchical clustering of the dataset used to enforce level-wise spanning tree construction. The clustering and class-color adjacencies in the dendrogram reveal how UPGMA spanning tree's constructed the correct path. **3.** Shows the CST built with the UPGMA and Centroid Point merging strategy. **4.** The MST built on this dataset. **5.** Is the diameter path of the CST which passes through all states in sequence. **6.** The diameter path of the MST which fails to pass through the light blue state. **7.** The diameter path of the data set with increasing noise with MST on top and CST below. The MSTs consistently fail to pass through the state coded in purple. (Color figure online)

reconstructed intact and in the proper temporal order; as shown in Fig. 5, the four cell lineages each branch off from the blue embryonic stem cells in the center of the tree. These results are comparable to those achieved by the Sample Progression Discovery method. The corresponding dendrogram confirms that the cell lineages are clustered in the clustering phase and the resulting reconstruction shows that temporal order is maintained within clusters.

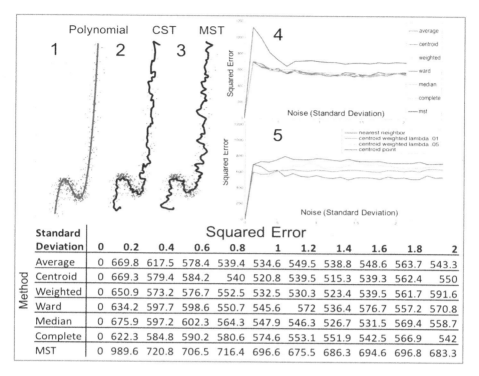

Fig. 4. CST and MST performance on a synthetic dataset sampled from the polynomial $y = x^3 + 3x^2 - 6x - 8$ with Gaussian noise. **1.** The original curve over the sampled points. **2.** The CST construction, **3.** The MST reconstruction. **4.** Is the squared reconstruction error of the six clustering methods with nearest neighbor merging, and the MST. Noise increases left to right. Cluster trees show consistently lower reconstruction error. **5.** Reconstruction error of UPGMA clustering with the three merging strategies described in Sect. 3.2.

4.3 Cell Cycle Reconstruction

Cellular reproduction is carried out in a well characterized and repeating sequence of biological phases. Specifically, a cell passes through the G1 phase, S phase, G2 phase, and then M phase to complete one iteration of the cell cycle, beginning again at G1 phase to repeat the process. Each phase has a number of genes that carry out the underlying biological function, these genes are often highly expressed during their associated phase. To capture the expression dynamics at each phase, cDNA microarray samples measure gene expression levels throughout the cycle. The gene expression profiles form natural clusters of genes that are associated with each phase [12].

To test our approach's ability to both capture the gene clusters and accurately reconstruct the sequence of phases in the process we applied the CST, MST, and SPD methods to the expression levels of the 1099 genes in the human tumor cell cycle dataset provided in [12]. Each gene is represented by a vertex in the tree with the color indicating its associated phase in Fig. 6. Visibly, the CST method performs better separation of the phases. Both the MST and SPD methods tend to merge the G2, M/G2 and G1/M gene

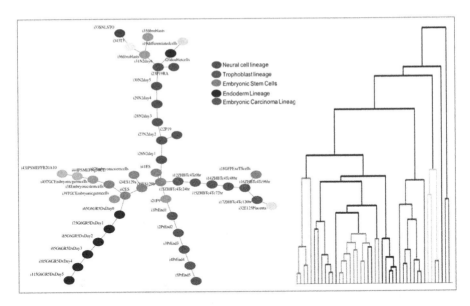

Fig. 5. Embryonic stem cell differentiation. Four cell differentiation lineages are reconstructed in order with sequential vertices representing increasing time. The dendrogram on the left shows the underlying hierarchical clustering that informed tree construction. The two images show that not only are the cell lineages generally clustered together, but their temporal order is maintained in the clusters.

groups. Because we know that the cell cycle is a repeating sequence with no branches, we observe the diameter path through the tree as a representation of the underlying biological sequence. To better represent phase regions of the diameter path, we performed neighbor smoothing whereby a vertex's phase assignment is determined by the majority vote of its raw phase and that of each of its neighbors. The smoothed diameter paths are shown in Fig. 6. The end points of the diameter are connected to show the cyclical nature of the process.

Observing these diameter paths, we see that the CST method correctly reconstructs the phase sequence with the minor exception of two G1/M phase nodes in the G1/S phase region, this can be explained by the implied overlap of G1 phase within the two regions. The MST method fails to represent the G1/M phase altogether while the SPD method combines M/G2, G1/M and G2 phase proteins.

Because most nodes do not appear on the diameter path and form clusters around the path, we measured the reconstruction error by counting the number of nodes whose phase assignment does not match the phase assignment of its nearest diameter node. The CST method had the lowest reconstruction error of the three methods followed by MST and SPD respectively.

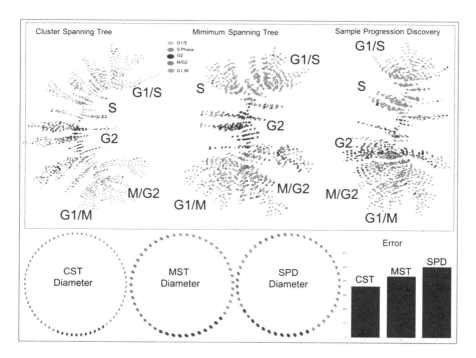

Fig. 6. Cell cycle gene reconstruction. The CST, MST and SPD methods were applied to the cell cycle gene expression microarray data. The cell cycle has a known sequence of phases: G1, S-phase, G2, M. Each gene is represented by a node in the tree colored by its associated phase in the cycle. The CST method properly separated the phases and reconstructed the sequence in the correct order. Phases were not sufficiently separated with the MST and SPD methods. The diameter paths of each tree with 1 neighbor smoothing are shown. The MST does not contain the G1/M phase. SPD mixes M/G2, G1/M and G2 proteins. Error is computed by summing the number of vertices that do not match the nearest diameter vertex's phase.

4.4   Reconstruction of Macro-parasite Phenotypic Screening Data

We consider phenotypic screening against parasites that cause the disease Schistosomiasis. Our data set consists of images of 95 *S. mansoni* somules taken on the first, second, third, and fourth day of exposure to a 10μM solution of the HMG-CoA reductase inhibitor Mevastatin which has been studied for its potential anthelmintic effects [14]. Each parasite is represented by 43 quantitative image features that describe the parasite's shape and texture. Parasites tend to show increasingly apparent deleterious effects as exposure time increases.

Like with the cell cycle example, this dataset contains a known linear progression (exposure duration) and natural clustering (images of parasite groups taken on specific days), so we seek to reconstruct the time progression of the clusters from the dataset. Error is measured using the same metric from the cell cycle dataset, namely mismatches along the smoothed diameter path. Figure 7 shows the trees resulting from the three algorithms along with parasite images across the CST. The CST result shows strong grouping and correct ordering of parasites from days one and four. It is not surprising that the intermediate exposure days are rather heterogeneously grouped due to the varying rate of response that individual parasites show to the drug. While days two and three are merged, we can interpret the results as showing three intuitive groupings, initial response, intermediate response, and maximal response. It is worth noting that, upon visual inspection of the underlying data, the three 'Day 3' parasites and one 'Day 1' parasite present in the 'Day 4' group all show significant effects and are properly placed, effect-wise, with the 'Day 4' parasites. Similarly, the two 'Day 4' parasites in the 'Day 1' group show idiosyncratic effects and are rightly not grouped with the other 'Day 4' parasites.

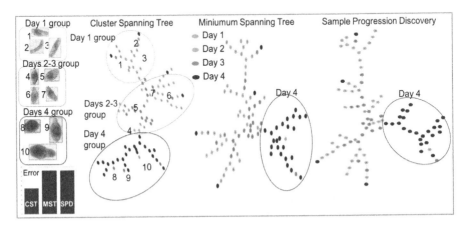

Fig. 7. Progression reconstruction of parasite phenotypic response after the first, second third and fourth day of exposure to 10 µM concentration of the drug Mevastatin. CST correctly groups the first and fourth day samples, while days two and three form a heterogeneous intermediate cluster. Example parasite images from various points on the tree are shown as well as the progression reconstruction error.

All three tree construction methods accurately grouped the 'Day 4' parasites, however only CST was able to group 'Day 1.' Both MST and SPD split the 'Day 1' group and placed them on opposite ends of the tree, significantly distorting the reconstruction. By reviewing the spatial organization of the underlying data through a lower dimensional projection (not shown) we observe that, while the parasites from Day 1 are near to each other in feature space, the MST and SPD algorithms do not take into account the local organization and one misplaced edge has significant effects on the overall graph topology. The local constraints enforced by CST help to ameliorate this problem and improve the overall reconstruction.

Acknowledgements. This research was funded in part by the National Science Foundation grant IIS-0644418 and the National Institutes of Health grant 1R01AI089896.

References

1. Amenta, N., Bern, M., Eppstein, D.: The crust and the B-skeleton: combinatorial curve reconstruction. Graph. Models Image Process. **60**(2), 125–135 (1998)
2. Hastie, T., Stuetzle, W.: Principal curves. J. Am. Stat. Assoc. **84**(406), 502–516 (1989)
3. Aigner, M., Ziegler, G.M., Erdos, P.: Proofs from THE BOOK, vol. 274. Springer, Berlin (2010)
4. Kruskal, B.: On the shortest spanning subtree of a graph and the traveling salesman problem. Proc. Am. Math. Soc. **7**(1), 48–50 (1956)
5. Boruvka, O.: Contribution to the solution of a problem of economical construction of electrical networks. Elektronický Obzor **15**, 153–154 (1926)
6. Sokal, R.R.: A statistical method for evaluating systematic relationships. Univ Kans Sci Bull. **38**, 1409–1438 (1958)
7. Székely, G.J., Rizzo, M.L.: Hierarchical clustering via joint between-within distances: extending ward's minimum variance method. J. Classif. **22**, 151–183 (2005)
8. Ward, J.H.: Hierarchical grouping to optimize an objective function. J. Am. Stat. Assoc. **58**(301), 236–244 (1963)
9. Magwene, P.M., Lizardi, P., Kim, J.: Reconstructing the temporal ordering of biological samples using microarray data. Bioinformatics **19**(7), 842–850 (2003)
10. Qiu, P., Gentles, A.J., Plevritis, S.K.: Discovering biological progression underlying microarray samples. PLoS Comput. Biol. **7**, 4 (2011)
11. Bochner, B.R.: Global phenotypic characterization of bacteria. FEMS microbiology Rev. **33**(1), 191–205 (2009)
12. Whitfield, M.L., et al.: Identification of genes periodically expressed in the human cell cycle and their expression in tumors. Mol. Biol. Cell **13**(6), 1977–2000 (2002)
13. Park, Y., Shackney, S., Schwartz, R.: Network-based inference of cancer progression from microarray data. IEEE/ACM Trans. Comput. Biol. Bioinform. **6**(2), 200–212 (2009)
14. Arreola, L.R., Long, T., Asarnow, D., Suzuki, B.M., Singh, R., Caffrey, C.: Chemical and genetic validation of the Statin drug target for the potential treatment of the Helminth disease. Schistosomiasis PLoS One **9**, 1 (2014)
15. Fitch, W.M.: Toward defining the course of evolution: minimum change for a specific tree topology. Syst. Zool. **20**, 406–416 (1971)
16. 1000 Genomes Project Consortium.: A map of human genome variation from population-scale sequencing. Nature **467**(7319), 1061–1073 (2010)
17. Behrends, S., Vehse, K., Scholz, H., Bullerdiek, J., Kazmierczak, B.: Assignment of GUCY1A3, a candidate gene for hypertension, to human chromosome bands 4q31. 1 → q31. 2 by in situ hybridization. Cytogenet. Genome Res. **88**(3–4), 204–205 (2000)
18. Yasuda, K., et al.: Variants in KCNQ1 are associated with susceptibility to type 2 diabetes mellitus. Nat. Genet. **40**(9), 1092–1097 (2008)
19. Platt, O.S., et al.: Pain in sickle cell disease: rates and risk factors. N. Engl. J. Med. **325**(1), 11–16 (1991)
20. Allison, A.C.: Protection afforded by sickle-cell trait against subtertian malarial infection. Br. Med. J. **1**(4857), 290–294 (1954)
21. Ehret, G.B., et al.: Genetic variants in novel pathways influence blood pressure and cardiovascular disease risk. Nature **478**(7367), 103–109 (2011)
22. Merrill, G.F.: Cell synchronization. Methods Cell Biol. **57**, 229–249 (1988)

Protein and RNA Structure

Consistent Visualization of Multiple Rigid Domain Decompositions of Proteins

Emily Flynn[1,2] and Ileana Streinu[2,3(✉)]

[1] Biomedical Informatics Program, Stanford University, Santa Clara, USA
erflynn@stanford.edu
[2] Department of Computer Science, Smith College, Northampton, USA
istreinu@smith.edu
[3] School of Computer Science, University of Massachusetts Amherst,
Amherst, USA
streinu@cs.umass.edu

Abstract. We describe an efficient method to facilitate the visual comparison of cluster decompositions obtained from multiple variations of a protein structure, as well as the results of using different computational and experimental methods for obtaining such decompositions. Implemented as a web server application, this tool is useful for gaining information about protein folding cores, the effect of mutations on a protein's stability, and for validation and better understanding of rigidity analysis.

1 Introduction

Collective Motions in Proteins. Understanding conformational motions of biological molecules is important for understanding their functions, yet experimental methods for observing such motions are expensive and limited in the information they provide. Specialized computer architectures [8] are capable now of performing molecular dynamics simulations in the range beyond microseconds, yet understanding the large, *slow conformational transitions* remains a difficult problem. Due to the often collective nature of macromolecular motions, where large groups of atoms move together in a coordinated fashion, their study can be approached with a variety of coarse-grained models. However, the gap between computation and experiment is wider at this time scale, as pointed out in [9]; thus model validation remains an important and challenging step in this area. Comparing the domain decompositions underlying these models with manual annotations, or with each other, is an important step in *validating* various decomposition methods.

E. Flynn—Research conducted while an undergraduate student at Smith College, with support from a Clare Boothe Luce Scholarship, NSF 4CBC Biomathematics Fellowship (through NSF UBM-1129194) and Goldwater Scholarship.

I. Streinu—Research supported by NSF CCF-1319366, NSF UBM-1129194 and NIH/NIGMS 1R01GM109456.

© Springer International Publishing Switzerland 2016
A. Bourgeois et al. (Eds.): ISBRA 2016, LNBI 9683, pp. 151–162, 2016.
DOI: 10.1007/978-3-319-38782-6_13

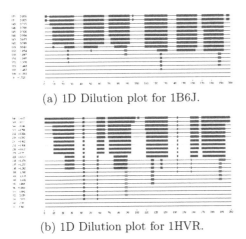

(a) 1D Dilution plot for 1B6J.

(b) 1D Dilution plot for 1HVR.

Fig. 1. Visualization of rigidity dilution profiles of two different structures of HIV-1 protease using the 1D method: (a) the profile for 1B6J (which can be compared with similar ones found in the literature [4]), and (b) of 1HVR.

Rigidity Analysis. Protein rigidity analysis is an efficient computational method for extracting flexibility and rigid domain information from static X-ray crystallography data. Atoms and bonds are modeled as a mechanical structure and analyzed with a fast graph-based algorithm, producing a decomposition of the flexible molecule into interconnected rigid clusters. Previous implementations have used 1D-visualization methods for comparing cluster decompositions, as in the dilution analysis application of the FlexWeb/FIRST server [12]. To provide a comparison with the 3D visualization method reported in this paper, we include in Fig. 1, a demonstration of the 1D method.

Comparing Cluster Decompositions. In this paper we focus on KINARI-Web [3], the web server for rigidity analysis developed in the senior author's lab, and on the new tools we developed for making such comparisons and visualizations as understandable and useful as possible. KINARI uses a much more intuitive 3D visualizer for rigid cluster decompositions (Fig. 2).

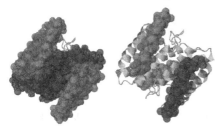

Fig. 2. Inconsistent coloring makes the comparisons of clusters confusing: the red and blue clusters switch location with each other from the undiluted (left) to the diluted (right) form. (Color figure online)

Visual comparison is an essential part of examining rigidity and flexibility results; however, comparisons of multiple decompositions is difficult due to the current coloring scheme used by the KINARI visualizer. This often leads to inconsistent colorings across different cluster decompositions of the same protein, as illustrated in Fig. 2. Here, the three largest clusters decrease in size from the undiluted form of cytochrome C to a diluted form (with hydrogen bonds weaker than $-4.0\,\text{kcal/mol}$ in energy removed). During this process, the largest rigid cluster changes from the central cluster to the cluster on the right. Since KINARI currently colors clusters based on ranked size, the top red color switches from the central cluster to the right cluster, which is confusing to the viewer. Visualizing the change in clusters from undiluted to diluted form would be more natural if the colors were consistent instead of swapping. *Our goal in this paper is to develop and demonstrate an improved system for automated consistent coloring of cluster decomposition for biological applications.* This adds to the usefulness of KINARI by assisting biologists' assessments of protein stability properties that are reveled in dilution and mutation applications.

Applications of Rigidity Analysis. The current KINARI-Web software is designed for analyzing and visualizing the rigidity of one macromolecule at a time. However, there are many biological applications which rely on comparing the rigidity of multiple structures or decompositions produced by different computational and experimental techniques. These applications include:

1. **Dilution analysis**: simulation of the unfolding pathway by generating and analyzing the rigidity of multiple structures along a modelled denaturation process
2. **Mutation analysis**: comparison of the rigidities of *in silico* point mutations with those of the original protein
3. **NMR models**: visualization of the rigidity of multiple NMR models of the same protein
4. **Conformations**: examination of the rigidity of multiple conformations of the same protein
5. **Computational methods**: comparison of multiple computational methods and experimental data on rigidity
6. **Domains**: comparison of crystallographer assigned domains of the same structure to each other and to computational methods

Types 1, 2, 3, and 4 all involve comparisons of rigidity analysis produced cluster decompositions, while types 5 and 6 involve comparisons of rigidity analysis to other computational and experimental decompositions. The last two types of comparisons (types 5 and 6) are particularly important for evaluating the biological relevance of rigidity analysis. Specifically, we would like to examine how the output produced by KINARI-Web [3] compares to both experimental data on molecular motions and the results of other computational methods for modeling bio-molecule flexibility. This analysis can lead to further validation of KINARI and greater understanding of the limitations of and differences between KINARI and other methods.

In Silico **Rigidity of Protein Dilutions.** Simulated protein unfolding can provides insight into protein folding, the steps by which a linear chain of amino acids folds into a complex three-dimensional structure. For some proteins, the unfolding pathway is reversible and therefore corresponds directly to the reverse of the folding process, while for other proteins, the unfolding pathway can still shed light into certain aspects of folding and intermediate structures that may form during the folding process. Additionally, understanding protein thermal stability and unfolding is important in protein design. Experimentally, chemical and thermal denaturants are used to examine protein stability, unfolding, and refolding.

Dilution Web Viewer

User ID: 99ZxM
Current PDB file: 1HHP

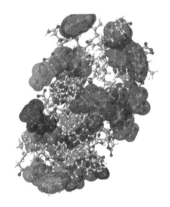

Number of hydrogen bonds remaining: **59**
Energy cutoff (kcal/mol):

Fig. 3. KINARI Dilution Viewer output for 1hhp. A scroll bar allows the user to examine the consistently colored rigid cluster decompositions of the successive dilutions and select particular dilutions for further inspection.

Rader et al. [12] simulated protein unfolding using rigidity analysis as implemented in the FIRST server. The weakening or dilution of non-covalent bonds during unfolding is modeled by removal of hydrogen bonds and salt bridges one by one in order from weakest to strongest. This response is similar to what is expected when a protein is denatured by increasing temperature. Hydrophobic interactions are not removed during dilution because they have been shown to have a stabilizing effect at higher temperatures. Rigidity analysis is performed after removal of each successive interaction.

Dilution analysis has been used in multiple computational studies to identify key transition states in the unfolding process [12] and the folding cores of particular proteins [11,12]. Computationally found folding cores and transition states match findings from experimental data, providing additional validation for rigidity analysis and dilution simulations. In particular, the unfolding of the transmembrane protein rhodopsin [11] was examined in detail and shown to correlate well with NMR data, GNM, and experimental mutation studies. More recently, dilutions have been used for examining other aspects of protein biology, including the unfolding patterns across multiple protein families [14], the relationship of thermostability to rigidity in thermophilic proteins [10], and the role of drug identity and binding on HIV-1 protease rigidity [4]. While FIRST provides a method for linear visualization of successive dilutions, this method obscures the three-dimensionality of the dilution; rigid regions are a consequence of 3D structure and as a result it is important to visualize them as such.

***In Silico* Rigidity of Proteins with Point Mutations.** Mutations that cause changes to the amino acid sequence of a protein can affect its function. Point mutations, which are changes to single amino acids, are implicated in many diseases including sickle cell anemia and cystic fibrosis. It is of interest to understand the effects of mutations on protein structures and rigidity, which has applications in drug design. In particular, we want to examine the effect of mutating a specific residue on a protein's rigidity, and identify which residues in a protein are important for its stability (i.e. have a destabilizing affect when mutated).

KINARI-Mutagen [7] is a server that generates *in silico* point mutations for a selected protein and then performs rigidity analysis on the mutated versions of the protein. This software can be used for examining the effect of single amino-acid substitutions on protein stability in order to help identify which residues may have a destabilizing effect on a protein if mutated. KINARI-Mutagen simulates *in silico* excision mutations of amino acid residues to glycine by removing the hydrogen bonds and hydrophobic interactions associated with the mutated residue, which has the same impact on rigidity as removing the amino acid's side chain. The software tool provides multiple features for examining the quantitative effect of excision mutations on a protein's rigidity; however, it does not provide a method for visualizing the effects of these mutations on the protein's rigidity.

Our New KINARI Dilution and Mutation Viewers. Consistently coloring clusters in two decompositions of the same protein, or in two different proteins, is computationally a complex problem. In this conference paper we report on solutions to this problem for the two biologically relevant, yet algorithmically simpler problems presented above. In these cases, rigidity analysis is performed multiple times on systematic changes applied to the same protein conformation. Since in both cases the resulting sequence of decompositions satisfies a hereditary property, the consistent coloring of the domains can be efficiently computed. To demonstrate their usefulness for practical applications, we implemented these methods in two web applications, Dilution Viewer and Mutation Viewer. Snapshots of the viewing interface appear in Figs. 3 and 6.

Further Steps: Multiple NMR Models and Multiple Conformations. We have also proposed a method for applications 3 and 4 above, namely *consistently coloring the rigid domain decompositions arising from multiple conformations of the same protein.* This part of our work, based on variations of weighted and stable matching algorithms, requires substantially more background for describing the algorithms and heuristics and is not included in this short conference submission. Rather, in this current version, we focus on the motivation for developing such software tools, and illustrate it with the easier-to-understand (and implement) Dilution and Mutation applications discussed above. We also present the general mathematical model underlying the consistent coloring problem and the software infrastructure required for making the visualization tools easily available to biologists. The detailed description of the underlying algorithms and the results obtained with NMR models and multiple conformations is deferred to the full version of the paper.

Kinari-2. The tools described above will become publicly available as part of Kinari-2 at the same url http://kinari.cs.umass.edu as the previous release, Kinari-1. Besides the improved visualization, Kinari-2 provides much more efficient algorithms and methods for curating and processing large data, in particular the sequences of diluted and mutated proteins described here. An overview of the new Kinari-2 design is given in [13]. The underlying infrastructure necessary for guaranteeing the reproducibility of the rigidity analysis results was described in [1]. *The work reported here focuses on the distinct problem of making the results of these calculations clear and intelligible for the biologists using the system, in the form of visualizations that are as consistent as computationally possible.* While the two cases described in this conference paper are amenable, after proper formulation, to simple solutions, the general problem of consistent visualization is not. However, we believe that we have made significant progress on several cases of practical relevance not just to the KINARI project, but to the general problem of visualizing distinct cluster decompositions of proteins (possibly obtained by other decomposition methods, not just rigidity analysis).

2 Methods

The problem of consistently coloring cluster decompositions can be formulated mathematically in terms of matchings between abstract sets of points. Let $U = \{1, ..., n\}$ be the *universe* of points representing the numbered atoms in a particular structure. A *cluster decomposition* is a *set system* R of the fixed point universe consisting of m subsets $\{C_1, ..., C_m\}$, where each subset C_i in the set system corresponds to a *cluster* containing those points. The set system is said to be *2-thin* when any two sets (clusters) in the set system have at most two points in common.

The consistent coloring problem can now be stated: *Given two or more set systems $R_1, \cdots, R_k$ over the same universe, assign colors to sets in each set system to maximize the overlap between similarly colored sets in the two systems.*

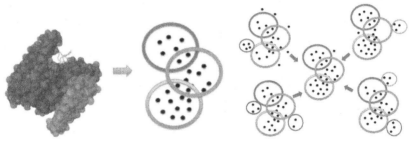

(a) A set system abstracting a protein cluster decomposition.　(b) Consistent coloring for a collection of set systems.

Fig. 4. The consistent coloring problem.

As stated, the problem focuses on colorings of two cluster decompositions from the same set of atoms. Figure 4(a) provides an example of converting the KINARI produced rigid cluster decomposition of 1BBH to a set system R_1 of the universe U, where each set C_i represents a cluster in the original decomposition. For simplicity, only a subset of the atoms in the original structure are included as points in the universe. Figure 4(b) formulates the consistent coloring problem in terms of set systems, showing the set system R_2 before and after coloring. R_2 is a set system of the same universe U as R_1 in (a) (In this particular case, R_1 and R_2 have a *hereditary relationship*, with R_1 as the parent of R_2).

Coloring cluster decompositions with different but overlapping sets of atoms has important applications in comparing the rigidity of proteins in the absence and presence of biological units, crystal structures [6], and nucleic acids [2]. This is a significantly more complex problem algorithmically and outside the scope of the work presented here. Even on the same set of atoms U, there is no easy-to-obtain optimal solution to the general problem of consistently coloring two cluster decompositions R_1 and R_2. Here we focus on particular subproblems directly motivated by biological applications. We examine how to best color R', the *new* cluster decomposition, based on a given coloring of R, the *reference* cluster decomposition, given that R and R' have a certain relationship.

Hereditary Decompositions. We examine two types of pairwise comparisons between set systems (cluster decompositions) R and R' of the same universe of U with m and m' clusters respectively where R and R' have a *hereditary relationship* where R is the *parent* of R'. Two clusters C_i and C'_j are said to have a *hereditary relationship*, with C_i the *parent* of C'_j if the set of points corresponding to C'_j is a subset of the set of points corresponding of C_i. Similarly two cluster decompositions R and R' of the same set of atoms A have a *hereditary relationship* with R the *parent* cluster decomposition of R' if every cluster C'_j in $\{C'_1 \ldots C'_{m'}\}$ of R' is a subset of exactly one cluster C_i in R (see Fig. 4 for an example of two cluster decompositions with this relationship)[1].

Given that R has been assigned a particular coloring, the problem of coloring a cluster decomposition R' where R is its *parent* is a relatively simple one. Algorithm 1 provides pseudocode for this approach. Briefly, each cluster C'_j in $\{C'_1 \ldots C'_{m'}\}$ of R' is matched to its corresponding parent cluster C_i in R based on the intersection between the two sets of points. Each C'_j is related to only one C_i, while one C_i may be the parent of multiple C'_j because R' is a subset of R. A cluster C'_j *inherits* the color of its parent cluster C_i if it is the largest *child* cluster of C_i. We call this largest child cluster of a particular parent the *favorite child* of that parent. If a cluster C'_j is not the favorite child of its parent C_i in R, it is assigned a new color.

Consistent Coloring Groups of Cluster Decompositions. The biological applications of rigidity analysis we are interested in, dilution and mutation, require consistent coloring of larger-scale *groups* of cluster decompositions rather than pairs. We define two types of group consistent coloring problems in order of increasing complexity, with each named after the biological application that inspires this type of coloring.

---

[1] The converse problem, where every C'_j in R' is a superset of at least one cluster C_i in R is not addressed because it does not correspond to any of the biological applications we are interested in and is only a slight variation of the original problem.

Algorithm 1. Consistent coloring algorithm for related clusters, where R is the reference cluster decomposition, and R' is the new cluster decomposition.

CONSISTENTCOLOR(R', R){

1: **for** C'_j in R' **do**
2:     $C_i = $ FINDPARENT(C'_k, R)
3:     **if** C_i already has a favorite child **then**
4:         Let C'_k be C_i's favorite child
5:         **if** C'_j is larger than C'_k **then**
6:             C'_k is no longer the favorte child of C_i
7:             Assign C'_k to a new color.
8:             Set C'_j to the favorite child of C_i. C'_j now inherits the color of C_i.
9:         **end if**
10:    **else**
11:        Set C'_j to the favorite child of C_i. C'_j now inherits the color of C_i.
12:    **end if**
13: **end for**

}

The Mutation Problem. The input is an unordered set of n cluster decompositions $S = \{R_1, R_2, \ldots, R_n\}$ such that each cluster decomposition R_i was created by splitting zero or more clusters in an original cluster decomposition R. The relationship between the cluster decompositions is star-like, where each of the n rigid cluster decompositions in $S = \{R_1, R_2, \ldots, R_n\}$ inherits coloring from the parent (or reference) cluster decomposition R. In terms of the biological application, R is the wild-type or normal form of the protein and each of the cluster decompositions in the set $S = \{R_1, R_2, \ldots, R_n\}$ is represents a particular point mutant.

The Dilution Problem. The input is an ordered list of n cluster decompositions $S = \{R_1, R_2, \ldots, R_n\}$ such that each cluster decomposition R_i was created by splitting zero or more clusters in the previous cluster decomposition R_{i+1}. The first cluster decomposition R_1 was created by splitting zero or more clusters in the original cluster decomposition R. The set of cluster decompositions then form a tree. The tree has a single root, and then the first level of the tree contains the original cluster decomposition R, with each cluster C in R represented by a node in this level of the tree. All of the original clusters C in R are children of the root.

Then each cluster decomposition R_i in S will be located at the $(i+1)$th level in the tree, with each cluster in $C_i = \{C_{i,1}, C_{i,2}, \ldots, C_{i,k}\}$ in R_i as a node in that level of the tree. The parent of a particular node $C_{i,j}$ in the $i+1$ level of the tree is the corresponding node in the ith level that it is a subset of. We color each level based on the previous level in the tree. A particular level i in the tree is *redundant* if every node in that level is also in the previous $(i-1)$ level of the tree. An *event* occurs when a level has more nodes than the previous level.

In terms of the biological application, R is the undiluted or normal form of the protein, while S contains the subsequent dilutions in the order in which they occur. For dilution analysis, we are only interested in steps of the dilution that produce events. As a result, during the process of coloring the dilutions we both identify dilution steps that correspond to events where the rigidity has changed and color based on the previous non-redundant dilution step in the tree.

Implementation. We developed the Dilution and Mutation Viewers based on the consistent coloring approach described above. The structure scheme of the software tools is diagrammed in Fig. 5. Both coloring tools share a common C++ back end that implements the algorithms described in Methods. The individual web applications each have their own user interfaces, which are variations of a common front end tool built in PHP, CSS, and JavaScript. Jmol, a Java viewer for chemical structures, is used for visualizing protein rigidity. XML configuration files are written by the front end and used to communicate with the back end. This structure of a web front end with Jmol visualization and a C++ back end that communicate via XML configuration files is similar to existing KINARI applications. Python and bash scripts are also used as part of the web back end to generate the dilutions and mutations and write configuration files for analyzing the rigidity of these structures with KINARI.

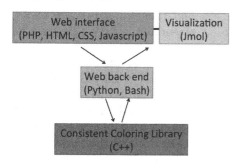

Fig. 5. Structure of comparison software tools. There are three layers: the web front end which consists of the interface and Jmol visualizers, the web back end, and the consistent coloring C++ library.

The interfaces for the dilution and mutation tools were designed to enable easy, intuitive visualization of the results of running each of the biological applications. In order to animate the unfolding pathway, the Dilution Viewer interface (displayed in Fig. 3) provides a scroll bar that the user can employ to view the successive events in the dilution process. The interface of the Mutation Viewer is designed to allow comparisons between the wild-type form of the protein and the mutated versions; the user can click each of the residues in the protein to visualize the rigidity with that residue mutated next to the original protein (Fig. 6). These initial visualizations in the Dilution and Mutation viewers provide snapshots of 3D Jmol images to allow the user to get an overview of the results of running dilution and mutation analysis. If further visualization is desired of a specific example, the user can elect to view and rotate a particular mutated or diluted version side-by-side with the original form in a 3D JMol visualizer.

3   Results and Discussion

While rigidity analysis has been used previously for simultating protein unfolding in dilution studies [12], until now, the main visualization method for dilution results was 1D rigidity line plots (as shown in Fig. 1). These plots lack information about the 3D spatial connectivity, shape, and cavities between the rigid clusters, and the eventual rigid core.

To validate our Dilution Viewer tool, we applied it to two proteins that have previously been examined with dilution analysis in the literature: HIV-1 protease and rhodopsin. For lack of space, we include here only a shortened version describing one of our studies.

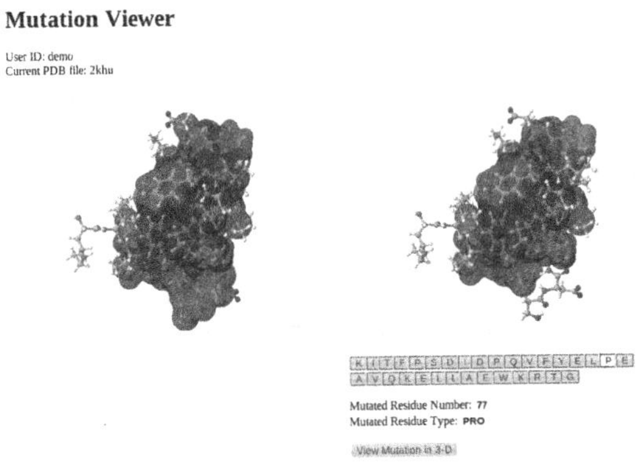

Mutation Viewer

User ID: demo
Current PDB file: 2khu

K L T F S D I D P Q V F V E L P E
A V Q K E L L A E W K P T Q

Mutated Residue Number: **77**
Mutated Residue Type: **PRO**

View Mutation in 3-D

Fig. 6. The new Mutation Viewer used in KINARI Mutation analysis, illustrated on 2 khu. The rigidity of the wild-type 2 khu is shown next of the rigidity of the same protein when a proline residue is mutated to glycine.

Case Study: HIV-1 Protease. We examined the change in rigidity during dilution of several PDB structures of HIV-1 protease, including 1B6J and 1HVR. HIV-1 protease is a protein involved in the replication cycle of HIV. It cleaves one of the viral proteins, performing a step that is necessary for the virus to mature and proliferate. Because of its essential role, HIV-1 protease is frequently used as a drug target, as inhibiting the enzyme prevents HIV from replicating. As of April 2014, there are 471 different crystal structures of this protein available in the PDB.

Heal et al. [4] examined a subset of these structures, performing dilution analysis on 206 high resolution structures in the presence and absence of known inhibitors and drugs. HIV-1 protease structures, specifically 1HVR, are also benchmark tests for KINARI and FIRST software.

Since we re-implemented dilution analysis, we wanted to first check whether our dilution results match those in the literature and then examine whether using our improved 3D dilution visualization leads to additional insights into the unfolding pathway. For this comparison, we ran dilution analysis on some of the HIV-1 protease examples and generated both 1D plots and 3D visualizations of the results. A Mathematica program was written to generate 1D plots from dilution results, and the consistent coloring framework was modified to output information on residue-level coloring in addition to the current atom-level cluster coloring output. While comparisons indicate that there is some similarity between our dilution results and those in the literature for HIV-1 protease, the correspondence between the methods was less than expected. A possible reason for these differences is the placement of hydrogen bonds by the two methods. Wells et al. [14] notes a weakness of dilution analysis is its high sensitivity to the placement of hydrogen bonds. Simple comparison of the numbers of hydrogen bonds between the two methods shows a large disparity: there are 250 hydrogen bonds in Heal et al.'s analysis and 153 in our analysis. One possible explanation for the difference in hydrogen bond identification and resulting dilutions is that Heal et al. did not directly identify interactions on the PDB structure [4]. The authors mention that they perturbed the structure by "representing the protein in a conformation which is part of the ensemble explored by the protein in its natural flexible motion" as done previously in [5], but it is not clear what the final structure used for analysis was. Other disparities in the dilution results could be due to the use of salt bridges in Heal et al.'s analysis.

The discrepancies between the results of the two software methods highlight a common issue with computational analyses; often two methods will come up with variable results and it is unclear which result is more accurate or closer to the biological "truth".

Although our dilution analysis of HIV-1 protease produced different results from the literature, our 3D results provide additional insight into the dilution pathway and rigidity during the process that is not apparent from examination of the 1D rigidity plots. Our Dilution Viewer provides a method to display and compare the sequential rigidity in a scrollable sequence. Figure 7 shows examples from the our dilution pathway of HIV-1 protease (1HVR). There are 20 steps in the dilution of 1HVR in which an *event* or change in rigidity occurs; four of them are displayed sequentially showing portions of the process. This visualization of the dilution pathway demonstrates how the shapes and locations of the rigid clusters change during the dilution pathway. The purpose of running dilution analysis is to simulate unfolding, which gives us insight into the folding process. Folding is intrinsically a 3D and not a 1D process and as such, methods for 3D visualization are essential for insight into the results of computational dilution simulations.

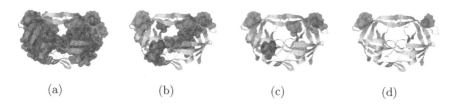

(a)   (b)   (c)   (d)

Fig. 7. 1HVR 3D dilution pathway produced by our Dilution Viewer: (a) the undiluted form of 1HVR; (b), (c), and (d) after the removal of 23, 38, and 138 hydrogen bonds respectively.

Conclusion. We developed methods for consistent coloring of rigid cluster decompositions produced by the KINARI software, and implemented these methods in two freely available web-based applications. We demonstrated the usefulness of the visualization methods through a comparative study of several structures of the HIV-1 protease, which were thoroughly compared with other studies of this nature reported in the literature. We found that the better visualization permitted the identification of differences between several methods for placing hydrogen bonds and performing rigidity analysis used by different implementations of the method. We also demonstrated the usefulness of our visualization method in the case of rigidity analysis for mutated proteins; no such tool was available until now.

Authors' Contributions. This paper is based on EF's honors thesis at Smith College under the supervision of IS. IS identified the problem of consistent coloring in the context of rigidity-analysis based applications, modeled it mathematically, designed and oversaw the project. EF implemented the prototype system described here and ran the experiments. Both authors contributed to writing the paper.

References

1. Bowers, J.C., John, R.T., Streinu, I.: Managing reproducible computational experiments with curated proteins in KINARI-2. In: Harrison, R., Li, Y., Măndoiu, I. (eds.) Bioinformatics Research and Applications. LNCS, vol. 9096, pp. 72–83. Springer, Heidelberg (2015)
2. Flynn, E., Jagodzinski, F., Santana, S.P., Streinu, I.: Rigidity and flexibility of protein-nucleic acid complexes. In: Proceedings of 3nd IEEE International Conference on Computational Advances in Bio and Medical Sciences (ICCABS 2013), June 2013
3. Fox, N., Jagodzinski, F., Li, Y., Streinu, I.: KINARI-Web: a server for protein rigidity analysis. Nucleic Acids Res. **39**(Web Server Issue), W177–W183 (2011)
4. Heal, J., Jimenez-Roldan, J., Wells, S., Freedman, R., Römer, R.: Inhibition of HIV-1 protease: the rigidity perspective. Bioinformatics **28**, 350–357 (2012)
5. Henzler-Wildmand, K., Kern, D.: Dynamic personalities of proteins. Nature **50**, 964–972 (2007)
6. Jagodzinski, F., Clark, P., Liu, T., Grant, J., Monastra, S., Streinu, I.: Rigidity analysis of periodic crystal structures and protein biological assemblies. BMC Bioinform. **14**(Suppl. 18), S2 (2013)
7. Jagodzinski, F., Hardy, J., Streinu, I.: Using rigidity analysis to probe mutation-induced structural changes in proteins. J. Bioinform. Comput. Biol. **10**(3), 1242010 (2012)
8. Klepeis, J.L., Lindorff-Larsen, K., Dror, R.O., Shaw, D.E.: Long-timescale molecular dynamics simulations of protein structure and function. Curr. Opin. Struct. Biol. **19**(2), 120–127 (2009)
9. Meuwly, M., Cui, Q.: Protein functional dynamics: from femtoseconds to milliseconds. Chem. Phys. **396**, 1–2 (2012)
10. Rader, A.: Thermostability in rubredoxin and its relationship to mechanical rigidity. Phys. Biol. **7**, 016002 (2010)
11. Rader, A., Anderson, G., Isin, B., Khorana, H., Bahar, I., Klein-Seetharaman, J.: Identification of core amino acids stabilizing rhodoposin. PNAS **101**, 7246–7251 (2004)
12. Rader, A.J., Hespenheide, B.M., Kuhn, L.A., Thorpe, M.F.: Protein unfolding: rigidity lost. Proc. Nat. Acad. Sci. **99**(6), 3540–3545 (2002)
13. Streinu, I.: Large scale rigidity-based flexibility analysis of biomolecules. Struct. Dyn. **3**, 012005 (2016)
14. Wells, S., Jimenez-Roldan, J., Römer, R.: Comparative analysis of rigidity across protein families. Phys. Biol. **6**, 046005 (2009)

A Multiagent *Ab Initio* Protein Structure Prediction Tool for Novices and Experts

Thiago Lipinski-Paes, Michele dos Santos da Silva Tanus,
José Fernando Ruggiero Bachega, and Osmar Norberto de Souza[✉]

Laboratório de Bioinformática, Modelagem e Simulação de Biossistemas - LABIO,
Pontifícia Universidade Católica do Rio Grande do Sul, Porto Alegre, Brazil
osmar.norberto@pucrs.br

Abstract. Proteins are vital to most biological processes by performing a variety of functions. Structure and function are intimately related,
thus highlighting the importance of predicting a proteins 3-D conformation. We propose GMASTERS, a multiagent tool to address the protein
structure prediction (PSP) problem. GMASTERS is a general-purpose
ab initio graphical program based on cooperative agents that explore the
protein conformational space using Monte Carlo and Simulated Annealing methods. The user can choose the abstraction level, energy function
and force field to perform simulations. Because bioinformatics demands
knowledge from diverse scientific fields, its tools are intrinsically complex.
GMASTERS abstracts away some of this complexity while still allowing
the user to learn and explore research hypotheses with the advantage
of an embedded graphical interface. Although this abstraction comes at
a cost, its performance is similar to state-of-the-art methods. Here, we
describe GMASTERS and how to use it to explore the PSP problem.

Keywords: PSP Problem · Multiagent system · Monte Carlo · AB
model

1 Introduction

Proteins are polymers of 20 different building blocks, called amino acids. These
building blocks interact physicochemically resulting in a unique spatial conformation for each protein [1]. Due to advances in the Genome Project, there is
a large number of protein sequences available in the GenBank [2]. Currently,
there are about 82 million non-redundant protein sequences. However, in the
Protein Data Bank or PDB [3], there are approximately 115,000 3-D structures
of proteins. Eliminating redundancy by filtering very similar structures (SCOP),
we get only 1,393 different folds or topologies. Under these circumstances, it is
evident the huge gap between our competence to produce protein sequences and
to determine 3-D structures of new proteins with yet unknown folds [4]. Computer Science, more specifically Structural Bioinformatics, has been a great ally
on reducing this gap.

© Springer International Publishing Switzerland 2016
A. Bourgeois et al. (Eds.): ISBRA 2016, LNBI 9683, pp. 163–174, 2016.
DOI: 10.1007/978-3-319-38782-6_14

The Protein Structure Prediction (PSP) problem emerged in the 60's and even today its solution remains a major challenge to molecular biology [5]. Limitations of 3-D structure experimental determination techniques, such as X-ray diffraction crystallography and nuclear magnetic resonance, highlight the importance of computational methods to predict the structure of proteins. Advances in handling the PSP problem will allow us to predict the 3-D structure of proteins with relevant applications in the biopharmaceutical industry. It will also improve our understanding of proteins involved in vital processes, including diseases such as cancer [6]. Considering the difficulties faced by traditional approaches (*in vitro* and *in vivo* experiments) concerning biological systems, the use of computers becomes attractive, making possible to execute low-cost and faster *in silico* experiments. An application that involves PSP must consider the system's real time adaptability, i.e., parameters modifications such as thermal bath temperature. There is a clear need for *in virtuo* experiments: computer simulations susceptible to perturbations during execution. While the easy modification of parameters is a typical property of all computer simulations (*in silico* experiments), the easy modification of the experiment itself is a property of multiagent systems (MAS), resulting in *in virtuo* experiments [7,8].

Here, we present a general tool that allows addressing PSP according to the user needs. The user is free to choose (via a Graphical User Interface) both abstraction level and force field to guide the simulation. The agents are organized in hierarchical levels. Optimization is done by Monte Carlo/Simulated Annealing and the user can modify parameters and optimization method. GMASTERS can be obtained at labio.org.

2   Background

2.1   Proteins and the PSP Problem

If we take a deep look into living organisms and observe their cellular level functions we will notice these functions are carried out by a variety of proteins. Zooming in our body toward the cells, we will realize that each cell has its own copy of the genome. The genome is what gives the cell functionality. From a computational point of view, we could state that the genome is a string or sequence composed by four kinds of letters (A,C,T and G) referring to four kinds of nucleotides. Scanning the genome from the left to the right certain substrings are found (called genes). Each triplet of nucleotides is a code that can be parsed to one of the 20 amino acids. The concatenation of parsed triplet nucleotides (amino acids) generates a protein sequence. A protein linear sequence of amino acid residues is called its primary structure [9]. Figure 1 shows a typical representation of a protein, as well as the abstraction level used in this work.

The PSP problem is the problem of predicting the 3-D structure of a protein starting from its primary structure or amino acid sequence. The physical process by which a polypeptide folds into a functional protein is an old question (reviewed by Snow [10]) and continues to be one of the biggest challenges in current structural bioinformatics [5].

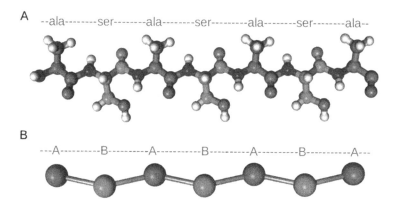

Fig. 1. (A) The representation of an extended hypothetical protein chain formed by alternated residues of alanines (hydrophobic) and serines (hydrophilic). (B) The representation of the abstraction level used in this work, where hydrophobic amino acids are marked as "A" while the hydrophilic amino acids are marked as "B".

The protein's 3-D structure is directly linked to its function. Determining its spatial conformation experimentally is expensive and time consuming. Bioinformatics has the important role of accelerating this knowledge discovery [11]. This paper's approach is based on Anfinsen's proposal which states that, at the environmental conditions (temperature, solvent concentration and composition) at which folding occurs, the native structure is a unique, stable and kinetically accessible minimum of the protein's free energy. However, finding this structure is not trivial and even simplified methods have NP-Complete complexity [12]. Figure 2 shows a hypothetical uni-dimensional energy function to illustrate the challenge of finding the lowest energy and achieving the native conformation.

Still regarding the inherent difficulty of the problem we can cite Levinthal's paradox [13], which states that for a 100-length chain there will be at least 2^{100} possible conformations (considering only two degrees of freedom), characterizing it as an intractable problem [14]. In the last five decades different algorithmic approaches have been tested and, although there has been progress, the problem remains unsolved even for small proteins. While the ultimate goal is to predict the 3-D or tertiary structure from the primary structure, the current knowledge and computing power is insufficient to handle a problem of such complexity [15].

2.2 Multiagent Systems

Multiagent systems (MAS) are part of the Artificial Intelligence field and refer to the modeling of autonomous agents in a common universe. MAS is a relatively new sub-field of Computer Science - it has been studied since about 1980 - and the field has gained widespread recognition around 1990 [16].

Agents are computational entities that interact with an environment and are goal-oriented, having a body and a location in time and space. An agent is capable of autonomous and flexible actions to reach its goals. According to

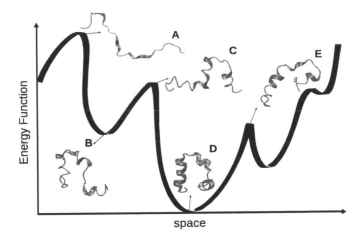

Fig. 2. Simplified protein structure prediction funnel. The figure shows an unidimensional component of a free energy hyper-surface where (A), (B), (C) and (E) represent local minima and (D) represents the global minimum (native structure).

Russel and Norvig [17], a rational agent is able to select the action that maximizes its performance, given the evidences accumulated by its perceptions and internal knowledge. According to Bradshaw [18], an agent is a software entity that works continuously and autonomously in a given environment. It should perceive and act in its environment in a flexible and intelligent way. It may learn from experience, communicate, cooperate with other agents and have the mobility required to satisfy its goals. The agents' autonomy means they have an existence independent of other agents, and have to achieve their own goals. Although there is no universally accepted definition of agent, some properties as autonomy, pro-activeness, reactivity and social ability are intrinsic of its behavior. A set of agents acting in an environment characterizes a MAS.

An agent needs an environment that can be of numerous types and complexities. The environment complexity is strongly linked to the agent's complexity.

Netlogo [19] is a very popular agent-based modeling tool and it is particularly suited for modeling complex systems that take time into account and where hundreds or thousands of agents can be programmed and interact independently, making possible to explore the connection between micro and macrolevels of behavioral patterns. One of the remarkable advantages of Netlogo is its embedded tools, and BehaviorSpace is one of them, offering the possibility of automatically perform a large set of experiments by changing parameters' values. Due to the BehaviorSpace capability, it is possible to explore more resourcefully the configuration space in PSP and tune parameters to improve results.

3 MASTERS

MASTERS [20] was built using Netlogo v5.0 (see Sect. 2.2 for details). A strong Netlogo's peculiarity is that it was developed for educational purposes, providing a rich modeling environment which allows experimental coding. *Ab initio* methods require three elements [21]: (i) a method for searching the energy landscape, (ii) an energy function (iii) a geometrical representation of the protein chain. As a pure *ab initio* method, MASTERS must comprise these elements. How MASTERS addresses these items is shown hereafter.

Searching Agents. The Searching Agents have the mission of exploring the conformational space. Usually, one or more searching agents are associated to each amino acid in the protein, depending on the selected abstraction. The agents' position is expressed as Cartesian coordinates, resulting in movements that are local, i.e., they do not affect the position of other agents.

Director Agent. The Director Agent has total knowledge about the protein's current spatial conformation and coordinates searching agents, aiming at a more efficient spatial exploration. The Director agent has no representation in the Cartesian space. It is an agent that acts on the 3-D space from outside. Director Agents perform global moves on the searching agents. It is mandatory to have at least one Director Agent in the simulation.

Environment Agent. There is only one Environment Agent. Its role is to control the simulation flow, simulated annealing scheme, number of movements per time/temperature step, real time plots, and outputs.

3.1   Hierarchical Cooperation

This is related to (i). A core MASTERS concept is the agents hierarchical organization (Fig. 3). Higher-level agents have the role of coordinating the actions of lower-level agents [22]. In a bottom-up hierarchical order.

Searching and Director agents cooperate to find the conformation that better suits their goals. They are autonomous, not depending on each other to perform their moves. The agents are reactive to their environment: the Searching agents can see their neighborhood, whereas the Director agent has full information on the environment, being able to influence the simulation in a broader manner.

MASTERS' environment is treated as a box with dimensions delimited by the user. In a multiagent perspective it can be considered both accessible (for Director and Environment agents) and inaccessible (for Searching agents, who have limited access to information), non deterministic, due to its stochastic nature, dynamic and discrete.

3.2   Sampling Technique

Also related to (i), MASTERS' movements are controlled by Monte Carlo (MC). MC [23] is one of the most used energy landscape exploration techniques. It has a probabilistic nature and it is a method for generating different configurations of a particles system, i.e., points in space compatible with external conditions.

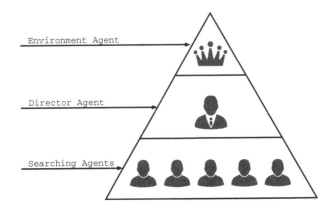

Fig. 3. MASTERS' hierarchy. The Environment Agent is responsible for the simulation flow. Searching agents explore the configuration space by performing move attempts. The Director agent is able to affect the position of all other agents in the environment.

Looking back to Fig. 2 we can pay attention to points (A) and (B). From there, MC defines two conformational states S_A and S_B, each one with its corresponding energy E_A and E_B. If $E_B < E_A$ the move is accepted. If $E_B > E_A$ there is still the possibility of accepting the move. However, in such cases, the probability of accepting a move from S_A to S_B follows the Eq. 1, where k is the Boltzmann constant and T the temperature. Once the system reaches (B) it is allowed to accept movements toward (C), hoping to someday find (D).

$$e^{-(EB-EA)kT} = e^{-(\Delta BA)kT} \tag{1}$$

Once an MC simulation is in progress it is necessary to constantly check its acceptance ratio, as it influences the agents' movements. Usually a ratio of 0.5 can be considered an optimal initial value for MC simulations involving PSP [24]. Since our method is based on different types of autonomous agents, each type of agent needs to have its acceptance ratio average assessed separately.

The MC method has drawbacks, though. In the PSP problem, the temperature of the system determines the size of energy barriers that could potentially be overcome. When dealing with temperatures that are too low, MC will not explore far from the minimum energy found, leading to local minima. Simulated Annealing (SA) is a simple MC modification that turns it into a global optimizer. At the beginning of the simulation the temperature is set high and fairly high-energy barriers are overcome. Then the system is gradually cooled, eventually being confined to a single energy. Due to the gradual cooling rate (logarithmic), the system ends up spending more time in low energy regions. This may increase the chances of finding the lowest energy state although there is no assurance [11]. Concerning SA performance, convergence is guaranteed only if the temperature is reduced to zero logarithmically. In MASTERS the temperature is gradually decreased according to Eq. 2, where $\alpha = 0.98$:

$$T_{k+1} = T_k * \alpha \tag{2}$$

Regarding the role of the agents on sampling, the simulation relies on accounting the number of movement attempts for each agent. To achieve minima at a given temperature, the system should explore the conformational space a large number of times. Counters are used to summarize the average number of move attempts per agent type (both Searching and Director Agents). Every time step is related to a specific temperature. The system will be stuck at each temperature until an average number of movement attempts has been attained. The simulation ends when the temperature reaches a value set by the user.

3.3 Choosing the Energy Function/Abstraction Level

Here the elements (ii) and (iii) are addressed. MASTERS currently incorporates MC/SA as sampling technique, not allowing the use of Molecular Dynamics or Genetic Algorithms. However, one of its particular characteristics is its orthogonality with respect to the simulation flow and the energy function/abstraction level used. The user must choose which energy function/force field to use and this is a primordial step within the framework's generality. MASTERS was not built exclusively to a particular energy function. The framework can be applied to a wide range of optimization problems that involve Cartesian coordinates (2-D or 3-D). Regardless of the problem, the energy function will directly affect the number of searching agents and their movements.

3.4 GMASTERS

The educational purpose is one of the main focuses of GMASTERS, providing a user-friendly interface where students that are not familiar with computer programming can explore the PSP problem in interactive ways. The MASTERS' version presented here includes a new graphical user interface named GMASTERS. The latter is an alternative to the old MASTERS' developed on the NetLogo environment. GMASTERS is written in the Python language and employs the GTK+ toolkit, which provides more sophisticated widgets and friendlier interface. We hope the interface to considerably assist users, something not commonly taken into consideration when simulating proteins. The results generated are plotted using the Matplotlib toolkit. For the visualization and 3D analysis of the obtained models GMASTERS connects with PyMOL [25] using a similar approach to GTKDynamo [26]. A typical session snapshot is shown in Fig. 4.

Creating Projects. In GMASTERS the user always works inside a given project. To create a new project the user provides information such as project directory, user name and protein sequence. Every project contains, beyond the information provided by the user, date and time of creation and the list of jobs.

Setup and Running Simulations. Once a project is created the user can setup and run Monte Carlo simulations inside GMASTERS. The user can set parameters such as box dimensions, temperature and maximum movement. Every new

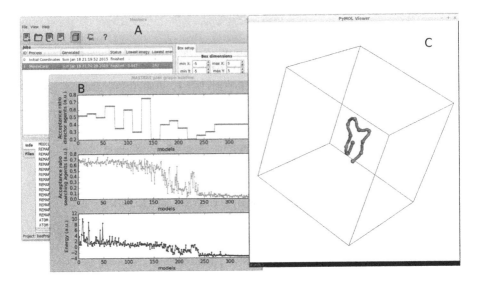

Fig. 4. GMASTER's GUI. (A) Main window, displaying simulation projects and parameters setup, (B) Plot window, showing energy and acceptance ratio, (C) PyMOL viewer, with the predicted conformation.

simulation result is a new item in the job list which provides time of creation, status, lowest obtained energy and the current step's energy. These data allow a preliminary comparison between different simulations and, at the same time, keep chronologically ordered steps performed by the user.

Data Analysis and Model Visualization. As a simulation proceeds its results are stored in a log file. GMASTERS is able to read and interpret these log files and display graphics of the relevant information, which can then be saved and manipulated by the user. Trajectories are also generated. These come in PDB format, and can hold a variety of information, including coordinates and energies.

4   Case Study

4.1   Geometry Representation and Energy Function

To examine GMASTERS' behavior and effectiveness, we adopted the same simplified model used in our last work [20], the AB Model (see Fig. 1).

The simplest and most conventional model among all applied abstractions used in PSP is the HP Model [27]. The HP model divides all 20 amino acids into two different groups, the hydrophobic (H) and the hydrophilic (P) ones. The amino acids are placed at an on-lattice grid, and the energy computation at each conformation takes into account only interactions between next-neighbored nonadjacent hydrophobic amino acids [27]. The energy of a conformation is the number of hydrophobic-hydrophobic contacts that are adjacent on the lattice, but not adjacent on the string (sequence). The main idea is to force the establishment of a compact hydrophobic core as observed in real proteins [28]. Lattice models have proven to be useful tools for reasoning about the PSP problems complexity [27] and despite its high abstraction level, the PSP problem with HP models is still an NP-Complete challenge [29]. The AB model is a lattice model in which the amino acids are once again divided into two groups: hydrophobic amino acids are marked as A while the hydrophilic ones are marked as B.

Many authors have been using this model as starting point for PSP understanding [30–32]. The AB model, in comparison to the HP model, has the additional capability of collecting information about local interactions that might be significant for the local structure of protein chains. This allows finding compact, well-defined native structures that would not be found if these local interactions were neglected [33]. Unlike the HP model, the interactions considered in the AB model include both sequence independent local interactions and the sequence dependent Lennard-Jones term that supports the energy convergence to a hydrophobic core [32,34].

The AB off-lattice energy model is described by Eq. 3, where θ is the bend angle between the two bonds defined by three consecutive residues and r_{ij} is the distance between residues i and j [32]. The first sum, the backbone bending potential, calculates the bending angle energy of the protein chain. The double sum is the Lennard-Jones potential. It calculates the long-range interaction energy, which is attractive for pairs of the same amino acids (AA or BB) and repulsive for AB pairs. The residue specific prefactor C is given by the Eq. 4.

$$E = \sum_{i=2}^{n-1} \frac{1}{4}(1 - cos\theta) + \sum_{i=1}^{n-2} \sum_{j=i+2}^{n-1} [r_{ij}^{-12} - C(\xi_i, \xi_j)r_{ij}^{-6}] \tag{3}$$

$$C(\xi_i, \xi_j) = \begin{Bmatrix} +1, & \xi_i\xi_j = A \\ +1/2, & \xi_i\xi_j = B \\ -1/2, & \xi_i \neq \xi_j \end{Bmatrix} \tag{4}$$

4.2   Target Sequence

Although several works in the literature use Fibonacci sequences as targets for their simulation [32,34], we chose to work on real sequence targets. The aminoacids A, C, G, I, L, M, P and V were set to hydrophobic (class A) and D, E, F, H, K, N, Q, R, S, T, W and Y were set to hydrophilic (class B). As means of demonstration we chose the PDB ID 1AGT protein, a recurrent target [35].

4.3   Simulation Setup and Running

Once the target sequence (and abstraction level) is chosen, a simulation setup is required. Here it is possible to parametrize the MC/SA scheme, selecting values to initial temperature, temperature decrease ratio, movement amplitudes, etc.

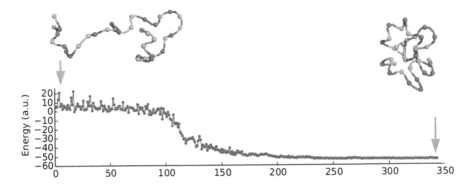

Fig. 5. The system starts with a high energy and the different kind of agents interact microscopically (MC/SA controlled moves) yielding a macroscopic behavior: the protein folding/energy convergence. The outputs are provided in a PyMOL canvas.

4.4   Data Analysis and Model Visualization

While the simulation runs it is possible to visualize the current energy profile and structure. Figure 5 shows that in the beginning of the simulation the protein is nearly unfolded and the first steps' energy is high. As the simulation progresses and the system cools down, lower levels of energy are reached and the protein starts to fold into more favorable conformations. If the system's energy doesn't start to decrease, it may be an indication to review the simulation setup parameters. In this case study, the simulation progresses normally and we can see the energy converging to lower levels. Furthermore it was observed an improvement in CPU time.

5   Conclusion and Future Works

This paper presented a tool for handling the protein structure prediction problem. More specifically GMASTERS, a GUI developed in GTK that runs over MASTERS core. The focus of the application is the user, making possible for students with different knowledge levels to learn about PSP and Monte Carlo and for experts to test their own methods.

To explore the new features available in GMASTERS we performed a case study with the protein whose PDB ID is 1AGT and, in addition to the new facilities provided, it was noticed a considerable improvement in CPU time.

We plan to evolve GMASTERS' architecture by parallelizing simulations, improving its overall performance. In addition, we intend to use different coarse grained abstraction levels and therefore different force fields.

Acknowledgements. This work was supported by grants CNPq-305984/20128 and FAPERGS-TO2054-2551/13-0 to ONS, a CNPq Research Fellow. TLP and MSS are supported by CAPES, JFRB by a CAPES/FAPERGS DOCFIX postdoctoral fellowship.

References

1. Lesk, A.M.: Introduction to Bioinformatics, 3rd edn. Oxford University Press, Oxford (2008)
2. Clark, K., Karsch-Mizrachi, I., Lipman, D.J., Ostell, J., Sayers, E.W.: Genbank. Nucleic Acids Res. **44**, D67–D72 (2016)
3. Berman, H.M., Westbrook, J., Feng, Z., Gilliland, G., Bhat, T.N., Weissig, H., Shindyalov, I.N., Bourne, P.E.: The protein data bank. Nucleic Acids Res. **28**(1), 235–242 (2000)
4. Pavlopoulou, A., Michalopoulos, I.: State-of-the-art bioinformatics protein structure prediction tools. Int. J. Mol. Med. **28**(3), 295–310 (2011)
5. Dill, K., MacCallum, J.: The protein-folding problem, 50 years on. Science **338**(6110), 1042–1046 (2012)
6. Duan, Y., Kollman, P.A.: Computational protein folding: from lattice to all-atom. IBM Syst. J. **40**(297–309), 0018–8670 (2001)
7. Amigoni, F., Schiaffonati, V.: Multiagent-based simulation in biology. In: Magnani, L., Li, P. (eds.) Model-based Reasoning in Science, Technology, and Medicine, SCI, vol. 64, pp. 179–191. Springer, Heidelberg (2007)
8. Tisseau, J.: Virtual reality, in virtuo autonomy. Ph.D. thesis, University of Rennes 1 (2001)
9. Nelson, D., Cox, M.: Lehninger Principles of Biochemistry, 5th edn. W. H. Freeman and Company, New York (2008)
10. Snow, C.D., Sorin, E.J., Rhee, Y.M., Pande, V.S.: How well can simulation predict protein folding kinetics and thermodynamics? Annu. Rev. Biophys. **34**, 43–69 (2005). Annual Reviews, Palo Alto
11. Zvelebil, M., Baum, J.: Understanding Bioinformatics. Garland Science, US (2007)
12. Crescenzi, P., Goldman, D., Papadimitriou, C., Piccolboni, A., Yannakakis, M.: On the complexity of protein folding. J. Comput. Biol. **5**, 597–603 (1998)
13. Levinthal, C.: Are there pathways for protein folding? J. Med. Phys. **65**(1), 44–45 (1968)
14. Tramontano, A.: Integral and differential form of the protein folding problem. Phys. Life Rev. **1**(2), 103–127 (2004)
15. Helles, G.: A comparative study of the reported performance of ab initio protein structure prediction algorithms. J. Roy. Soc. Interface **5**(21), 387–396 (2008)
16. Wooldridge, M., Jennings, N.: Intelligent agents - theory and practice. Knowl. Eng. Rev. **10**(2), 115–152 (1995)
17. Russell, S.J., Norvig, P.: Artificial Intelligence: A Modern Approach, 2nd edn. Prentice Hall, Englewood Cliffs (2003)
18. Bradshaw, J.M.: An introduction to software agents. In: Bradshaw, J.M. (ed.) Software Agents, pp. 3–46. AAAI Press / The MIT Press, Cambridge (1997)
19. Tisue, S., Wilensky, U.: Netlogo: a simple environment for modeling complexity (2004)
20. Lipinski-Paes, T., Norberto de Souza, O.: MASTERS: a general sequence-based multiagent system for protein tertiary structure prediction. Electron. Notes Theor. Comput. Sci. **306**, 45–59 (2014)
21. Osguthorpe, D.J.: Ab initio protein folding. Curr. Opin. Struct. Biol. **10**(2), 146–152 (2000)
22. Roli, A., Milano, M.: MAGMA: a multiagent architecture for metaheuristics. IEEE Trans. Syst. Man Cybern. Part B (Cybernetics) **34**, 925–941 (2004)

23. Metropolis, N., Rosenbluth, A.W., Rosenbluth, M.N., Teller, A.H., Teller, E.: Equation of state calculations by fast computing machines. J. Chem. Phys. **21**(6), 1087–1092 (1953)
24. Allen, P., Tildesley, D.: Computer Simulation of Liquids. Oxford science publications, Clarendon Press (1987)
25. Schrödinger, L.L.C.: The PyMOL molecular graphics system, February 2016
26. Bachega, J.F.R., Timmers, L.F.S.M., Assirati, L., Bachega, L.R., Field, M.J., Wymore, T.: GTKDynamo: A PyMOL plug-in for QC/MM hybrid potential simulations. J. Comput. Chem. **34**(25), 2190–2196 (2013)
27. Dill, K.: Theory for the folding and stability of globular-proteins. Biochemistry **24**(6), 1501–1509 (1985)
28. William, E.H., Alantha, N.: Protein structure prediction with lattice models. Chapman & Hall/CRC Computer & Information Science Series, pp. 30-1–30-24. Chapman and Hall/CRC (2005)
29. Berger, B., Leighton, T.: Protein folding in the hydrophobic-hydrophilic (HP) model is NP-complete. J. Comput. Biol. **5**(1), 27–40 (1998)
30. Bachmann, M., Arkin, H., Janke, W.: Multicanonical study of coarse-grained off-lattice models for folding heteropolymers. Phys. Rev. E Stat. Nonlin Soft Matter Phys. **71**(3), 031906 (2005). doi:10.1103/PhysRevE.71.031906
31. Hsu, H.P., Mehra, V., Grassberger, P.: Structure optimization in an off-lattice protein model. Physical Review E **68**(3), 037703 (2003). doi:10.1103/PhysRevE.68.037703
32. Stillinger, F., Head-Gordon, T.: Collective aspects of protein-folding illustrated by a toy model. Phys. Rev. E **52**(3), 2872–2877 (1995)
33. Irback, A., Peterson, C., Potthast, F., Sommelius, O.: Local interactions and protein folding: a three-dimensional off-lattice approach. J. Chem. Phys. **107**(1), 273–282 (1997)
34. Stillinger, F., Head-Gordon, T., Hirshfeld, C.: Toy model for protein-folding. Phys. Rev. E **48**(2), 1469–1477 (1993)
35. Krezel, A.M., Kasibhatla, C., Hidalgo, P., Mackinnon, R., Wagner, G.: Solution structure of the potassium channel inhibitor agitoxin 2: Caliper for probing channel geometry. Protein Sci. **4**(8), 1478–1489 (1995)

Filling a Protein Scaffold with a Reference

Letu Qingge[1], Xiaowen Liu[2], Farong Zhong[3], and Binhai Zhu[1($\boxtimes$)]

[1] Department of Computer Science, Montana State University,
Bozeman, MT 59717, USA
letu.qingge@msu.montana.edu, bhz@montana.edu
[2] School of Informatics and Computing, Indiana University-Purdue
University Indianapolis, Indianapolis, IN 46202, USA
xwliu@iupui.edu
[3] School of Mathematics, Physics and Informatics, Zhejiang Normal University,
Jinhua 324001, Zhejiang, China
zfr@zjnu.cn

Abstract. In mass spectrometry-based de novo protein sequencing, it is hard to complete the sequence of the whole protein. Motivated by this we study the (one-sided) problem of filling a protein *scaffold* S with some missing amino acids, given a sequence of contigs none of which is allowed to be altered, with respect to a complete reference protein P of length n, such that the BLOSUM62 score between P and the filled sequence S' is maximized. We show that this problem is polynomial-time solvable in $O(n^{26})$ time. We also consider the case when the contigs are not of high quality and they are concatenated into an (incomplete) sequence I, where the missing amino acids can be inserted anywhere in I to obtain I', such that the BLOSUM62 score between P and I' is maximized. We show that this problem is polynomial-time solvable in $O(n^{22})$ time. Due to the high running time, both of these algorithms are impractical, we hence present several algorithms based on greedy and local search, trying to solve the problems practically. The empirical results show that the algorithms can fill protein scaffolds almost perfectly, provided that a good pair of scaffold and reference are given.

1 Introduction

In mass spectrometry-based de novo protein sequencing, it is hard to complete the sequence of the whole protein [2]. An incomplete sequence contains one or several contigs, each is a protein segment. When the order of the contigs is known (e.g., ordered contigs reported by top-down mass spectrometry-based de novo protein sequencing [11]), these contigs are called a *scaffold*. In many applications, it is more desirable to obtain complete protein sequences. We comment that a similar phenomena occurs in the sequencing of genomes, firstly noticed by Mũnoz et al. [14], and have resulted in a series of interesting algorithmic studies [3,5–9].

Hence, a natural combinatorial problem is to fill the missing amino acids into scaffolds. As one must find a biologically meaningful way of filling scaffolds, it makes sense to use a complete homologous protein sequence as a reference. Here

© Springer International Publishing Switzerland 2016
A. Bourgeois et al. (Eds.): ISBRA 2016, LNBI 9683, pp. 175–186, 2016.
DOI: 10.1007/978-3-319-38782-6_15

we consider two kinds of scaffolds. One kind is lists of contigs which are computed with a good confidence and should not be altered. Throughout this paper, we use $\mathcal{S}$ to denote such a scaffold, composed of contigs $C_1, C_2, ..., C_m$. The other kind are sequences, which usually arise when contigs are computed without a very high success confidence — in which case it would be unrealistic not to alter the contigs. We then simply concatenate the contigs into $\mathcal{I}$, the incomplete sequence. We try to carry out this idea to fill a scaffold (resp. sequence) $\mathcal{S}$ (resp. $\mathcal{I}$) to have $\mathcal{S}'$ (resp. $\mathcal{I}'$), such that the similarity between $\mathcal{S}'$ (resp. $\mathcal{I}'$) and a given (complete) protein $\mathcal{P}$ is maximized. These problems are called *Contig-Preserving Protein Scaffold Filling* (CP-PSF for short) and *Protein Scaffold Filling* (PSF for short) respectively.

The BLOSUM62 matrix [4] is the most popular similarity measure to align a pair of protein sequences. (The matrix can be found in the appendix.) We will use BLOSUM62 to measure the accuracy of our scaffold filling algorithms. However, for empirical results it is not very indicative to interpret BLOSUM62 scores for measuring the similarity between sequences. Therefore, we will use the number of *matched pairs* as a similarity measure (this can be considered a rounded BLOSUM62 matrix: if the score between two amino acids in BLOSUM62 is less than 4, then set the score as 0; otherwise, set the score as 1). Of course, when a space (denoted as a $\star$ in this paper) is used, with BLOSUM62 we get a negative score; and this is not reflected in the number of matched pairs.

Firstly, we show that the PSF problem is solvable in $O(n^{22})$ time. As this method is impractical, we make use of the standard Needleman-Wunsch algorithm to align $\mathcal{I}$ with $\mathcal{P}$, we then design two heuristic algorithms, based on greedy and local search, to insert the amino acids in $X = \mathcal{P} - \mathcal{I}$ into $\mathcal{P}$. We test our algorithm using some real datasets (i.e., 4 chains from two antibody proteins). Secondly, for the CP-PSF problem of filling $\mathcal{S}$, we also present a polynomial time solution which takes $O(n^{26})$ time. We then study the problem of aligning the contigs in $\mathcal{S}$ to $\mathcal{P}$ to achieve the maximum BLOSUM62 score, and show that it is solvable in $O(n^5)$ time. Then, based on this, we design and implement two algorithms using greedy and local search methods. We test our algorithms on the same datasets.

We comment that our problems are related to but different from the *one-sided* scaffold filling problem for genomes (with gene repetitions), the main difference is that the similarity measure between genomes is different from that between two protein sequences. For protein sequences, the order of its amino acids is even more critical compared with genomes. Given a complete genome $\mathcal{G}$ and a genomic scaffold $\mathcal{H}$, the one-sided scaffold filling problem, i.e., filling $\mathcal{H}$ into $\mathcal{H}'$ such that the number of adjacencies between $\mathcal{G}$ and $\mathcal{H}'$ is maximized, is NP-hard [5,6] and the best approximation algorithm has a factor of 1.20 [7].

This paper is organized as follows. In Sect. 2, we give necessary definitions. In Sect. 3, we present an $O(n^{22})$ time solution and two practical algorithms, and we present some empirical results using some real datasets. In Sect. 4, we first show that CP-PSF can be solved in $O(n^{26})$ time and a special case for the CP-PSF problem is solvable in $O(n^5)$ time, and then, based on it, we design two practical algorithms for CP-PSF. We then present some empirical results using the same datasets. In Sect. 5, we conclude the paper.

2   Preliminaries

We first present some necessary definitions.

We denote the set of 20 amino acids as $\Sigma = \{A, C, D, E, F, G, H, I, K, L, M, N, P, Q, R, S, T, V, W, Y\}$. A protein sequence $\mathcal{P}$ is a sequence over Σ. We also use $c(\mathcal{P})$ to denote the multiset of elements in $\mathcal{P}$. For example, $\mathcal{P} = \langle A, D, C, I, K, W, Y, C, I \rangle$ (or simply, $\mathcal{P} =$ ADCIKWYCI) with $c(\mathcal{P}) = \{A, C, C, D, I, I, K, W, Y\}$. In bottom-up mass spectrometry based de novo protein sequencing, we first derive peptide sequences from tandem mass spectra of the target protein, then build up longer contigs by spectral or peptide assembly [1, 10]. For real datasets, the length of a peptide is typically between 5 and 40.

Given two protein sequences $\mathcal{P} = \langle p_1, p_2, ..., p_n \rangle$ and $\mathcal{Q} = \langle q_1, q_2, ..., q_n \rangle$, (p_i, q_i) forms a *matched pair* if $p_i = q_i$, i.e., p_i and q_i are the same amino acid. We use $\mathcal{P}[i]$ to represent p_i, $\mathcal{P}[i, j]$ to represent the substring $\langle p_i, ..., p_j \rangle$ and $\mathcal{P}[i..]$ to represent the substring $\langle p_i, ..., p_n \rangle$. The length of $\mathcal{P}$ is denoted as $|\mathcal{P}|$, which is n here. We use $mp(\mathcal{P}, \mathcal{Q})$ to denote the corresponding number of matched pairs. Let T_1 and T_2 denote two sets of protein sequences of the same length, we use $mp(T_1, T_2)$ to denote the maximum number of matched pairs $mp(\mathcal{P}, \mathcal{Q})$, for any $\mathcal{P} \in T_1, \mathcal{Q} \in T_2$. This definition also holds when $\mathcal{P}$ and $\mathcal{Q}$ are aligned from initial protein sequences, possibly of different lengths, using the standard sequence alignment algorithm by Needleman and Wunsch [15], i.e., when $\mathcal{P}$ and $\mathcal{Q}$ contain $\star$ (the gap, or space) letters. (For the ease of presentation we use $\star$ instead of $-$, as the latter is used as subtraction as well.) Note that a $\star$ does not form a matched pair with any amino acid.

BLOSUM62 (B62 for short), coming from BLOck SUbstitution Matrix [4], is based on comparisons of multiply (locally) aligned ungapped segments corresponding to the most highly conserved regions of proteins in the Blocks database [16]. The number of 62 means the comparisons are based on ungapped sequence alignments with $< 62\%$ identity. BLOSUM62 is the default matrix for the standard protein-BLAST program. The integer values in BLOSUM62 vary from -4 to 11, e.g., $B62[W, D] = -4$ and $B62[W, W] = 11$. It should be noted that a score -8 is applied for introducing a $\star$, which is a restriction for using too many $\star$'s. Throughout this paper, we use B62 to measure the similarity of protein sequences.

A *scaffold* $\mathcal{S}$ is a list of contigs $C_1, C_2, ..., C_m$, where each contig is a sequence of amino acids. For example, $\mathcal{S} = \langle C_1, C_2, C_3 \rangle$, where $C_1 = \boxed{\text{AEFGIA}}$, $C_2 = \boxed{\text{CDIKLNTVW}}$, and $C_3 = \boxed{\text{PQAWYA}}$. These contigs are usually computed with some weight constraint, say, the total weight of amino acids in C_1 is roughly $71+129+147+57+113+71=598$ Dalton. Hence, we should be more careful in inserting some missing amino acid into a contig. In fact, throughout this paper, we do not alter the amino acids in a given contig at all. A *sequence scaffold* (or just sequence) $\mathcal{I}$ is an incomplete protein sequence, i.e., with some unknown missing amino acids. This sequence is obtained usually when the contigs are not of high quality, and we just concatenate C_i's to obtain a sequence. We are allowed to insert missing amino acids anywhere in $\mathcal{I}$.

Given a multiset X of amino acids, and a scaffold $\mathcal{S} = \langle C_1, ..., C_m \rangle$, $\mathcal{S} + X$ is the set of all protein sequences obtained by inserting the amino acids in X in between C_i's (i.e., C_i's are not altered).

In practice, if the total mass of the target protein is known, which can be measured by top-down mass spectrometry, we would know the total mass of missing amino acids. In this case, one way for handling this is to enumerate the sets of amino acids which sum to a certain mass — corresponding to that of the missing amino acids. Of course, when this mass is decently large we could have an exponential number of such sets. In this paper, we only focus on a given set X of missing amino acids, which can be computed as the difference between the reference protein and given scaffold (or sequence).

Given two protein sequences $\mathcal{P}, \mathcal{Q}$, we use $B62(\mathcal{P}, \mathcal{Q})$ to denote the maximum BLOSUM62 score when aligning them. Let $\mathcal{Z}$ be a set of protein sequences. Then $B62(\mathcal{P}, \mathcal{Z})$ is the maximum BLOSUM62 score when aligning $\mathcal{P}$ and any sequence $z \in \mathcal{Z}$. Given a protein sequence $\mathcal{I}$ and a multiset X of amino acids, we denote $\mathcal{I} + X$ as the set of all protein sequences obtained by filling all the amino acids in X into $\mathcal{I}$. We use $|X|$ to denote the size of the set X.

The contig-preserving protein scaffold filling (CP-PSF) problem is defined as follows.

Contig-Preserving Protein Scaffold Filling to Maximize the B62 Score (CP-PSF)

Input: a complete protein sequence $\mathcal{P}$, a protein scaffold $\mathcal{S} = \langle C_1, ..., C_m \rangle$, and a multiset X of amino acids.
Question: maximize the B62 score $B62(\mathcal{P}, \mathcal{S} + X)$.

For most of the practical instances we could assume that X is given as $X = c(\mathcal{P}) - \cup_i c(C_i)$. Note that we have no restriction on the length of $\mathcal{P}$ and the filled sequences, as with B62 we could use $\star$'s.

When the scaffold is a sequence $\mathcal{I}$, the problem can be simplified as follows.

Protein Scaffold Filling to Maximize the B62 Score (PSF)

Input: a complete protein sequence $\mathcal{P}$ with $|\mathcal{P}| = n$, an incomplete protein sequences $\mathcal{I}$ and a multiset of amino acids X.
Question: maximize the B62 score $B62(\mathcal{P}, \mathcal{I} + X)$.

Note that, again, for most of the practical instances we could assume that X is given as $X = c(\mathcal{P}) - c(\mathcal{I})$.

As solutions for PSF could be used as subroutines for CP-PSF, in the next section we first show that the PSF problem is polynomially solvable by giving a complex dynamic programming solution which runs in $O(n^{22})$ time. We then present two practical methods and show some empirical results.

3   Algorithms and Empirical Results for PSF

3.1   PSF is in P

We present a dynamic programming solution for solving the PSF problem, given $\mathcal{P}, \mathcal{I}$ and X. The objective is to insert X into $\mathcal{I}$ to obtain $\mathcal{I}'$ such that $B62(\mathcal{P}, \mathcal{I}')$ is maximized.

The idea of our algorithm is as follows. In an optimal solution, when scanning $\mathcal{P}$ from left to right, there are 4 cases: (1) $\mathcal{P}[i]$ is aligned to an inserted amino acid in X, (2) $\mathcal{P}[i]$ is aligned to some amino acid $\mathcal{I}[j]$, (3) $\mathcal{P}[i]$ is inserted with a $\star$ and it is aligned to some amino acid $\mathcal{I}[j]$, and (4) $\mathcal{P}[i]$ is aligned to an inserted $\star$ in $\mathcal{I}[j]$. To denote the set X (or any of its subset), we use a 20-dimension vector $X = \langle a_1, a_2, ..., a_{20} \rangle$, where a_k is the number of amino acid k ($k = 1..20$). Define $|X| = a_1 + a_2 + \cdots + a_{20}$. It is clear that the number of distinct subsets of X is bounded by

$$(a_1 + 1) \times (a_2 + 1) \times \cdots \times (a_{20} + 1) \le (\frac{a_1 + a_2 + \cdots + a_{20} + 20}{20})^{20} = (\frac{|X| + 20}{20})^{20} = O(n^{20}).$$

We now define $S_1[i, k]$ as the maximum B62 score when aligning $\mathcal{P}[1..i]$ and part of $\mathcal{I}'$, where $\mathcal{P}[i]$ is aligned to an inserted amino acid k from X and some $\star$'s could be inserted in $\mathcal{P}[1..i]$ (the resulting sequence is denoted as $\mathcal{P}'[1..i'']$). $T_1[i, k]$ contains all subsets of X, each of a uniform size ℓ and can be inserted to part of $\mathcal{I}$ such that the maximum value of $S_1[i, k]$ is achieved. If $t \in T_1[i, k]$, then t is a 20-vector. Hence at this point there are still $|X| - |t|$ amino acids to be inserted. Note that here k is a positive integer bounded by 20. If a specific k has been used up then $S_1[i, k]$ is undefined. At the end, $|\mathcal{P}'|$ could be larger than n (due to the insertion of $\star$'s), but it is still bounded by $O(n)$. We denote the largest index i'' in $\mathcal{P}'$ as n_{max}, where $n_{max} - n$ is the number of $\star$'s inserted in $\mathcal{P}$.

Likewise, we define $S_2[i, i']$ as the maximum B62 score when aligning $\mathcal{P}[1..i]$ and part of $\mathcal{I}'$, where $\mathcal{P}[i]$ is aligned to $\mathcal{I}[i']$. Define $T_2[i, i']$ as all subsets of X, each can be inserted to part of $\mathcal{I}$ such that the maximum value of $S_1[i, i']$ is achieved. Note that here we have $i, i' = O(n)$.

$S_3[i, j]$ is defined as the maximum B62 score when aligning $\mathcal{P}[1..i]$ and part of $\mathcal{I}'$, where $\mathcal{P}[i]$ is inserted by a $\star$ and the $\star$ is aligned to $\mathcal{I}[j]$. $S_4[i, j]$ is defined as the maximum B62 score when aligning $\mathcal{P}[1..i]$ and part of $\mathcal{I}'$, where $\mathcal{I}[j]$ is inserted by a $\star$ and the $\star$ is aligned to $\mathcal{P}[i]$. $T_3[i, j], T_4[i, j]$ are defined similarly.

We show the update of these tables as follows.

$$S_1[i+1, k] = \max \begin{cases} \max_{k'}\{S_1[i, k'] + B62((\mathcal{P}[i+1], k))\}, & \text{if } t[k] > 0, \ t \in T_1[i, k'] \ \text{(case 1)} \\ S_2[i, i'] + B62(\mathcal{P}[i+1], k), & \text{if } t[k] > 0, \ t \in T_2[i, i'] \ \text{(case 2)} \\ S_3[i, j] + B62(\mathcal{P}[i+1], k), & \text{if } t[k] > 0, \ t \in T_3[i, j] \ \text{(case 3)} \\ S_4[i, j] + B62(\mathcal{P}[i+1], k), & \text{if } t[k] > 0, \ t \in T_4[i, j] \ \text{(case 4)} \end{cases}$$

$$T_1[i+1, k] = \begin{cases} T_1[i, k'], \text{ and for } t \in T_1[i+1, k], \text{ update } t[k] \leftarrow t[k] - 1, & \text{(case 1)} \\ T_2[i, i'], \text{ and for } t \in T_1[i+1, k], \text{ update } t[k] \leftarrow t[k] - 1, & \text{(case 2)} \\ T_3[i, j], \text{ and for } t \in T_1[i+1, k], \text{ update } t[k] \leftarrow t[k] - 1, & \text{(case 3)} \\ T_4[i, j], \text{ and for } t \in T_1[i+1, k], \text{ update } t[k] \leftarrow t[k] - 1, & \text{(case 4)} \end{cases}$$

$$S_2[i+1, i'] = \max \begin{cases} \max_{k'}\{S_1[i, k'] + B62(\mathcal{P}[i+1], \mathcal{I}[i'])\}, & \text{(case 5)} \\ S_2[i, i'-1] + B62(\mathcal{P}[i+1], \mathcal{I}[i']), & \text{(case 6)} \\ S_3[i, i'-1] + B62(\mathcal{P}[i+1], \mathcal{I}[i']), & \text{(case 7)} \\ S_4[i, i'-1] + B62(\mathcal{P}[i+1], \mathcal{I}[i']), & \text{(case 8)} \end{cases}$$

$$T_2[i+1, i'] = \begin{cases} T_1[i, k'], & \text{(case 5)} \\ T_2[i, i'-1], & \text{(case 6)} \\ T_3[i, i'-1], & \text{(case 7)} \\ T_4[i, i'-1], & \text{(case 8)} \end{cases}$$

$$S_3[i+1,j] = \max \begin{cases} \max_{k'}\{S_1[i,k']\} - 8, & \text{(case 9)} \\ S_2[i,j-1] - 8, & \text{(case 10)} \\ S_3[i,j-1] - 8, & \text{(case 11)} \\ S_4[i,j-1] - 8, & \text{(case 12)} \end{cases}$$

$$T_3[i+1,j] = \begin{cases} T_1[i,k'], & \text{(case 9)} \\ T_2[i,j-1], & \text{(case 10)} \\ T_3[i,j-1], & \text{(case 11)} \\ T_4[i,j-1], & \text{(case 12)} \end{cases}$$

$$S_4[i+1,j] = \max \begin{cases} \max_{k'}\{S_1[i,k']\} - 8, & \text{(case 13)} \\ S_2[i,j-1] - 8, & \text{(case 14)} \\ S_3[i,j-1] - 8, & \text{(case 15)} \\ S_4[i,j-1] - 8, & \text{(case 16)} \end{cases}$$

$$T_4[i+1,j] = \begin{cases} T_1[i,k'], & \text{(case 13)} \\ T_2[i,j-1], & \text{(case 14)} \\ T_3[i,j-1], & \text{(case 15)} \\ T_4[i,j-1], & \text{(case 16)} \end{cases}$$

The optimal solution value is

$$\max\{S_1[n_{\max},k], 1 \leq k \leq 20; S_2[n_{\max},m]; S_3[n_{\max},m]; S_4[n_{\max},m]\}.$$

As the number of subsets of X is bounded by $O(n^{20})$, it is obvious that the algorithm takes $O(n^{22})$ time and space. We thus have the following theorem.

Theorem 1. *PSF can be solved in $O(n^{22})$ time and space.*

Even though PSF is polynomially solvable, the high running time makes the solution practically infeasible to implement. We next present two practical methods.

3.2   Practical Algorithms for PSF

In the following, we assume that a complete protein sequence $\mathcal{P}$ is always given. Our first algorithm is called *Align+Greedy*. We first align $\mathcal{I}$ to $\mathcal{P}$ using the standard Needleman-Wunsch algorithm, and then we insert the missing amino acids at the $\star$ positions in the aligned $\mathcal{I}$ with a greedy method (i.e., according to the maximum of their B62 scores).

Algorithm 1. *Align+Greedy($\mathcal{P},\mathcal{I},X$)*

1   Align $\mathcal{I}$ with $\mathcal{P}$ to obtain $\mathcal{I}_1$ with the maximum B62 score.
2   Find a $\star$ position in the aligned $\mathcal{I}_1$ such that inserting an amino acid in X would incur the maximum B62 score among all $\star$ positions.
3   Repeat Step 2 until all elements in X are inserted into $\mathcal{I}_1$.
4   Return the filled $\mathcal{I}_1$ as $\mathcal{I}'$, with the total alignment score between $\mathcal{I}'$ and $\mathcal{P}$ being $b62(\mathcal{P},\mathcal{I},X)$.

An improved version of our first algorithm is based on a local search method. The idea is that if Algorithm 1 does not give us the best return, there must be a way to locally update the solution to have a better result. The algorithm is as follows.

Algorithm 2. *LocalSearch($\mathcal{P}$,$\mathcal{I}$,X)*
1   Compute $b62(\mathcal{P},\mathcal{I},X)$ using Algorithm 1. Assign $U \leftarrow b62(\mathcal{P},\mathcal{I},X)$.
2   Insert an amino acid $x \in X$ into a position i of $\mathcal{I}$ to obtain $\mathcal{I}(i,x)$ such that the score $b62(\mathcal{P},\mathcal{I}(i,x),X-\{x\})$ is maximum, for all $x \in X$ and for all $i \in [0,|\mathcal{I}|]$.
3   Run Algorithm 1 to obtain $b62(\mathcal{P},\mathcal{I}+x,X-\{x\})$.
     If $b62(\mathcal{P},\mathcal{I}+x,X-\{x\}) \leq U$, then return the solution $\mathcal{I}''$ incurring U; else update $U \leftarrow b62(\mathcal{P},\mathcal{I}(i,x),X-\{x\})$, $\mathcal{I} \leftarrow \mathcal{I}(i,x)$, $X \leftarrow X-\{x\}$, and repeat Step 2.

Our datasets are based on MabCampath (or Alemtuzumab) and Humira (or Adalimumab), which are two similar antibody proteins. Both of them contain a light chain and a heavy chain, the lengths for them are 214 and 449 for MabCampath, and 214 and 453 for Humira respectively. The pairwise alignments of the two light chains and two heavy chains display 91.1 % and 86.6 % identity respectively. For each protein sequence we compute a set of peptides from bottom up tandem mass spectra using PEAKS [12,13], which is a de novo peptide sequencing software tool. Then we simply select a maximal set of disjoint peptides for each protein sequence. For the light chain of MabCampath (MabCampath-L for short): we have two (disjoint) peptides of lengths 12 and 19. For the heavy chain of MabCampath (MabCampath-H for short): we have eight (disjoint) peptides of lengths 9, 7, 13, 12, 14, 15, 12 and 19. For the heavy chain of Humira (Humira-H for short): we have six (disjoint) peptides of lengths 7, 7, 9, 9, 10 and 8. For the light chain of Humira (Humira-L for short), PEAKS is not able to obtain any peptides of decent quality. So we will only use Humira-L for reference purpose. Due to that the amino acids I and L have the same mass, in the peptides and all our comparisons, all I's have been converted to L's. The datasets can be found in
http://www.cs.montana.edu/~qingge.letu/ResearchData.html.
Our code was written in Matlab and Java.

3.3 Empirical Results

Let $x \in \{$MabCampath-H, Humira-H, MabCampath-L$\}$ and let the corresponding references of x be Humira-H, MabCampath-H, and Humira-L respectively. For each x, we use Algorithm 1 to compute the filled sequence $\mathcal{I}'_x$, and we use Algorithm 2 to compute the filled sequence $\mathcal{I}''_x$. As the B62 scores are not very indicative, in the following we use the number of matched pairs resulting from the computed B62 scores. The empirical results are shown in the following two tables (Tables 1 and 2). Note that in most cases, Algorithm 2 produces slightly better results. Of course, that should be considered as normal.

Table 1. Empirical results for the three computed $\mathcal{I}'_x$ using Algorithm 1. In all the tables, **ref_length, tar_length** represents the length of the reference and target protein sequence respectively.

	x=MabCampath-H	x=Humira-H	x=MabCampath-L
mp($\mathcal{I}'_x$,ref)	406	442	214
mp($\mathcal{I}'_x$,ref)/ref_length	406/453=89.62 %	442/449=98.44 %	214/214=100 %
mp($\mathcal{I}'_x$,target)	355	385	195
mp($\mathcal{I}'_x$,target)/tar_length	355/449=79.06 %	385/453=84.99 %	195/214=91.12 %

Table 2. Empirical results for the three computed $\mathcal{I}''_x$ using Algorithm 2.

	x=MabCampath-H	x=Humira-H	x=MabCampath-L
mp($\mathcal{I}''_x$,ref)	415	442	214
mp($\mathcal{I}''_x$,ref)/ref_length	415/453=91.61 %	442/449=98.44 %	214/214=100 %
mp($\mathcal{I}''_x$,target)	370	385	195
mp($\mathcal{I}''_x$,target)/tar_length	370/449=82.41 %	385/453=84.99 %	195/214=91.12 %

4   Algorithms for Empirical Results for CP-PSF

4.1   CP-PSF is Polynomially Solvable

In this section, we first show that CP-PSF is also solvable in polynomial time. As the running time of this algorithm is too high (hence infeasible for implementation), we just sketch a solution without necessarily trying to obtain the best running time.

Define $B[i, j, k, \ell]$ as the maximum B62 score when the last element of $S_{\ell-1}$ is aligned with $\mathcal{P}[i]$, and S_ℓ is aligned with $\mathcal{P}[j..k]$. Define $C[i, j, k, \ell]$ as the set of amino acids inserted to obtain $B[i, j, k, \ell]$. Let $INS(\mathcal{P}[i+1..j-1], C[i, j, k, \ell])$ be the maximum B62 score obtained by inserting the amino acids in $C[i, j, k, \ell]$ to align with the substring $\mathcal{P}[i+1..j-1]$. Apparently, $INS(-, -)$ can be computed using the dynamic programming algorithm summarized in Theorem 3.1. Note that $*$'s could be inserted in $\mathcal{P}$ and in between the scaffolds in $\mathcal{S}$. The optimal solution is $\max_{i,j,k}\{B[i, j, k, m] + INS(\mathcal{P}[k+1..n], X - C[i, j, k, m])\}$, which might not be unique. As we have $O(mn^3) = O(n^4)$ cells in the table $B[i, j, k, \ell]$, the following theorem is straightforward.

Theorem 2. *The Contig-Preserving PSF can be solved in $O(n^{26})$ time and space.*

Unfortunately, the running time of this algorithm is too high. Hence, we first design some practical algorithms and then show some empirical results for the CP-PSF problem using these practical algorithms.

4.2   Practical Algorithms CP-PSF-B62

We show some property for CP-PSF.

Lemma 1. *The problem of computing an alignment of $\mathcal{P}$ and $\mathcal{S}$ by only inserting spaces between the contigs in $\mathcal{S}$ such that $B62(\mathcal{P}, \mathcal{S})$ is maximized, is polynomially solvable.*

Proof. Let $C[1], C[2], ..., C[m]$ be the sequence of contigs in $\mathcal{S}$. Let $c[i]$ be the size of $C[i]$, i.e., $c[i] = |C[i]|$, for $i = 1, .., m$. Let $d[i]$ be the last letter of $C[i]$, for $i = 1, .., m$. Note that $B62^*(\mathcal{P}[i..j], C[k])$ is the maximum B62 score when aligning $\mathcal{P}[i..j]$ and $C[k]$, where no $\star$ can be inserted into $C[k]$. This can be pre-computed in $O(c[k]^2)$ time by setting a large penalty for introducing a $\star$ in $C[k]$ (say $-c[k]$). Therefore, the problem is to align $C[k]$'s, without any gap in each $C[k]$, to disjoint positions at $\mathcal{P}$.

Define $A[i, j, k]$ as the maximum total B62 score obtained when $\mathcal{P}[i..j]$ is aligned with $C[k]$, without using any gap in $C[1], ..., C[k]$.

$$A[i, j, k + 1] = \max_{i', j'}\{A[i', j', k] + B62^*(\mathcal{P}[i..j], C[k + 1])\},$$

where $i' \leq i - c[k], \sum_{\ell=k+1..m} c[\ell] \leq n - j + 1$.

The initialization is done as follows. $A[i, j, 1] = B62^*(\mathcal{P}[i..j], C[1]), \sum_{\ell=2..m} c[\ell] \leq n - j + 1$. The final solution can be found at $\max_{1 \leq i < j \leq n}\{A[i, j, m]\}$. The algorithms takes $O(mn^4) = O(n^5)$ time and $O(mn^2) = O(n^3)$ space.   □

The above lemma, though does not solve CP-PSF, does give us a heuristic algorithm. After the contigs in $\mathcal{S}$ are aligned at $\mathcal{P}$, we could use a greedy method to form a feasible solution. Among all the gap positions out of any contig in the aligned $\mathcal{S}$, we insert the elements in X at all the $\star$'s positions, from left to right, to maximize the total B62 score. We could use the lemma as a subroutine to have Algorithm 3, which runs in $O(mn^4)$ time. In fact, we could implement a simplified heuristic version for the above lemma; namely, we could scan from left to right in $\mathcal{P}$ to find the best locations to locate $C_1, C_2, ..., C_m$. In the worst case, that would only take $O(mn^2)$ time.

Algorithm 3. *ContigAlign+Greedy($\mathcal{P}, \mathcal{S}, X$)*

1   Align $\mathcal{S}$ with $\mathcal{P}$ to obtain $\mathcal{S}_1$ with the maximum B62 score based on Lemma 4.
2   Find a $\star$ position in the aligned $\mathcal{S}_1$ such that inserting an amino acid in X would incur the maximum B62 score among all $\star$ positions.
3   Repeat Step 2 until all elements in X are inserted into $\mathcal{S}_1$.
4   Return the filled $\mathcal{S}_1$ as $\mathcal{S}'$, with the total alignment score between $\mathcal{S}'$ and $\mathcal{P}$ being $b62(\mathcal{P}, \mathcal{S}, X)$.

Similar to the idea in Sect. 3.2, we could use a local search idea to try to improve Algorithm 3. Here, we could augment the contigs by appending some amino acids at its two ends, but the initial contigs are never altered.

Algorithm 4. *ContigLocalSearch(P,S,X)*

1   Compute $b62(\mathcal{P},\mathcal{S},X)$ using Algorithm 3. Assign $U \leftarrow b62(\mathcal{P},\mathcal{S},X)$.

2   Insert an amino acid $x \in X$ at the beginning (resp. end) of contig $S_i \in \mathcal{S}$ to obtain $S_i + x$ such that the score $b62(\mathcal{P},\mathcal{S} - \{S_i\} \cup \{S_i + x\}, X - \{x\})$ is the maximum, for all $x \in X$ and all $S_i \in \mathcal{S}$.

3   Run Algorithm 3 to obtain $b62(\mathcal{P},\mathcal{S} - \{S_i\} \cup \{S_i + x\}, X - \{x\})$. If $b62(\mathcal{P},\mathcal{S} - \{S_i\} \cup \{S_i + x\}, X - \{x\}) \leq U$, then return the solution $\mathcal{S}''$ incurring U; else update $U \leftarrow b62(\mathcal{P},\mathcal{S} - \{S_i\} \cup \{S_i + x\}, X - \{x\})$, $\mathcal{S} \leftarrow \mathcal{S} - \{S_i\} \cup \{S_i + x\}$, $X \leftarrow X - \{x\}$ and repeat Step 2.

Table 3. Empirical results for the three computed $\mathcal{S}'_x$ using Algorithm 3.

	x=MabCampath-H	x=Humira-H	x=MabCampath-L
mp($\mathcal{S}'_x$,ref)	369	421	214
mp($\mathcal{S}'_x$,ref)/ref_length	369/453=81.46%	421/449=93.76%	214/214=100%
mp($\mathcal{S}'_x$,target)	347	363	195
mp($\mathcal{S}'_x$,target)/tar_length	347/449=77.28%	363/453=80.13%	195/214=91.12%

Table 4. Empirical results for the three computed $\mathcal{S}''_x$ using Algorithm 4.

	x=MabCampath-H	x=Humira-H	x=MabCampath-L
mp($\mathcal{S}''_x$,ref)	377	422	214
mp($\mathcal{S}''_x$,ref)/ref_length	377/453=83.22%	422/449=93.97%	214/214=100%
mp($\mathcal{S}''_x$,target)	345	364	195
mp($\mathcal{S}''_x$,target)/tar_length	345/449=76.84%	364/453=80.35%	195/214=91.12%

4.3   Empirical Results

Similar to Sect. 3.3, let $x \in \{$MabCampath-H, Humira-H, MabCampath-L$\}$ and let the corresponding references of x be Humira-H, MabCampath-H, and Humira-L respectively. For each x, we use Algorithm 3 to compute the filled sequence $\mathcal{S}'_x$, and we use Algorithm 4 to compute the filled sequence $\mathcal{S}''_x$. Instead of directly using the B62 scores, we again use the number of matched pairs resulting from the computed B62 scores. The empirical results are shown in the following two tables. Note that in all cases, compared with Algorithm 3, with respect to a reference Algorithm 4 produces slightly better results. This can be seen in the first two lines in Tables 3 and 4. However, with respect to the corresponding target, it is not always the case that Algorithm 4 performs better than Algorithm 3 (though the difference is small). This can be checked in the last two lines of Tables 3 and 4 (and when $x =$MabCampath-H).

5   Concluding Remarks

In this paper, we study the protein scaffold filling problem when a reference protein is given and we solve the two corresponding versions in $O(n^{22})$ and

$O(n^{26})$ time, which are both impractical, respectively. We then design two practical methods, for each version, and obtain some empirical results using some datasets from four antibody protein sequences. Our empirical results are very promising: as long as the right reference protein and a high-quality scaffold are given, the algorithms can fill the scaffold with 76%-91% accuracy.

In practice, with top-down mass spectrometry, we might know the length of the target protein sequence. In this case, we might only need to insert a subset $X' \subset X$ of the missing amino acids. This will be an interesting direction for future research.

Acknowledgments. This research is partially supported by NSF of China under grant 60928006 and by the Opening Fund of Top Key Discipline of Computer Software and Theory in Zhejiang Provincial Colleges at Zhejiang Normal University. We also thank anonymous reviewers for several useful comments.

Appendix

See Table 5.

Table 5. The BLOSUM62 score matrix.

	C	S	T	P	A	G	N	D	E	Q	H	R	K	M	I	L	V	F	Y	W
C	9																			
S	-1	4																		
T	-1	1	5																	
P	-3	-1	-1	7																
A	0	1	0	-1	4															
G	-3	0	-2	-2	0	6														
N	-3	1	0	-2	-2	0	6													
D	-3	0	-1	-1	-2	-1	1	6												
E	-4	0	-1	-1	-1	-2	0	2	5											
Q	-3	0	-1	-1	-1	-2	0	0	2	5										
H	-3	-1	-2	-2	-2	-2	1	-1	0	0	8									
R	-3	-1	-1	-2	-1	-2	0	-2	0	1	0	5								
K	-3	0	-1	-1	-1	-2	0	-1	1	1	-1	2	5							
M	-1	-1	-1	-2	-1	-3	-2	-3	-2	0	-2	-1	-1	5						
I	-1	-2	-1	-3	-1	-4	-3	-3	-3	-3	-3	-3	-3	1	4					
L	-1	-2	-1	-3	-1	-4	-3	-4	-3	-2	-3	-2	-2	2	2	4				
V	-1	-2	0	-2	0	-3	-3	-3	-2	-2	-3	-3	-2	1	3	1	4			
F	-2	-2	-2	-4	-2	-3	-3	-3	-3	-3	-1	-3	-3	0	0	0	-1	6		
Y	-2	-2	-2	-3	-2	-3	-2	-3	-2	-1	2	-2	-2	-1	-1	-1	-1	3	7	
W	-2	-3	-2	-4	-3	-2	-4	-4	-3	-2	-2	-3	-3	-1	-3	-2	-3	1	2	11

References

1. Bandeira, N., Pham, V., Pevzner, P., Arnott, D., Lill, J.: Beyond Edman degradation: automated de novo protein sequencing of monoclonal antibodies. Nat. Biotechnol. **26**(12), 1336–1338 (2008)
2. Bandeira, N., Tang, H., Bafna, V., Pevzner, P.: Shotgun protein sequencing by tandem mass spectra assembly. Anal. Chem. **76**, 7221–7233 (2004)
3. Bulteau, L., Carrieri, A.P., Dondi, R.: Fixed-parameter algorithms for scaffold filling. Theo. Comput. Sci. **568**, 72–83 (2015)
4. Henikoff, S., Henikoff, J.: Amino acid substitution matrices from protein blocks. PNAS **89**(22), 10915–10919 (1992)
5. Jiang, H., Zhong, F., Zhu, B.: Filling scaffolds with gene repetitions: maximizing the number of adjacencies. In: Giancarlo, R., Manzini, G. (eds.) CPM 2011. LNCS, vol. 6661, pp. 55–64. Springer, Heidelberg (2011)
6. Jiang, H., Zheng, C., Sankoff, D., Zhu, B.: Scaffold filling under the breakpoint and related distances. IEEE/ACM Trans. Comput. Biol. Bioinf. **9**(4), 1220–1229 (2012)
7. Jiang, H., Ma, J., Luan, J., Zhu, D.: Approximation and nonapproximability for the one-sided scaffold filling problem. In: Xu, D., Du, D., Du, D. (eds.) COCOON 2015. LNCS, vol. 9198, pp. 251–263. Springer, Heidelberg (2015)
8. Liu, N., Jiang, H., Zhu, D., Zhu, B.: An improved approximation algorithm for scaffold filling to maximize the common adjacencies. IEEE/ACM Trans. Comput. Biol. Bioinf. **10**(4), 905–913 (2013)
9. Liu, N., Zhu, D., Jiang, H., Zhu, B.: A 1.5-approximation algorithm for two-sided scaffold filling. Algorithmica **74**(1), 91–116 (2016)
10. Liu, X., Han, Y., Yuen, D., Ma, B.: Automated protein (re)sequencing with MS/MS and a homologous database yields almost full coverage and accuracy. Bioinformatics **25**, 2174–2180 (2009)
11. Liu, X., Dekker, L., Wu, S., Vanduijn, M., Luider, T., Tolic, N., Kou, Q., Dvorkin, M., Alexandrova, S., Vyatkina, K., Pasa-Tolic, L., Pevzner, P.: De Novo protein sequencing by combining top-down and bottom-up tandem mass spectra. J. Proteome Res. **13**, 3241–3248 (2014)
12. Ma, B., Zhang, K., Hendrie, C., Liang, C., Li, M., Doherty-Kirby, A., Lajoie, G.: PEAKS: powerful software for peptide de novo sequencing by tandem mass spectrometry. Rapid Commun. Mass Spectrom. **17**(20), 2337–2342 (2003)
13. Ma, B., Zhang, K., Liang, C.: An effective algorithm for peptide de novo sequencing from MA/MS spectra. J. Comput. Syst. Sci. **70**(3), 418–430 (2005)
14. Muñoz, A., Zheng, C., Zhu, Q., Albert, V., Rounsley, S., Sankoff, D.: Scaffold filling, contig fusion and gene order comparison. BMC Bioinf. **11**, 304 (2010)
15. Needleman, S., Wunsch, C.: A general method applicable to the search for similarities in the amino acid sequence of two proteins. J. Mol. Biol. **48**(3), 443–453 (1970)
16. Pietrokovski, S., Henikoff, J., Henikoff, S.: The Blocks database - a system for protein classification. Nucl. Acids Res. **24**(1), 197–200 (1996)

Phylogenetics

Mean Values of Gene Duplication and Loss Cost Functions

Paweł Górecki[⊠], Jarosław Paszek, and Agnieszka Mykowiecka

Faculty of Mathematics, Informatics and Mechanics,
University of Warsaw, Warsaw, Poland
{gorecki,jpaszek,a.mykowiecka}@mimuw.edu.pl

Abstract. Reconciliation based cost functions play crucial role in comparing gene family trees with their species tree. To provide a better understanding of tree reconciliation we derive mean formulas for gene duplication, gene loss and gene duplication-loss cost functions, for a fixed species tree under the uniform model of gene trees. We then analyse the time complexity and study mathematical properties of these formulas. Finally, we provide several computational experiments on empirical datasets for the duplication, duplication-loss and deep coalescence means under the uniform model.

Keywords: Tree reconciliation · Duplication-loss model · Deep coalescence · Speciation · Gene duplication · Gene loss · Bijectively labelled tree · Uniform model of trees · Mean value

1 Introduction

Species phylogeny that represent evolutionary history of species is usually inferred from gene family trees. However, gene and species trees are usually incongruent which can be due to data selection, sequencing errors, inference methods or evolutionary events such as speciation, gene duplication, gene loss, lineage sorting or horizontal gene transfer [14] events. Studies on gene and species phylogeny have been conducted since 1980s. Goodman et al. [6] introduced a model of *tree reconciliation* in which gene duplication and loss events are invoked to address the differences between a gene tree and its species tree. This concept was formalized by Page and others [4,10,21]. The crucial notion in this model is the *duplication-loss cost* defined as the minimal number of gene duplications and losses required to explain all differences between a given gene tree with its species tree.

The model of reconciled trees is closely related to a stochastic coalescent model [1,22] in which a gene tree is evolving over time in a process that is dependent on the second tree, called a species tree. Maddisson [17,18] proposed a new type of lineage sorting event, called *deep coalescence*, that occurs when a common ancestor of two genes extends deeper than their corresponding speciation event. Minimizing the number of deep coalescences leads to the notion of the

© Springer International Publishing Switzerland 2016
A. Bourgeois et al. (Eds.): ISBRA 2016, LNBI 9683, pp. 189–199, 2016.
DOI: 10.1007/978-3-319-38782-6_16

deep coalescence cost, which is a measure, similar to duplication-loss cost, that can be used to compare gene and species trees. Mathematical properties the deep coalescence cost have been intensively studied in the recent years [8,9,23,26–29]. Furthermore, all reconciliation based costs including the deep coalescence have been successfully applied in many classical problems such as species tree inference, error correction or rooting an unrooted gene tree [2,7,11,16].

In this article we focus on mean values of gene duplication based cost functions where a species tree is a parameter. Similar results were established for the deep coalescence cost by Than and Rosenberg in 2012 [29] under two standard probability models of trees: the uniform model and the Yule-Harding model [15,19,25]. From the mathematical point of view, the gene loss cost function is a linear combination of gene duplication and deep coalescence [11,32], therefore, any property derived for these two functions can naturally be translated into gene loss and gene duplication-loss cost functions.

Our contribution. In this article we derive mean formulas for gene duplication, gene loss and gene duplication-loss cost functions for a fixed species tree under the uniform model of gene trees. We show that the mean values for the duplication based cost functions depends on the cardinalities of sibling clusters present in the species tree. Next, we analyse the time complexity of these formulas. Finally, by using our computationally efficient formulas, we provide a comparative study for the mean values under the uniform model performed on two empirical datasets for three standard reconciliation cost functions such as gene duplication, gene duplication-loss and deep coalescence.

2   Basic Definitions

We introduce several notions from phylogenetic theory [11,21]. Let X be a non-empty set of n species (taxons). By $R(X)$ we denote the set of all rooted binary trees whose leaves are bijectively labeled by the species from X. Let $T \in R(X)$. A *cluster*[1] of v is the set of all leaf labels present in the subtree of S rooted at v. Let $\mathsf{lca}_T(v, w)$ denote the least common ancestor of nodes v and w in T. The root of a tree T is denoted by $\mathsf{root}(T)$ and the parent of a non-root node v is denoted by $\mathsf{par}(v)$. By V_T and E_T we denote the sets of all nodes and all edges in T, respectively.

In the model of reconciled trees a gene tree is compared with its species tree. In this article both types of trees have the same bijective labelling of leaves, therefore, we assume that every gene tree and every species tree is an element of $R(X)$. For a (gene) tree $G \in R(X)$ and a (species) tree $S \in R(X)$ *the least common ancestor mapping between G and S, or lca-mapping*, $\mathsf{M}: V_G \to V_S$, is defined as

$$\mathsf{M}(g) := \begin{cases} s & g \text{ and } s \text{ are leaves having the same label,} \\ \mathsf{lca}_S(\mathsf{M}(g'), \mathsf{M}(g'')) & g \text{ is an internal node having two children } g' \text{and } g''. \end{cases}$$

---

[1] Called sometimes a clade in the literature.

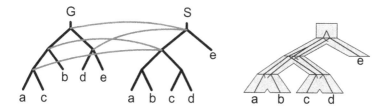

Fig. 1. *An example of rooted reconciliation.* The lca-mapping between a gene tree G and a species tree S and the embedding of G into S, i.e., an informal representation of an evolutionary scenario explaining differences between G and S by using gene duplications and gene losses. Here the DL cost is 9 (2 duplications + 7 gene losses) and the DC cost is 3. The lca-mapping is shown only for the internal nodes of G.

An internal node g is a *duplication* if $M(g) = M(a)$ for a child a of g. Every internal non-duplication node is called a *speciation*. *The duplication cost*, denoted by $D(G, S)$, equals the total number of duplications in G when reconciling G with S [20]. Formally,

$$D(G, S) := |\{g \in V_G : M(g) = M(a) \text{ and } \mathsf{par}(a) = g\}|.$$

The formula for counting deep coalescence events [17,18], that occur when a common ancestor of two genes extends deeper than their corresponding speciation event, can be expressed by [32]:

$$DC(G, S) := \sum_{g \in V_G \setminus \{\mathsf{root}(G)\}} (\|M(g), M(\mathsf{par}(g))\| - 1).$$

where $\|a, b\|$ is the number of edges on the path connecting a and b in S. See also [29] for alternative definitions of DC. The total number of *gene losses* required to reconcile G and S is defined by [32]

$$L(G, S) := 2D(G, S) + DC(G, S). \tag{1}$$

Finally, the duplication-loss cost we define as

$$DL(G, S) := D(G, S) + L(G, S). \tag{2}$$

We denote trees by using the standard nested parenthesis notation. For instance, in Fig. 1, $G = (((a, c), b), (d, e))$ is a five-leaf gene tree over a species set $\{a, b, c, d, e\}$.

3   Results

In the uniform model of binary trees an equal probability is assigned to each possible leaf labeled binary tree with n leaves. Unrooted trees in this model can be generated by uniform and random insertions of one edge to any edge at each step. For rooted trees the insertion can be additionally performed on the rooting edge as indicated in Fig. 2. We analyse the mean of the duplication cost in the uniform model of rooted leaf-labeled trees. Recall that $R(X)$ denote the set of all bijectively labeled rooted trees over X and $|X| = n > 0$. The mean of duplication cost for a fixed species tree $S \in R(X)$ under a probabilistic model of gene trees is defined as:

$$\overline{\mathsf{D}}_u(S) = \sum_{G \in R(X)} \mathbb{P}(G)\mathsf{D}(G, S). \tag{3}$$

The size of $R(X)$ is given by the following classical formula:

$$b(n) = (2n - 3)!!,$$

where $k!!$ is the double factorial, i.e., $k!! = k \cdot (k - 2)!!$ and $0!! = (-1)!! = 1$. Thus, in the uniform model every (gene) tree $G \in R(X)$ has the same probability $\mathbb{P}(G) = \frac{1}{b(n)}$.

Fig. 2. Uniform model for leaf-labeled rooted trees. A new edge with $n + 1$-th label is uniformly added to any edge including the rooting edge. Every tree among the four-leaf trees on the right can be created from $(a, (b, c))$ with equal probability.

In the duplication model a type of a node from a gene tree depends on the clusters of its children, therefore, we introduce a notion of a (rooted) split. Every internal node s of S, induces the set $\{A, B\}$, called a *split* and denoted by $A|B$, such that A and B are the clusters of children of s. The set of all splits in S we denote by $\mathsf{Spl}(S)$. For example, $\mathsf{Spl}(((a, b), (c, d)))$ is equal to $\{\{\{a, b\}, \{c, d\}\}, \{\{a\}, \{b\}\}, \{\{c\}, \{d\}\}\}$, or by using a simplified split notation: $\{ab|cd, a|b, c|d\}$.

For $A, B \subset X$, by $e_n^{\mathsf{Dup}}(A, B)$ we denote the number of duplication nodes present in trees from $R(X)$ mapped into a node whose split is $A|B$. Similarly, we define $e_n^{\mathsf{Spec}}(A, B)$ for speciation nodes.

Lemma 1. *For a species tree S with n leaves,*

$$\sum_{A|B \in \mathsf{Spl}(S)} (e_n^{\mathsf{Dup}}(A, B) + e_n^{\mathsf{Spec}}(A, B)) = b(n) \cdot (n - 1).$$

Proof. The result follows from the fact that there are $b(n)$ trees each having $n - 1$ internal nodes. Next, for a fixed species tree, every internal node present in a tree from $R(X)$ is either a speciation or a duplication.

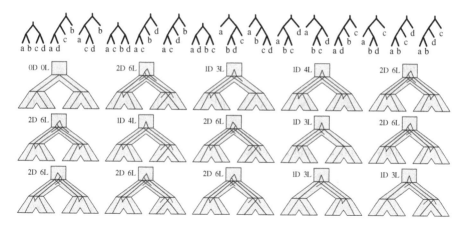

Fig. 3. *Duplication-loss scenarios for $n = 4$ and the species tree $S = ((a, b), (c, d))$. Top:* All 15 bijectively labeled gene trees with four leaves. *Bottom:* Embeddings (scenarios) of every gene tree into the species tree S [10]. Each scenario is summarized with two numbers denoting the number of gene duplications (D) and the number of gene losses (L). We omit leaf labels for brevity. We have 22 duplications, 23 speciation nodes and 68 gene losses in total. In this example, $\overline{D}_u(S) = 22/15$, $\overline{L}(S) = 68/15$ and $\overline{DL}_u(S) = 90/15$.

Hence, the mean (3) is equivalent to

$$\overline{D}_u(S) = \frac{1}{b(n)} \sum_{A|B \in \mathsf{Spl}(S)} e_n^{\mathsf{Dup}}(A, B) = n - 1 - \frac{1}{b(n)} \sum_{A|B \in \mathsf{Spl}(S)} e_n^{\mathsf{Spec}}(A, B). \quad (4)$$

The mappings of children of a speciation node induce disjoint clusters in a species tree, therefore, it is more convenient to count directly the number of speciation nodes rather then duplications.

Lemma 2. *For a species tree S with n leaves and a split $A|B$ present in S*

$$e_n^{\mathsf{Spec}}(A, B) = \sum_{i=1}^{|A|} \sum_{j=1}^{|B|} \binom{|A|}{i} \binom{|B|}{j} b(i)b(j)b(n - i - j + 1).$$

Proof. Assume that $s \in S$ has the split $A|B$. A gene tree having a speciation node mapped into s can be constructed as follows. Let z be an element not in X, let A' and B' be nonempty subsets of A and B, respectively. Then, a gene tree $G \in R(X)$ which has a speciation node with split $A'|B'$ can be constructed by replacing the leaf z in a tree $R((X \setminus (A' \cup B')) \cup \{z\})$ by a tree (G_A, G_B) such that $G_A \in R(A')$ and $G_B \in R(B')$. Clearly, the root of (G_A, G_B) is a speciation mapped into s. It should be clear that the above method counts every speciation node from $R(X)$ exactly once (even if a single tree from $R(X)$ may have more than one speciation mapped into s). □

Finally, we have the main result.

Theorem 1 (Mean of D under the uniform model). *For a given species tree S with n leaves*

$$\overline{D}_u(S) = n - 1 - \frac{1}{b(n)} \sum_{A|B \in \mathsf{Spl}(S)} \sum_{i=1}^{|A|} \sum_{j=1}^{|B|} \binom{|A|}{i}\binom{|B|}{j} b(i)b(j)b(n-i-j+1).$$

Proof. It follows from Lemma 2 and Eq. 4.   □

To obtain the result for the duplication-loss cost we need the mean of the deep coalescence cost.

Theorem 2 (Mean of DC under the uniform model; adopted from Cor. 5 [29]). *For a species tree S with n leaves:*

$$\overline{DC}_u(S) = -2n(2n-2) + 2\,\mathsf{epl}(S) + \frac{(2n-2)!!}{b(n)} \sum_{A \in C(S)} \frac{(2n-2|A|-1)!!}{(2n-2|A|-2)!!},$$

where $C(S)$ is the set of all clusters in S excluding the cluster of the root, and $\mathsf{epl}(S)$ is the external path length of S [3, 29] equaling $\sum_{v \in L_S} \|v, \text{root}(S)\|$ (or equivalently $\sum_{A \in C(S)} |A|$).

Finally, we have the result for the duplication-loss and loss costs.

Theorem 3 (Mean of DL and L under the uniform model). *For a species tree S we have*

$$\overline{DL}_u(S) = 3 \cdot \overline{D}_u(S) + \overline{DC}_u(S)$$

and

$$\overline{L}_u(S) = 2 \cdot \overline{D}_u(S) + \overline{DC}_u(S).$$

Proof. The result follows easily from the definition of gene loss (see Eq. (1)) and duplication-loss (see Eq. (2)) cost functions and the properties of mean values.□

Now it is straightforward to combine the formulas from Theorems 1, 2 and 3 to obtain the final formula for the mean of both cost functions. We omit easy details. An example is depicted in Fig. 3.

From the computational point of view computing the mean of deep coalescence for a fixed species tree can be completed in $O(n)$ steps under assumption that double factorials are memorized and the required size of clusters is stored with the nodes of the standard pointer-like implementation of trees. For the mean of the remaining cost functions, however, we need two additional loops. Therefore, the time complexity of computing $\overline{D}_u(S)$, $\overline{L}_u(S)$ and $\overline{DL}_u(S)$ is $O(n^3)$.

4   Experimental Evaluation

4.1   Mean Values for Tree Shapes

Table 1 depicts mean values of four cost functions for all tree shapes with $3, 4, \ldots 9$ leaves. Note that the mean formula (Theorem 2) for deep coalescence depends on the size of clusters from a species tree. On the other hand the mean of the duplication cost (Theorem 1) depends on the cardinality of splits from a species tree. The same applies for the duplication-loss and loss costs. Therefore the same cardinality of splits in tree shapes induces the same mean values. The smallest n, for which we can observe this property is 9 as indicated in the last row of Table 1, where the trees marked in red have the following split cardinalities: $4|5, 2|2, 1|4, 1|3, 1|2$ and three times $1|1$. In Table 1 the mean of gene duplication cost grows with the Furnas rank [5] of a species tree (a general statement of this type for any n, however, needs to be proved), while the deep coalescence and other cost functions have rather the opposite property (see also [29]).

For a more general view, we computed mean values for all tree shapes with up to 20 leaves. Note that for $n = 20$ there are 293547 shapes. The result is summarized in Fig. 5. We observe that mean of the duplication cost has the smallest variability, which can partially be explained by the fact that the maximal number of gene duplications is $n - c - 1$ [12], where c is the number of cherries, i.e., subtrees consisting of exactly two leaves, in a gene tree. In consequence, other cost functions such as duplication-loss, deep coalescence or loss (not depicted here), being a linear combination of gene duplication and deep coalescence functions, share a similar structure of values as indicated in Fig. 5.

4.2   Empirical Study

In our experimental evaluation we used two publicly available datasets.

Dataset I. We downloaded a collection of curated unrooted gene family trees from *TreeFam v7.0* [24]. From the TreeFam species tree based on the NCBI taxonomy, we selected 24 species out of 29 species in order to obtain gene trees with a bijective labelling. The following species were removed: *Oryzias latipes, Fugu rubripes, Ciona savignyi, Drosophila pseudoobscura, Caenorhabditis remanei*. Next, after contracting gene trees to the set of 24 species, we obtained 11 bijectively labelled gene trees. Finally, these gene trees were rooted by using Urec [13].

Dataset II. We downloaded gene families of nine yeast genomes from *Génolevures* dataset [30]. From the whole dataset, we selected 1690 gene families having exactly one gene sampled from each species. Finally, for each gene family we inferred an unrooted gene tree having bijective leaf labelling by using PhyML with the standard parameter setting. Similarly, to the first dataset, the gene trees were rooted by Urec [13]. To reconcile the gene trees we used the original Génolevures species phylogeny [31].

Data processing. For both datasets we conducted an experiment accordingly to the following procedure. For a dataset with the species tree S, for every

Table 1. *Mean values for all tree shapes with* $n \in \{3, 4, \ldots, 9\}$ *leaves.* The shapes are shown ordered by Furnas rank [5]. The table is patterned after [29]. Below every shape we present the corresponding mean values. The two shapes marked in red have the splits with the same cardinality, which implies equal values of the corresponding means. Note that for $n = 1, 2$ the means are 0.

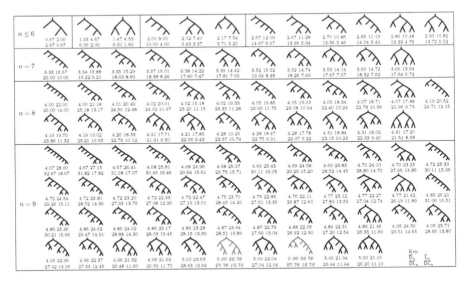

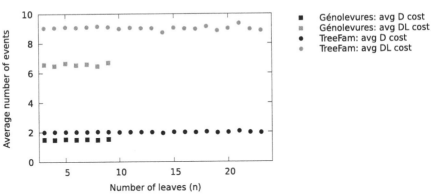

Fig. 4. Average number of gene duplication and gene duplication-loss events located in gene trees sampled from TreeFam and Génolevures.

$k = 0, 1, \ldots, |L_S| - 3$, and for $i = 1, 2, \ldots, 100$ we randomly chose a set R_k^i of k species present in S. Then, we created $(|L_S| - 2) \cdot 100$ sample datasets by contracting the species tree S and every gene tree to the set of species from $L_S \setminus R_k^i$. Next, for every resulting sample dataset consisting of a species tree S' and a collection of gene tree $\mathscr{G}$ we calculated the average number of gene duplication and gene duplication-loss events by $\frac{1}{|\mathscr{G}|} \sum_{G \in \mathscr{G}} \mathsf{D}(G, S')$ and $\frac{1}{|\mathscr{G}|} \sum_{G \in \mathscr{G}} \mathsf{DL}(G, S')$,

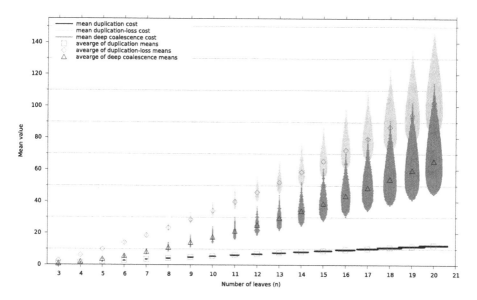

Fig. 5. *Frequency diagram of mean values of duplication, duplication-loss and deep coalescence costs for all fixed species tree shapes for $n = 3, 4, \ldots 20$ under the uniform model of gene trees.* For each n, mean values for every cost were grouped into bins of size 0.01. The width of each bin is proportional to $\log_2 K$, where K is the number of species tree shapes having the mean value in this bin.

respectively. Finally, we averaged these values in the 100 sample datasets having the same size of contracted trees.

The results are depicted in Fig. 4. We observe that the average number of gene duplication in empirical datasets can be well approximated by a constant function. For instance, every gene tree in TreeFam, whether contracted or not, have on average 2 gene duplications. On the other hand, the mean values over the whole set of gene trees are monotonically growing with n as indicated in Fig. 5. We conclude that the properties of empirical datasets are significantly different than uniformly random ones.

5   Conclusion

In this article we derived mean formulas for gene duplication, gene loss and gene duplication-loss cost functions for a fixed species tree under the uniform model of gene trees. Our empirical study shows that the properties of mean values of the duplication cost are different than properties of the deep coalescence cost. For instance, we shown that only the mean of duplications tend to grow with the Furnas rank. As expected our study confirmed that the structure of mean values is similar between the gene loss, the deep coalescence and the duplication-loss cost functions. Similar results can be obtained for the Yule-Harding model for bijectively labelled trees [15,19,25].

Mean values can naturally be used to normalize costs in comparative studies with gene and species trees [29]. For example, in [12], we demonstrated how a species tree can be inferred from a set of gene trees under the normalized duplication cost, where the normalizing factor was a diameter of a duplication cost defined as the maximal number of gene duplications for a fixed species tree (see also Sect. 4.1). Further investigation into the normalization with mean values provides an important direction for future work.

Acknowledgements. We would like to thank the four reviewers for their detailed comments that allowed us to improve our paper. JP was supported by the DSM funding for young researchers of the Faculty of Mathematics, Informatics and Mechanics of the University of Warsaw.

References

1. Aldous, D.J.: Stochastic models and descriptive statistics for phylogenetic trees, from Yule to today. Stat. Sci. **16**, 23–34 (2001)
2. Bansal, M.S., Burleigh, J.G., Eulenstein, O.: Efficient genome-scale phylogenetic analysis under the duplication-loss and deep coalescence cost models. BMC Bioinf. **11**(Suppl 1), S42 (2010)
3. Blum, M.G., François, O.: On statistical tests of phylogenetic tree imbalance: the sackin and other indices revisited. Math. Biosci. **195**(2), 141–153 (2005)
4. Bonizzoni, P., Della Vedova, G., Dondi, R.: Reconciling a gene tree to a species tree under the duplication cost model. Theoret. Comput. Sci. **347**(1–2), 36–53 (2005)
5. Furnas, G.W.: The generation of random, binary unordered trees. J. Classif. **1**(1), 187–233 (1984)
6. Goodman, M., Czelusniak, J., Moore, G.W., Romero-Herrera, A.E., Matsuda, G.: Fitting the gene lineage into its species lineage, a parsimony strategy illustrated by cladograms constructed from globin sequences. Syst. Zool. **28**(2), 132–163 (1979)
7. Górecki, P., Eulenstein, O.: Deep coalescence reconciliation with unrooted gene trees: linear time algorithms. In: Gudmundsson, J., Mestre, J., Viglas, T. (eds.) COCOON 2012. LNCS, vol. 7434, pp. 531–542. Springer, Heidelberg (2012)
8. Górecki, P., Eulenstein, O.: Gene tree diameter for deep coalescence. IEEE-ACM Trans. Comput. Biol. Bioinf. **12**(1), 155–165 (2015)
9. Górecki, P., Paszek, J., Eulenstein, O.: Unconstrained gene tree diameters for deep coalescence. In: Proceedings of the 5th ACM Conference on Bioinformatics, Computational Biology, and Health Informatics. BCB 2014, NY, USA, pp. 114–121. ACM, New York (2014)
10. Górecki, P., Tiuryn, J.: DLS-trees: a model of evolutionary scenarios. Theoret. Comput. Sci. **359**(1–3), 378–399 (2006)
11. Górecki, P., Eulenstein, O., Tiuryn, J.: Unrooted tree reconciliation: a unified approach. IEEE-ACM Trans. Comput. Biol. Bioinf. **10**(2), 522–536 (2013)
12. Górecki, P., Paszek, J., Eulenstein, O.: Duplication cost diameters. In: Basu, M., Pan, Y., Wang, J. (eds.) ISBRA 2014. LNCS, vol. 8492, pp. 212–223. Springer, Heidelberg (2014)
13. Górecki, P., Tiuryn, J.: URec: a system for unrooted reconciliation. Bioinformatics **23**(4), 511–512 (2007)

14. Hallett, M.T., Lagergren, J.: Efficient algorithms for lateral gene transfer problems. In: Proceedings of the Fifth Annual International Conference on Computational Biology. RECOMB 2001, NY, USA, pp. 149–156. ACM, New York (2001)

15. Harding, E.F.: The probabilities of rooted tree-shapes generated by random bifurcation. Adv. Appl. Probab. **3**(1), 44–77 (1971)

16. Ma, B., Li, M., Zhang, L.: From gene trees to species trees. SIAM J. Comput. **30**(3), 729–752 (2000)

17. Maddison, W.P.: Gene trees in species trees. Syst. Biol. **46**, 523–536 (1997)

18. Maddison, W.P., Knowles, L.L.: Inferring phylogeny despite incomplete lineage sorting. Syst. Biol. **55**(1), 21–30 (2006)

19. McKenzie, A., Steel, M.: Distributions of cherries for two models of trees. Math. Biosci. **164**(1), 81–92 (2000)

20. Page, R.: From gene to organismal phylogeny: reconciled trees and the gene tree/species tree problem. Mol. Phylogenet. Evol. **7**(2), 231–240 (1997)

21. Page, R.D.M.: Maps between trees and cladistic analysis of historical associations among genes, organisms, and areas. Syst. Biol. **43**(1), 58–77 (1994)

22. Pamilo, P., Nei, M.: Relationships between gene trees and species trees. Mol. Biol. Evol. **5**(5), 568–583 (1988)

23. Rosenberg, N.A.: The probability of topological concordance of gene trees and species trees. Theoret. Popul. Biol. **61**(2), 225–247 (2002)

24. Ruan, J., Li, H., Chen, Z., Coghlan, A., Coin, L.J., Guo, Y., Hériché, J.K., Hu, Y., Kristiansen, K., Li, R., Liu, T., Moses, A., Qin, J., Vang, S., Vilella, A.J., Ureta-Vidal, A., Bolund, L., Wang, J., Durbin, R.: TreeFam: 2008 update. Nucleic Acids Res. **36**, D735–D740 (2008)

25. Steel, M.A., Penny, D.: Distributions of tree comparison metrics – some new results. Syst. Biol. **42**(2), 126–141 (1993)

26. Than, C., Nakhleh, L.: Species tree inference by minimizing deep coalescences. PLoS Comput. Biol. **5**(9), e1000501 (2009)

27. Than, C.V., Rosenberg, N.A.: Consistency properties of species tree inference by minimizing deep coalescences. J. Comput. Biol. **18**(1), 1–15 (2011)

28. Than, C.V., Rosenberg, N.A.: Mathematical properties of the deep coalescence cost. IEEE-ACM Trans. Comput. Biol. Bioinf. **10**(1), 61–72 (2013)

29. Than, C.V., Rosenberg, N.A.: Mean deep coalescence cost under exchangeable probability distributions. Discrete Appl. Math. **174**, 11–26 (2014)

30. Sherman, D.J., Martin, T., Nikolski, M., Cayla, C., Souciet, J.L., Durrens, P.: Génolevures: protein families and synteny among complete hemiascomycetous yeast proteomes and genomes. Nucleic Acids Res. **37**(suppl 1), D550–D554 (2009)

31. The Génolevures Consortium, et al.: Comparative genomics of protoploid saccharomycetaceae, Genome Res. **19**(10), 1696–1709 (2009)

32. Zhang, L.: From gene trees to species trees II: Species tree inference by minimizing deep coalescence events. IEEE-ACM Trans. Comput. Biol. Bioinf. **8**, 1685–1691 (2011)

The SCJ Small Parsimony Problem
for Weighted Gene Adjacencies

Nina Luhmann[1,2(✉)], Annelyse Thévenin[2], Aïda Ouangraoua[3],
Roland Wittler[1,2], and Cedric Chauve[4]

[1] International Research Training Group "Computational Methods
for the Analysis of the Diversity and Dynamics of Genomes",
Bielefeld University, Bielefeld, Germany
[2] Genome Informatics, Faculty of Technology and Center for Biotechnology,
Bielefeld University, Bielefeld, Germany
nina.luhmann@uni-bielefeld.de
[3] Department of Computer Science, Université de Sherbrooke, Sherbrooke, Canada
[4] Department of Mathematics, Simon Fraser University, Burnaby, Canada
cedric.chauve@sfu.ca

Abstract. Reconstructing ancestral gene orders in a given phylogeny
is a classical problem in comparative genomics. Most existing methods compare conserved features in extant genomes in the phylogeny to
define potential ancestral gene adjacencies, and either try to reconstruct
all ancestral genomes under a global evolutionary parsimony criterion,
or, focusing on a single ancestral genome, use a scaffolding approach to
select a subset of ancestral gene adjacencies. In this paper, we describe
an exact algorithm for the small parsimony problem that combines both
approaches. We consider that gene adjacencies at internal nodes of the
species phylogeny are weighted, and we introduce an objective function
defined as a convex combination of these weights and the evolutionary
cost under the Single-Cut-or-Join (SCJ) model. We propose a Fixed-
Parameter Tractable algorithm based on the Sankoff-Rousseau dynamic
programming algorithm, that also allows to sample co-optimal solutions. An implementation is available at http://github.com/nluhmann/
PhySca.

1 Introduction

Reconstructing ancestral gene orders is a long-standing computational biology
problem with important applications, as shown in several recent large-scale
projects [8,17,18]. Informally, the problem can be defined as follows: Given a
phylogenetic tree representing the speciation history leading to a set of extant
genomes, we want to reconstruct the structure of the ancestral genomes corresponding to the internal nodes of the tree.

Existing ancestral genome reconstruction methods concentrate on two main
strategies. *Local* approaches consider the reconstruction of one specific ancestor
at a time independently from the other ancestors of the tree. Usually, they do

© Springer International Publishing Switzerland 2016
A. Bourgeois et al. (Eds.): ISBRA 2016, LNBI 9683, pp. 200–210, 2016.
DOI: 10.1007/978-3-319-38782-6_17

not consider an evolutionary model and proceed in two stages: (1) comparing gene orders of ingroup and outgroup species to define potential ancestral gene adjacencies, and (2) selecting a conflict-free subset of ancestral gene adjacencies to obtain a set of Contiguous Ancestral Regions (CARs) [3,7,14,15]. *Global* approaches on the other hand simultaneously reconstruct ancestral gene orders at all internal nodes of the considered phylogeny, generally based on a parsimony criterion within an evolutionary model. This *small parsimony problem* has been studied with several underlying genome rearrangement models, such as the breakpoint distance or the Double-Cut-and-Join (DCJ) distance [1,11,23]. While rearrangement scenarios based on complex rearrangement models can give insights into underlying evolutionary mechanisms, from a computational point of view, the small parsimony problem is NP-hard for most rearrangement distances [21]. One exception is the Single-Cut-or-Join (SCJ) distance, for which linear/circular ancestral gene orders can be found in polynomial time [9], however constraints required to ensure algorithmic tractability yield fragmented ancestral gene orders.

The work we present is an attempt to reconcile both approaches. We introduce a variant of the small parsimony problem based on an optimality criterion that accounts for both an evolutionary distance and the difference between the initial set of potential ancestral adjacencies and the final consistent subset of adjacencies. More precisely we consider that each potential ancestral gene adjacency can be provided with a (prior) non-negative weight at every internal node. These adjacency weights can e. g. be obtained as probabilities computed by sampling scenarios for each potential adjacency independently [6] or can be based on ancient DNA (aDNA) sequencing data providing direct prior information assigned to certain ancestral nodes. It follows that the phylogenetic framework we present can then also assist in scaffolding fragmented assemblies of aDNA sequencing data [12,19]. We describe an exact exponential time algorithm for reconstructing consistent ancestral genomes under this optimality criterion, and show that the small parsimony problem variant we introduce is Fixed-Parameter Tractable (FPT), with a parameter linked to the amount of conflict in the data. Moreover, this also allows us to provide a FPT sampling algorithm for co-optimal solutions. We evaluate our method on two data sets: mammalian genomes spanning roughly one million years of evolution, and bacterial genomes (pathogen *Yersinia*) spanning 20,000 years of evolution. See [13] for an extended preprint of this paper.

2   Background

Genomes and adjacencies. Genomes consist of chromosomes and plasmids. Each such component can be represented as a linear or circular sequence of oriented markers over a marker alphabet. Markers correspond to homologous sequences between genomes, e. g. genes or synteny blocks. We assume that each marker appears exactly once in each genome, so our model does not consider duplications or deletions. To account for its orientation, each marker x is encoded as a pair of marker extremities (x_h, x_t) or (x_t, x_h). An *adjacency* is an unordered pair of marker extremities, e. g. $\{x_t, y_h\}$. The order of markers in a genome can be encoded by a set of adjacencies. Two distinct adjacencies are said to be *conflicting* if they share a common marker extremity. If a set of adjacencies contains conflicting adjacencies, it is not *consistent* with a mixed linear/circular genome model.

The small parsimony problem and rearrangement distances. In a global phylogenetic approach, we are given a phylogenetic tree with extant genomes at its leaves and internal nodes representing ancestral genomes. We denote by $\mathcal{A}$ the set of all adjacencies present in at least one extant genome and assume that every ancestral adjacency belongs to $\mathcal{A}$. Then the goal is to find a labeling of the internal nodes by consistent subsets of $\mathcal{A}$ minimizing a chosen genomic distance over the tree. This is known as the *parsimonious labeling problem*. It is NP-hard for most rearrangement distances. The only known exception is the set-theoretic Single-Cut-or-Join (SCJ) distance [9]. It defines a rearrangement distance by two operations: the *cut* and *join* of adjacencies. Given two genomes defined by consistent sets of adjacencies A and B, the SCJ distance between these genomes is $d_{SCJ}(A, B) = |A - B| + |B - A|$.

The small parsimony problem under the SCJ model can be solved by computing a parsimonious gain/loss history for each adjacency separately with the dynamic programming Fitch algorithm [10]. Consistent labelings can be ensured with the additional constraint that in case of ambiguity at the root of the tree, the absence of the adjacency is chosen [9]. As each adjacency is treated independently, this constraint might automatically exclude all adjacencies being part of a conflict to ensure consistency and thus results in an unnecessarily sparse reconstruction.

Generalization by weighting adjacencies. When considering an internal node v, we define node u as its parent node in T. We assume that a specific adjacency graph is associated to each ancestral node v, whose edges are annotated by a weight $w_{v,a} \in [0, 1]$ representing a confidence measure for the presence of adjacency a in species v. Then in a global reconstruction, cutting an adjacency of a higher weight has higher impact in terms of the optimization criterion, than cutting an adjacency of lower weight.

Formally, we define two additional variables for each adjacency $a \in \mathcal{A}$ at each internal node $v \in V$: The presence (or absence) of a at node v is represented by $p_{v,a} \in \{0, 1\}$, while $c_{v,a} \in \{0, 1\}$ indicates a change for the status of an adjacency along an edge (u, v), i.e., $p_{u,a} \neq p_{v,a}$. We consider the problem of optimizing the following objective function, where $\alpha \in [0, 1]$ is a convex combination factor.

Definition 1 (Weighted SCJ labeling problem). *Let $T = (V, E)$ be a tree with each leaf l labeled with a consistent set of adjacencies $\mathcal{A}_l \subseteq \mathcal{A}$ and each adjacency $a \in \mathcal{A}$ is assigned a given weight $w_{v,a} \in [0, 1]$ for each node $v \in V$. A labeling λ of the internal nodes of T with $\lambda(l) = \mathcal{A}_l$ for each leaf is an optimal weighted SCJ labeling if none of the internal nodes $v \in V$ contains a conflict and it minimizes the criterion*

$$D(\lambda, T) = \sum_{a,v} \alpha(1 - p_{a,v})w_{a,v} + (1 - \alpha)c_{a,v}$$

Further, we can state the corresponding co-optimal sampling problem.

Definition 2 (Weighted SCJ sampling problem). *Given the setting of the weighted SCJ labeling problem, sample uniformly from all labelings λ of the internal nodes of T that are solutions to the weighted SCJ optimal labeling problem.*

Existing results. There exist a few positive results for the weighted SCJ labeling problem with specific values of α. If $\alpha = 0$, the objective function corresponds to the small parsimony problem under the SCJ distance and hence a solution can be found in polynomial time [9]. A generalization towards multifurcating, edge-weighted trees including prior information on adjacencies at exactly one internal node of the tree is given in [12]. Recently, Miklós and Smith [16] proposed a Gibbs sampler for sampling optimal labelings under the SCJ model with equal branch lengths. This method addresses the issue of the high fragmentation of internal node labelings, but convergence is not proven, and so there is no bound on the computation time. If $\alpha = 1$, i.e., we do not take evolution in terms of SCJ distance along the branches of the tree into account, we can solve the problem by applying independently a maximum-weight matching algorithm at each internal node [15]. So the extreme cases of the problem are tractable, and it remains open to see if the general problem is hard.

3 Methods

In order to find a solution to the weighted SCJ labeling problem, we first show that we can decompose the problem into smaller independent subproblems. Then, for each subproblem containing conflicting adjacencies, we show that, if it contains a moderate level of conflict, it can be solved using the Sankoff-Rousseau algorithm [20] with a complexity parameterized by the size of the subproblem. For a highly conflicting subproblem, we show that it can be solved by an Integer Linear Program (ILP).

Decomposition into independent subproblems. We first introduce a graph that encodes all adjacencies present in at least one internal node of the considered phylogeny (Definition 3). As introduced previously, we consider a tree $T = (V, E)$ where each node is augmented by an adjacency graph.

Definition 3 (Global adjacency graph). *The set of vertices V_{AG} of the global adjacency graph AG consists of all marker extremities present in at least one of the adjacency graphs. There is an edge between two vertices $a, b \in V_{AG}$ that are not extremities of a same marker, if there is an internal node in the tree T whose adjacency graph contains the adjacency $\{a, b\}$. The edge is labeled with the list of all internal nodes v that contain this adjacency.*

Each connected component C of the global adjacency graph defines a subproblem composed of the species phylogeny, the set of marker extremities equal to the vertex set of C and the set of adjacencies equal to the edge set of C. According to the following lemma, whose proof is straightforward, it is sufficient to solve each such subproblem independently.

Lemma 1. *The set of all optimal solutions of the weighted SCJ labeling problem is the set-theoretic Cartesian product of the sets of optimal solutions of the instances defined by the connected components of the global adjacency graph.*

To solve the problem defined by a connected component C of the global adjacency graph containing conflicts, we rely on an adaptation of the Sankoff-Rousseau algorithm with exponential time complexity, parameterized by the size and nature of conflicts of C, and thus can solve subproblems with moderate amount of conflict.

Application to the weighted SCJ optimal labeling problem. In order to use the Sankoff-Rousseau algorithm to solve the problem defined by a connected component C of the global adjacency graph, we define a label of an internal node of the phylogeny as the assignment of at most one adjacency to each marker extremity. More precisely, let x be a marker extremity in C, v an internal node of T, and $e_1, \ldots, e_{d_x}$ be all edges in the global adjacency graph that are incident to x and whose label contains v (i.e., represent adjacencies in the adjacency graph of node v). We define the set of possible labels of v as $L_{x,v} = \{\emptyset, e_1, \ldots, e_{d_x}\}$. The set of potential labels L_v of node v is then the Cartesian product of the label sets $L_{x,v}$ for all $x \in V(C)$ resulting in a set of discrete labels for v of size $\prod_{x \in V(C)}(1 + d_x)$. Note that not all of these joint labelings are valid as they can assign an adjacency $a = (x, y)$ to x but not to y, or adjacency $a = (x, y)$ to x and $b = (x, z)$ to z thus creating a conflict (see [13] for an example).

For an edge (u, v) in the tree, we can then define a cost matrix that is indexed by pairs of labels of L_u and L_v, respectively. The cost is infinite if one of the labels is not valid, and defined by the objective function otherwise. We can then apply the Sankoff-Rousseau approach to find an optimal labeling of all internal nodes of the tree for connected component C. Note that, if C is a connected component with no conflict, it is composed of two vertices and a single edge, and can be solved in space $O(n)$ and time $O(n)$.

Solving a general instance. Given a general instance, i.e., an instance not limited to a single connected component of the global adjacency graph, we can consider each connected component independently (Lemma 1). For a set of N markers and c connected components in the global adjacency graph defining a conflicting instance, we define D as the maximum degree of a vertex and M as the maximum number of vertices in all such components. Then, the complexity analysis in the appendix shows that the problem is Fixed-Parameter Tractable (FPT).

Theorem 1. *The weighted SCJ labeling problem can be solved in worst-case time $O(nN(1 + D)^{2M})$ and space $O(nN(1 + D)^M)$.*

In practice, the exponential complexity of our algorithm depends on the structure of the conflicting connected components of the global adjacency graph. The dynamic programming algorithm will be effective on instances with either small conflicting connected components or small degrees within such components, and will break down with a single component with a large number of vertices of high degree. For such components, the time complexity is provably high and we propose an ILP (see [13]) to solve such components.

Sampling co-optimal labelings. The Sankoff-Rousseau DP algorithm can easily be modified to sample uniformly from the space of all optimal solutions to the weighted SCJ labeling problem in a forward-backward fashion. The principle is to proceed in two stages: first, for any pair (v, a) we compute the number of optimal solutions under this label for the subtree rooted at v. Then, when computing an optimal solution, if a DP equation has several optimal choices, one is randomly picked according to the distribution of optimal solutions induced by each choice (see [13] for more details). This classical dynamic programming approach leads to the following result.

Theorem 2. *The weighted SCJ sampling problem can be solved in worst-case time $O(nN(1 + D)^{2M})$ and space $O(nN(1 + D)^M)$.*

For subproblems that are too large for being handled by the Sankoff-Rousseau algorithm, the SCJ small parsimony Gibbs sampler recently introduced [16] can easily be modified to incorporate prior weights, although there is currently no proven property regarding its convergence.

4   Results

We evaluated our reconstruction algorithm on two datasets: *mammalian* and *Yersinia* genomes. The mammalian dataset was used in the studies [7,16]. Our second dataset contains eleven *Yersinia* genomes, an important human pathogen. This dataset contains contigs from the recently sequenced extinct agent of the Black Death pandemic [4] that occurred roughly 650 years ago. We refer to [13] for the species phylogenies of these two datasets and extended information on how adjacency weights have been obtained for both datasets.

4.1   Mammalian Dataset

Unique and universal markers were computed as synteny blocks with different resolution in terms of minumum marker length. Note that all rearrangement breakpoints are therefore located outside of marker coordinates. It results in five different datasets varying from 2, 185 markers for a resolution of 100 kb to 629 markers for a resolution of 500 kb.

We considered all adjacencies present in at least one extant genome as potentially ancestral. To weight an adjacency at all internal nodes of the tree, we relied on evolutionary scenarios for each single adjacency, in terms of gain/loss, independently of the other adjacencies (i. e. without considering consistency of ancestral marker orders). We obtain these weights using the software DeClone [6], and we refer to them as *DeClone weights*. We considered two values of the DeClone parameter kT, 0.1 and 1, the former ensuring that only adjacencies appearing in at least one optimal adjacency scenario have a significant DeClone weight, while the latter samples adjacencies outside of optimal scenarios. For the analysis of the ancestral marker orders obtained with our algorithm, we considered the data set at 500 kb resolution and sampled 500 ancestral marker orders for all ancestral species under different values of α.

The complexity of our algorithm is dependent on the size of the largest connected component of the global adjacency graph. In order to restrict the complexity, we kept only adjacencies whose weights are above a given threshold x. In most cases, all connected components are small enough to be handled by our exact algorithm in reasonable time except for very large components in the marker sets with higher resolution under a low threshold x. For the 500 kb dataset with $x = 0.2$ and $kT = 1$, the computation of one solution takes on average 200 s on a 2.6 GHz i5 with 8 GB of RAM. It can be reduced to 30 s when DeClone weights are based on $kT = 0.1$. This illustrates that our algorithm, despite an exponential worst-case time complexity, can process realistic datasets in practice. Next, we analyzed the 500 optimal SCJ labelings obtained for $\alpha = 0$, i. e. aiming only at minimizing the SCJ distance, and considered the fragmentation of the ancestral gene orders (number of CARs) and the total evolutionary distance. Note that, unlike the Fitch algorithm used in [9], our algorithm does not favor fragmented assemblies by design but rather considers all optimal labelings. Sampling of co-optimal solutions shows that the pure SCJ criterion leads to some significant variation in terms of number of CARs (Fig. 1). The optimal SCJ distance in the tree for $\alpha = 0$ is 1,674, while the related DCJ distance in the sampled reconstructions varies between 873 and 904 (Fig. 2). In comparison, we obtained a DCJ distance of 829 with GASTS [22], a small parsimony solver directly aiming at minimizing the DCJ distance. This illustrates both a lack of robustness of the pure SCJ optimal labelings, and some significant difference between the SCJ and DCJ distances.

For $\alpha > 0$, our method minimizes a combination of the SCJ distance with the DeClone weights of the adjacencies discarded to ensure valid ancestral gene orders. We distinguish between DeClone parameter $kT = 0.1$ and $kT = 1$. Figures 2 and 3 show the respective observed results in terms of evolutionary distance and fragmentation. For $kT = 0.1$, the optimal SCJ and DCJ distance over the whole tree hardly depends on α. Including the DeClone weights in the objective actually results in the same solution, independent of $\alpha > 0$. In fact, while applying a low weight threshold of $x = 0.2$, the set of potential adjacencies is already consistent at all internal nodes except for a few conflicts at the root that are solved unambiguously for all values of α. This indicates that building DeClone weights on the basis of mostly optimal adjacency scenarios (low kT) results in a weighting scheme that agrees with the evolution along the tree for this

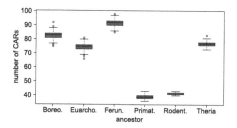

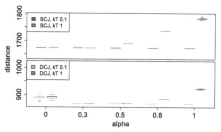

Fig. 1. Number of reconstructed CARs at each internal node in 500 samples for the mammalian dataset with 500 kb resolution, $x = 0.2$ and $\alpha = 0$.

Fig. 2. SCJ distance (upper half) and DCJ (lower half) distance in the whole tree for all samples and selected values of α in the mammalian dataset.

dataset. More importantly, Figs. 2 and 3 show that the combination of DeClone weights followed by our algorithm, leads to a robust set of ancestral gene orders.

In comparison, for $kT = 1$, we see an increase in SCJ and DCJ distance for higher α, while the number of CARs at internal nodes decreases, together with a loss of the robustness of the sampled optimal results when α gets close to 1. It can be explained by the observation that the weight distribution of ancestral adjacencies obtained with DeClone and $kT = 1$ is more balanced than with $kT = 0.1$ as it considers suboptimal scenarios of adjacencies with a higher probability.

4.2 *Yersinia Pestis* Dataset

We started from fully assembled DNA sequences of seven *Yersinia pestis* and four *Yersinia pseudotuberculosis* genomes. In addition, we included aDNA single-end reads and 2 134 contigs of length >500bp assembled from these reads for the Black Death agent, considered as ancestral to several extant strains [4]. We refer to this augmented ancestral node as the *Black Death (BD) node*. The marker sequences for all extant genomes were computed as described in [19], restricting the set of markers to be unique and universal. We obtained a total of 2, 207 markers in all extant genomes and 2, 232 different extant adjacencies. As for the mammalian dataset, we considered as potentially ancestral any adjacency that appears in at least one extant genome. However for this dataset, reducing the complexity by applying a weight threshold x was not necessary. For the BD node, adjacency weights can be based on the given aDNA reads for a given potential ancestral adjacency as follows. First, we used FPSAC [19] to compute DNA sequences filling the gaps between any two adjacent marker extremities. Then we computed the weights as a likelihood of this putative gap sequence given the aDNA reads, using the GAML probabilistic model described in [5].

Again we sampled 500 solutions for this dataset. We computed the weights at the BD node based on the aDNA data, while adjacencies at all other nodes were given weight 0. Hence we can investigate the influence of including the aDNA sequencing data in the reconstruction while for the rest of the tree, the

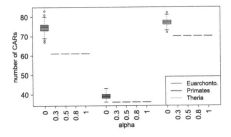

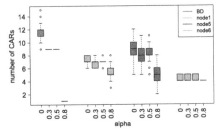

Fig. 3. Number of CARs in all samples at selected internal nodes for different values of α reconstructed with DeClone weights under $kT = 0.1$.

Fig. 4. Reconstructed number of CARs in the *yersinia* dataset with a DNA weights at the BD node and 0 otherwise, for four ancestral nodes.

weights do not impact the objective function. As shown in Fig. 4, for selected internal nodes of the phylogeny, the pure SCJ solutions at $\alpha = 0$ result in the highest fragmentation, while the number of CARs decreases as we increase the importance of the adjacency weights in the objective of our method. For the BD node, when including the aDNA weights, the fragmentation is decreasing while the reconstructions for each $\alpha > 0$ are robust. At the other nodes, the applied sequencing weights also reduce the fragmentation except for node6 which is located in the pseudotuberculosis subtree and hence more distant to the BD node. This shows that the aDNA weights not only influence the reconstructed adjacencies at the BD node, but also other nodes of the tree.

5   Conclusion

Our main contributions are the introduction of the small parsimony problem under the SCJ model with adjacency weights, together with an exact parameterized algorithm for the optimization and sampling versions of the problem. The motivation for this problem is twofold: incorporating sequence signal from aDNA data when it is available, and recent works showing that the reconstruction of ancestral genomes through the independent analysis of adjacencies is an interesting approach [2,6,9,16].

Regarding the latter motivation, we address a general issue of these approaches that either ancestral gene orders are not consistent or are quite fragmented if the methods are constrained to ensure consistency. The main idea we introduce is to take advantage of sampling approaches recently introduced in [6] to weight potential ancestral adjacencies and thus direct, through an appropriate objective function, the reconstruction of ancestral gene orders. Our results on the mammalian dataset suggest that this approach leads to a robust ancestral genome structure. However, we can observe a significant difference with a DCJ-based ancestral reconstruction, a phenomenon that deserves to be explored further. Our sampling algorithm improves on the Gibbs sampler introduced in [16] in terms of computational complexity and provides a useful tool to study ancestral genome reconstruction from a Bayesian perspective.

There are several research avenues opened by our work. From a theoretical point of view, we know the problem we introduced is tractable for $\alpha = 0$ and $\alpha = 1$, but it remains to see whether it is hard otherwise. Further, given that the considered objective is a combination of two objectives to be optimized simultaneously, Pareto optimization is an interesting aspect that should be considered. From a more applied point of view, one would like to incorporate duplicated and deleted markers into our small parsimony problem. There exist efficient algorithms for the case of a single adjacency [2,6] that can provide adjacency weights, and natural extensions of the SCJ model to incorporate duplicated genes.

Acknowledgements. NL and RW are funded by the International DFG Research Training Group GRK 1906/1. CC is funded by NSERC grant RGPIN-249834.

References

1. Alekseyev, M., Pevzner, P.A.: Breakpoint graphs and ancestral genome reconstructions. Genome Res. **19**, 943–957 (2009)
2. Bérard, S., Gallien, C., et al.: Evolution of gene neighborhoods within reconciled phylogenies. Bioinformatics **28**, 382–388 (2012)
3. Bertrand, D., Gagnon, Y., Blanchette, M., El-Mabrouk, N.: Reconstruction of ancestral genome subject to whole genome duplication, speciation, rearrangement and loss. In: Moulton, V., Singh, M. (eds.) WABI 2010. LNCS, vol. 6293, pp. 78–89. Springer, Heidelberg (2010)
4. Bos, K.I., Schuenemann, V., et al.: A draft genome of yersinia pestis from victims of the black death. Nature **478**, 506–510 (2011)
5. Boža, V., Brejová, B., Vinař, T.: GAML: genome assembly by maximum likelihood. Algorithms Mol. Biol. **10**, 18 (2015)
6. Chauve, C., Ponty, Y., Zanetti, J.: Evolution of genes neighborhood within reconciled phylogenies: an ensemble approach. BMC Bioinform. **16**(Suppl. 19), S6 (2015)
7. Chauve, C., Tannier, E.: A methodological framework for the reconstruction of contiguous regions of ancestral genomes and its application to mammalian genomes. PLoS Comput. Biol. **4**, e1000234 (2008)
8. Denoeud, F., Carretero-Paulet, L., et al.: The coffee genome provides insight into the convergent evolution of caffeine biosynthesis. Science **345**, 125527 (2013)
9. Feijão, P., Meidanis, J.: SCJ: a breakpoint-like distance that simplifies several rearrangement problems. IEEE/ACM TCBB **8**, 1318–1329 (2011)
10. Fitch, W.: Toward defining the course of evolution: minimum change for a specific tree topology. Syst. Biol. **20**, 406–416 (1971)
11. Kováč, J., Brejová, B., Vinař, T.: A practical algorithm for ancestral rearrangement reconstruction. In: Przytycka, T.M., Sagot, M.-F. (eds.) WABI 2011. LNCS, vol. 6833, pp. 163–174. springer, Heidelberg (2011)
12. Luhmann, N., Chauve, C., Stoye, J., Wittler, R.: Scaffolding of ancient contigs and ancestral reconstruction in a phylogenetic framework. In: Campos, S. (ed.) BSB 2014. LNCS, vol. 8826, pp. 135–143. Springer, Heidelberg (2014)
13. Luhmann, N., Thevenin, A., et al.: The SCJ small parsimony problem for weighted gene adjacencies (extended version) (2016). Preprint http://arxiv.org/abs/1603.08819

14. Ma, J., Zhang, L., et al.: Reconstructing contiguous regions of an ancestral genome. Genome Res. **16**, 1557–1565 (2006)
15. Maňuch, J., Patterson, M., et al.: Linearization of ancestral multichromosomal genomes. BMC Bioinform. **13**(Suppl. 19), S11 (2012)
16. Miklós, I., Smith, H.: Sampling and counting genome rearrangement scenarios. BMC Bioinform. **16**(Suppl 14), S6 (2015)
17. Ming, R., Van Buren, R., et al.: The pineapple genome and the evolution of CAM photosynthesis. Nature Genet. **47**, 1435 (2015)
18. Neafsey, D.E., Waterhouse, R.M., et al.: Highly evolvable malaria vectors: the genome of 16 Anopheles mosquitoes. Science **347**, 1258522 (2015)
19. Rajaraman, A., Tannier, E., Chauve, C.: FPSAC: fast phylogenetic scaffolding of ancient contigs. Bioinformatics **29**, 2987–2994 (2013)
20. Sankoff, D., Rousseau, P.: Locating the vertices of a steiner tree in an arbitrary metric space. Math. Program. **9**, 240–246 (1975)
21. Tannier, E., Zheng, C., Sankoff, D.: Multichromosomal median and halving problems under different genomic distances. BMC Bioinform. **10**(1), 120 (2009)
22. Xu, A.W., Moret, B.M.E.: GASTS: parsimony scoring under rearrangements. In: Przytycka, T.M., Sagot, M.-F. (eds.) WABI 2011. LNCS, vol. 6833, pp. 351–363. Springer, Heidelberg (2011)
23. Zheng, C., Sankoff, D.: On the pathgroups approach to rapid small phylogeny. BMC Bioinform. **12**(Suppl. 1), S4 (2011)

Path-Difference Median Trees

Alexey Markin and Oliver Eulenstein[✉]

Department of Computer Science, Iowa State University, Ames, IA 50011, USA
{amarkin,oeulenst}@iastate.edu

Abstract. Synthesizing large-scale phylogenetic trees is a fundamental problem in evolutionary biology. Median tree problems have evolved as a powerful tool to reconstruct such trees. Given a tree collection, these problems seek a median tree under some problem-specific tree distance. Here, we introduce the median tree problem for the classical path-difference distance. We prove that this problem is NP-hard, and describe a fast local search heuristic that is based on solving a local search problem exactly. For an effective heuristic we devise a time efficient algorithm for this problem that improves on the best-know (naïve) solution by a factor of n, where n is the size of the input trees. Finally, we demonstrate the performance of our heuristic in a comparative study with other commonly used methods that synthesize species trees using published empirical data sets.

Keywords: Phylogenetic trees · Median trees · Supertrees · Path-difference distance · Local search

1 Introduction

Large-scale phylogenetic trees that represent the evolutionary relationships, or genealogy, among thousands of species offer enormous promise for society's advancements. While such species trees are fundamental to evolutionary biology, they are also benefiting many other disciplines, such as agronomy, biochemistry, conservation biology, epidemiology, environmental sciences, genetics, genomics, medical sciences, microbiology, and molecular biology [13,14,21]. However, despite these promises, synthesizing large-scale species trees is confronting us with one of the most difficult computational challenges in evolutionary biology today. Here, we are focusing on synthesizing large species trees from a given collection of typically smaller phylogenetic trees.

Traditionally, a species tree for a set of species is inferred by first selecting a gene that is common to them, and then inferring the evolutionary history for this gene, which is called a *gene tree*. Gene trees describe partial evolutionary histories of the species genomes, and therefore, it is often assumed that gene trees have evolved along the edges of the species tree, imitating it. However, a major shortcoming of the traditional approach is that different gene trees for the same set of species can describe discordant evolutionary histories. Such discordance is frequently caused by erroneous gene trees, or can be the result

© Springer International Publishing Switzerland 2016
A. Bourgeois et al. (Eds.): ISBRA 2016, LNBI 9683, pp. 211–223, 2016.
DOI: 10.1007/978-3-319-38782-6_18

of genes which have evolved differently due to complex evolutionary processes that have shaped the species genomes [23]. To confront these challenges, median tree problems (also called supertree problems [4]) have emerged as a powerful tool for inferring species trees from a collection of discordant gene trees. These problems seek a tree, called a median tree, that is minimizing the overall distance to the input trees based on some problem-specific distance measure. Typically, measures that have been well-established in comparative phylogenetics are used to compute median trees [4], and one of the oldest measures to compare trees is the path-difference distance. However, despite the tradition and popularity of the path-difference distance, median trees under this measure and their computation have not been analyzed.

In this work we are studying the computation of median trees under the path-difference distance. We show that computing median trees under the path-difference distance is an NP-hard problem for rooted as well as unrooted input trees. While most median tree problems used in practice are NP-hard, they have been effectively addressed by standard local search heuristics that solve a local search problem thousands of times. Encouraged by these promising results we introduce a novel local search heuristic to compute median trees under the path-difference distance. The heuristic is based on our $\Theta(kn^3)$ time algorithm (introduced here) that solves the corresponding local search problem exactly, where n and k is the size and number of trees in a given instance of the median tree problem respectively. Our new local search heuristic allows to compute the first large-scale median trees under the path-difference distance. Finally, we demonstrate the performance of our heuristic in a comparative study on several published empirical data sets, and demonstrate that it outperforms other standard heuristics in minimizing the overall path-difference. Software implementing our local search heuristic is freely available from the authors.

Related Work. *Median tree problems* are a popular tool to synthesize large-scale species trees from a collection of smaller trees. Given a collection of input trees, such problems seek a tree, called a *median tree*, that minimizes the sum of its distances to each of the input trees. Since the ultimate goal of median tree problems is to synthesize accurately species trees of enormous scale, a large body of work has focussed on the biological, mathematical, and algorithmic properties of median tree problems adopting numerous definitions of distance measures from comparative phylogenetics [4]. One of the oldest such measures, however, is the path-difference distance [5,12,26,30], and median trees under this distance have not been analyzed. The *path-difference distance* between two trees is defined through the Euclidean distance between their path-length vectors. Each such vector represents the pairwise distances between all leaves of the corresponding tree (i.e., the number of edges on a simple path between leaves). Steel and Penny [30] have studied the distribution of the path-difference distance for un-rooted trees. Complementing this work, Mir and Rosello [19] computed the mean value of this distance for fully resolved unrooted trees with n leaves, and showed that this mean value grows in $O(n^3)$. Variants of the path-difference distance are the Manhattan distance of the path-length vectors [33], and their correlation [24].

Median tree problems are typically NP-hard [4], and therefore are, in practice, approached by using local search heuristics [1,9,17,18,32] that make truly large-scale phylogenetic analyses feasible [18,32]. Effective local search heuristics have been proposed and analyzed [1,9,17,18,32], and provided various credible species trees [18,32]. Given an instance I of a median tree problem, such heuristics start with some initial candidate species tree T and find a minimum cost tree for I under the tree distance measure of the problem in the (local) neighborhood of T, and so on, until a local minima is reached. At each local search step, the heuristic solves an instance of a local search problem. The time complexity of this local search problem depends on the tree edit operation that defines the neighborhood, as well as on the computation time of the tree distance measure that is used. A classical and well-studied tree edit operation is the *subtree prune and regraft (SPR)* operation [27] where a subtree of the edited tree is pruned and regrafted back into the tree at another location. The *SPR neighborhood* of T is the set of all trees into which T can be transformed by one SPR operation, and this neighborhood contains $\Theta(n^2)$ trees. Further, the best-known algorithm to compute the path-difference distance between two trees with n leaves requires $\Theta(n^2)$ time [30]. Therefore, given an instance of k trees over n different taxa of the SPR based local search problem, this problem can be naïvely solved by complete enumeration in $\Theta(kn^4)$ time, which is the best-known algorithm. However, when faced with heuristically estimating larger median trees this runtime becomes prohibitive.

Our Contribution. We introduce the *path-difference median tree problem* under the classical path-difference distance to synthesize large-scale phylogenetic trees. To prove its NP-hardness for rooted and unrooted input trees we are using polynomial time mapping-reductions from the maximum triplet consistency problem and from the quartet compatibility problem respectively. To solve large-scale instances of the path-difference median tree problem, we have devised a standard SPR local search heuristic. For time efficiency, we design a $\Theta(kn^3)$ time algorithm for an instance of the local search problem that improves on the best-known (naïve) solution by a factor of n, where n and k is the size and number of the input trees of the median tree problem respectively. Finally, we demonstrate the performance of our new local search heuristic through comparative studies using empirical data sets.

2 Basics and Preliminaries

Basic Definitions. A *(phylogenetic) tree* T is a rooted full binary tree. We denote its node set, edge set, leaf set, and root, by $V(T)$, $E(T)$, $\mathsf{L}(T)$, and $\mathsf{Rt}(T)$ respectively. Given a node $v \in V(T)$, we denote its parent by $\mathsf{Pa}_T(v)$, its set of children by $\mathsf{Ch}_T(v)$, its sibling by $\mathsf{Sb}_T(v)$, the subtree rooted at v by $T(v)$, and $T|v$ is the phylogenetic tree that is obtained by pruning $T(v)$ from T. Note that we identify the leaf set with the respective set of leaf-labels (taxa).

Let $L \subseteq \mathsf{L}(T)$ and T' be the minimal subtree of T with leaf set L. We define the *leaf-induced subtree* $T[L]$ of T to be the tree obtained from T' by successively removing each node of degree two (except for the root) and adjoining its two neighbors.

Path-difference Distance. Given a tree T and two leaves $u, v \in \mathsf{L}(T)$, let $\mathsf{d}_{u,v}(T)$ denote the length in edges of the unique path between u and v in T. Let $\mathsf{d}(T)$ be an associated vector obtained by a fixed ordering of pairs i, j [30], e.g., $\mathsf{d}(T) = (\mathsf{d}_{1,2}(T), \mathsf{d}_{1,3}(T), \ldots, \mathsf{d}_{n-1,n}(T))$, where n is the number of leaves. Then the *path-difference distance (PDD)* between two trees G and S over the same leaf set is defined as $\mathsf{d}(G, S) := || \mathsf{d}(G) - \mathsf{d}(S)||_2$.

We also define $\mathsf{PLM}(T)$ to be the matrix of path-lengths between each two leaves in T. That is, a matrix of size $|\mathsf{L}(T)| \times |\mathsf{L}(T)|$, where rows and columns represent leaves of T, and $\mathsf{PLM}_{u,v}(T) = \mathsf{d}_{u,v}(T)$. Let G and S be trees over the same leaf set, then we define $\Delta(G, S) := \mathsf{PLM}(G) - \mathsf{PLM}(S)$ to be the *matrix of path-length differences*.

3   Path-Difference Median Tree Problem

Let $\mathcal{P}$ be a set of trees $\{G_1, \ldots, G_k\}$. We define $\mathsf{L}(\mathcal{P}) := \cup_{i=1}^{k} \mathsf{L}(G_i)$ to be the *leaf set* of $\mathcal{P}$. A tree S is called a *supertree* of $\mathcal{P}$, if $\mathsf{L}(S) = \mathsf{L}(\mathcal{P})$. Further, we extend the definition of the path-difference distance to a set of trees. Note, we defined PDD only for two trees over the same leaf set. However, we do not want to enforce such a restriction on the set of input trees, since it is generally not the case for real world data. Therefore, in order to compare two trees S and G, where $\mathsf{L}(G) \subseteq \mathsf{L}(S)$ we use the *minus method* [11]. That is, we calculate a distance between G and the subtree of S induced by $\mathsf{L}(G)$: $\mathsf{d}(S, G) = \mathsf{d}(S[\mathsf{L}(G)], G)$. We now define PDD for an input set $\mathcal{P}$ and a supertree S as a sum $\mathsf{d}(\mathcal{P}, S) := \sum_{i=1}^{k} \mathsf{d}(G_i, S[\mathsf{L}(G_i)])$, which is used to establish the following problem.

Problem 1 (PD median tree (supertree) – decision version).
Instance: a set of input trees $\mathcal{P}$ and a real number p;
Question: determine whether there exist a supertree S, such that $\mathsf{d}(\mathcal{P}, S) \leq p$.

3.1   The PD Median Tree Problem is NP-hard

We show this by a polynomial time reduction from the MaxRTC problem.

Problem 2 (Maximum Compatible Subset of Rooted Triplets – MaxRTC).
Instance: a set of rooted triplets R and an integer $0 \leq c \leq |R|$;
Question: Is there a subset $R' \subseteq R$, such that R' is compatible and $|R'| \geq c$.

A *rooted triplet* is a (rooted full binary) tree with exactly three leaves. A set of trees $\mathcal{P}$ is called *compatible* if there exist a supertree T consistent with every tree in $\mathcal{P}$, and a tree T is *consistent* with a tree G if $T[\mathsf{L}(G)] \equiv G$.

Theorem 1. *The PD median tree problem is NP-hard.*

Proof. We map an instance $\langle R, c \rangle$ of the MaxRTC problem to an instance $\langle R, \sqrt{2}(|R| - c) \rangle$ of the PD median tree problem. The MaxRTC problem is known to be NP-complete [6]. This transformation works due to the following property. Assume that S is a supertree of a set of rooted triplets $R = \{T_1, \ldots, T_k\}$. Then we observe that $\mathsf{d}(S[\mathsf{L}(T_i)], T_i)$ is 0, when S is *consistent* with T_i, and is $\sqrt{2}$ otherwise. Therefore, $\mathsf{d}(R, S) = \sqrt{2}(|R| - c')$, where c' is the number of triplets in R, which are consistent with S. That is, there are at least c' compatible triplets in R. Now, we can conclude the proof.

(i) If $\langle R, c \rangle$ is a *yes*-instance of the MaxRTC problem, then there exist a tree S, such that S is consistent with $|R'| \geq c$ triplets. As we shown above, in that case $\mathsf{d}(P, S) \leq \sqrt{2}(k - c)$. Therefore, $\langle R, \sqrt{2}(|R| - c) \rangle$ is a *yes*-instance of the PD median tree problem.
(ii) Clearly, the same argument works in the other direction.   □

In practice median trees are sometimes computed for multi-sets of trees. However, our results, shown for sets of input trees, easily extend to multi-sets.

4   Local Search for PD Median Tree Problem

As stated in the introduction, we address the NP-hardness by devising a new SPR based local search heuristic. Next, we introduce needed definitions.

4.1   SPR-Based Local Search

Definition 1. *Given a node $v \in V(S) \setminus \{\mathsf{Rt}(S)\}$, and a node $u \in V(S) \setminus (V(S(v)) \cup \{\mathsf{Pa}(v)\})$, $SPR_S(v, u)$ is a tree obtained as follows:*

(i) Prune the subtree $S(v)$ by (i) removing the edge $\{\mathsf{Pa}(v), v\}$, and (ii) removing $\mathsf{Pa}(v)$ by adjoining its parent and child.
(ii) If u is a root of $S|v$, then a new root w' is introduced, so that u is a child of w'. Otherwise, an edge $(\mathsf{Pa}(u), u)$ is subdivided by a new node w'.
(iii) Connect the subtree $S(v)$ to the node w'.

In addition, we introduce the following useful notation
$$SPR_S(v) := \bigcup_u SPR_S(v, u); \quad SPR_S := \bigcup_{v,u} SPR_S(v, u). \; SPR_S \text{ is called}$$
an *SPR-neighborhood* of a tree S, and $|SPR_S| = O(n^2)$, where $n = |\mathsf{L}(S)|$.

Given a set of input trees $\mathcal{P} = \{G_1, \ldots, G_k\}$, the search space in the median tree problem can be viewed as a graph $\mathcal{T}$, where nodes represent supertrees of $\mathcal{P}$. There is an edge $\{S_1, S_2\}$ in $\mathcal{T}$, if S_1 could be transformed to S_2 with a single SPR operation. As was mentioned in the introduction, local search is designed to terminate at a local minimum of $\mathcal{T}$. More formally, at each iteration the following problem is solved

Problem 3 (PD local search).
Instance: An input set $\mathcal{P}$ and a supertree S;
Find: $S' = \underset{S' \in SPR_S}{\arg\min} \, \mathsf{d}(\mathcal{P}, S')$.

Next we describe an algorithm for the PD local search problem that improves on the best-known naïve algorithm (see Introduction) by a factor of n.

4.2   Local Search Based on an SPR Semi-structure

Let $\mathbf{G} \in \mathcal{P}$ be a fixed input tree, and let S_i be a supertree in the i-th iteration of the local search. Throughout this section we refer to the restricted tree $S_i[\mathsf{L}(G)]$ as $\mathbf{S}$. To reduce the complexity of the naïve algorithm, we exploit a *semistructure* of an SPR-neighborhood initially introduced in [8]. Let $N := SPR_S(v, \mathsf{Rt}(S))$ for some $v \in V(S)$, then $SPR_N(v)$ is equivalent to $SPR_S(v)$. This property is essential for the further analysis, which is motivated by the following theorem.

Theorem 2. *Given $\Delta(N, G)$, $\mathsf{d}(T, G)$ is computable in $\Theta(n)$ time for any $T \in SPR_S(v)$ with a single precomputation step of time complexity $\Theta(n^2)$.*

This theorem implies that for a fixed input tree G and a fixed prune node $v \in V(S)$ we can compute the PD distance for every $T \in SPR_S(v)$ in $\Theta(n^2)$ time. Therefore, computing $\mathsf{d}(T, G)$ for **all** $T \in SPR_S$ and **all** $G \in \mathcal{P}$ takes $\Theta(\mathbf{n^3 k})$ time, where $k = |\mathcal{P}|$. This is the time complexity of our algorithm for the PD local search problem. In the remainder of this section we detail the precomputation idea and prove Theorem 2.

Consider a tree $T := SPR_S(v, y)$, where $y \in V(S|v)$, and let $(u_0, ..., u_t)$ be a simple path in $S|v$, where $u_0 = \mathsf{Rt}(S|v)$ and $u_t = y$. Note, this path is also a path in N, since $S|v$ is a subtree of N.

For convenience, let C_u denote $\mathsf{L}(N(u))$ for any $u \in V(N)$. Thus, we have $C_v = \mathsf{L}(N(v)) = \mathsf{L}(S(v))$. Table 1 shows a path-length difference matrix $\mathsf{PLM}(T) - \mathsf{PLM}(N)$. Using this table, it is possible to derive the difference between $\mathsf{d}(T, G)$ and $\mathsf{d}(N, G)$. It was constructed by partitioning the leaf set of S as follows (see also Fig. 1): $\mathsf{L}(S) = C_v \cup (C_{\mathsf{Sb}(u_1)} \cup \ldots \cup C_{\mathsf{Sb}(u_t)}) \cup C_{u_t}$.

In order to explain Table 1 we need to explore how the path between two leaves changes when regrafting node v. We consider all possibilities for a pair of leaves i and j (except for the cases, when i and j belong to the same subset from the table, since the path does not change in that case).

(i) $i \in C_v$, $j \in C_{u_t}$. In N the path between i and j could be denoted by $A_i \sqcup (\mathsf{Pa}(v), u_0, \ldots, u_t) \sqcup B_j$. Note that partial paths A_i and B_j are not changed by the regrafting operation. In T the path between i and j is $A_i \sqcup (\mathsf{Pa}_T(v), u_t) \sqcup B_j$. The number of edges in the path is decreased by t.

(ii) $i \in C_v$, $j \in C_{\mathsf{Sb}(u_p)}$, where $1 \le p \le t$. Again, we denote the path between i and j in N by $A_i \sqcup (\mathsf{Pa}(v), u_0, \ldots, u_{p-1}, \mathsf{Sb}(u_p)) \sqcup B_j$. Then the corresponding path in T is $A_i \sqcup (\mathsf{Pa}_T(v), u_{t-1}, \ldots, u_p, u_{p-1}, \mathsf{Sb}(u_p)) \sqcup B_j$. It is easy to see that the path length increased by $(t - p) - (p - 1)$.

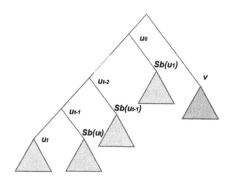

Fig. 1. Scheme of the $N = SPR_S(v, \mathrm{Rt}(S))$ tree, depicting how the leaf set was partitioned to create Table 1.

Table 1. Note that $1 \le p \le t$. Values inside the table indicate the difference in path lengths between leaves from different subsets, i.e., for $i \in C_v$ and $j \in C_{u_t}$: $\mathsf{d}_{i,j}(T) = \mathsf{d}_{i,j}(N) - t$.

	C_v	C_{u_t}	$C_{\mathsf{Sb}(u_p)}$
C_v	0	$-t$	$-2p+1$ $+t$
C_{u_t}	$-t$	0	$+1$
$C_{\mathsf{Sb}(u_p)}$	$-2p+1$ $+t$	$+1$	0

(iii) $i \in C_{u_t}$, $j \in C_{\mathsf{Sb}(u_p)}$, where $1 \le p \le t$. We denote a path between i and j in N by $A_i \sqcup (u_{t-1}, \ldots, u_p, u_{p-1}, \mathsf{Sb}(u_p)) \sqcup B_j$. Then the corresponding path in T is $A_i \sqcup (\mathsf{Pa}_T(v), u_{t-1}, \ldots, u_p, u_{p-1}, \mathsf{Sb}(u_p)) \sqcup B_j$. Exactly one edge was added to the path (as a result of regrafting v above u_t).

(iv) $i \in C_{\mathsf{Sb}(u_p)}$, $j \in C_{\mathsf{Sb}(u_q)}$, where $1 \le p, q \le t$. Clearly, the path between i and j is not affected by the regrafting operation.

Let A and B be two elements from $\{C_v, C_{u_t}, C_{\mathsf{Sb}(u_t)}, \ldots, C_{\mathsf{Sb}(u_1)}\}$ (set of disjoint subsets), and $dif_{A,B}$ be the corresponding value according to Table 1. For convenience we will refer to $\Delta_{i,j}(N, G)$ as simply $\Delta_{i,j}$.

$$
\begin{aligned}
\mathsf{d}^2(T, G) - \mathsf{d}^2(N, G) &= \sum_{\forall \{A,B\}} \sum_{\substack{i \in A \\ j \in B}} (\Delta_{i,j} + dif_{A,B})^2 - (\Delta_{i,j})^2 \\
&= \sum_{\forall \{A,B\}} \left(|A||B| dif_{A,B}^2 + 2 dif_{A,B} \sum_{\substack{i \in A \\ j \in B}} \Delta_{i,j} \right).
\end{aligned}
\tag{1}
$$

Precomputation. The above equation shows that in order to efficiently calculate $\mathsf{d}(T, G)$ for an arbitrary $T \in SPR_v(S)$ with a fixed v, we need to know $\sum_{\substack{i \in A \\ j \in B}} \Delta_{i,j}$ for every pair of distinct A, B, such that $dif_{A,B} \ne 0$. Note that there are only $O(t)$ such pairs. Further, we observe that those sums can be exhaustively precomputed as the following values for each $u \in V(N)$.

– $\mathsf{BDist}(u) := \sum_{\substack{i \in C_v \\ j \in C_u}} \Delta_{i,j}$, for any $u \in V(S|v)$. This sum is called the *Base Dis-*

tance (BDist): sum of path-length differences between all pairs of leaves from a subset C_u and C_v.

- RDist$(u) := \sum\limits_{\substack{i \in C_u \\ j \in L(S|v) \backslash C_u}} \Delta_{i,j}$, for any $u \in V(S|v)$. This sum is called the

  Remaining Distance (RDist): sum of path-length differences between all pairs of leaves from a subset C_u, and all the other leaves in N (excluding C_v).

Consider any node $u \in V(S|v)$. If u is not a leaf, then we denote its children as c_1 and c_2. $BDist(u)$ and $RDist(u)$ can be equivalently computed as follows.

- BDist$(u) = \begin{cases} \text{BDist}(c_1) + \text{BDist}(c_2), & p \text{ is not a leaf;} \\ \sum\limits_{i \in C_v} \Delta_{i,u}, & \text{otherwise.} \end{cases}$

- RDist$(u) = \begin{cases} \text{RDist}(c_1) + \text{RDist}(c_2) - 2 \cdot \textbf{SDist}(\textbf{c}_1), & p \text{ is not a leaf;} \\ \sum\limits_{i \in C_{u_0}} \Delta_{i,u}, & \text{otherwise.} \end{cases}$

- SDist$(c_1) = $ SDist$(c_2) := \sum\limits_{i \in C_{c_1}, j \in C_{c_2}} \Delta_{i,j}$ (*sibling distance*).

Time Complexity. The precomputation step is divided into three sub-steps according to the relations for the distances BDist, SDist and RDist. Below we assess their computation time separately.

- BDist is calculated in constant time for internal nodes and in $O(|C_v|)$ time for leaves. Therefore, it requires $\Theta(n^2)$ operations to claculate BDist for all $u \in V(N)$.
- RDist is similar to BDist: it is calculated in constant time for internal nodes and in $O(|C_{u_0}|)$ for leaves. Once again, the overall time complexity is $\Theta(n^2)$.
- SDist is calculated across all siblings, and thus requires the computation of sums over multiple sub-matrices of Δ. However, these sub-matrices do not overlap for different pairs of siblings. Hence, the overall time complexity to compute SDist is $O(n^2)$.

After the precomputation step, the sums in Eq. 1 can be substituted with BDist and RDist values in order to calculate $d^2(T, G) - d^2(N, G)$ in time $O(t)$, where $t \leq n$ for any $G \in SPR_S(v)$. This concludes the proof of Theorem 2.

Unrooted Case. The PD median tree problem for unrooted trees is NP-hard, which follows from a straightforward polynomial time reduction from the NP-hard quartet compatibility problem [29] (as in the rooted case, we observe that the PD distance is 0 when all input trees are consistent with a supertree). Moreover, our local search algorithm for rooted trees can be extended to work with unrooted trees as well. We are describing the key ideas of this algorithm, and omitting details for brevity. The semi-structure of the SPR-neighborhood can be exploited in the unrooted case as well: one can root a supertree at an edge $(Pa(v), v)$, where v is the "prune" node, and traverse the SPR-neighborhood in the same way as in the rooted case. Table 1 would be slightly changed to account for the artificial root, though the same precomputation idea is still applicable.

5   Experimental Evaluation

Median tree methods under the path-difference objective have never been studied before. Therefore, we adhere to a classical evaluation approach by comparing our median tree heuristic against standard supertree methods with different objectives [2,20]. We processed two published baseline phylogenetic datasets, the Marsupials dataset [7] and the Cetartiodactyla dataset [25]. These datasets have been actively used for experimental supertree evaluations throughout the evolutionary community (see, for example, [2,10,16,28]).

Following the experiments presented in one of the recent supertree papers [16], we compare our PD local search method against the following supertree methods: the maximum representation with parsimony (MRP) heuristic [31], the modified min-cut (MMC) algorithm [22], and the triplet supertree heuristic [16]. MRP heuristics are addressing the NP-hard MRP problem [20], and are among the most popular supertree methods in evolutionary biology [4]. For our evaluation we use the MRP local search heuristic implemented in PAUP* [31] with Tree Bisection and Reconnection (TBR) branch swapping [16]. The *TBR edit operation* is an extension of the SPR operation, where the pruned subtree is allowed to be re-rooted before regrafting it. The MMC algorithm computes supertrees (that satisfy certain desirable properties) in polynomial time, which makes this method especially attractive for large-scale phylogenetic analysis [22]. The triplet supertree heuristic is a local search heuristic that is addressing the well-studied NP-hard triplet supertree problem [16]. We are using the triplet heuristic based on SPR and TBR local searches, called TH(SPR) and TH(TBR) respectively.

Hybrid Heuristic. In a classical local search scenario there are two major steps. In the first step a supertree is constructed incrementally. Typically, the process is initiated with some t taxa, and an optimal supertree over the chosen taxa is computed exactly. Here, t is typically small, e.g., three. Next, on each iteration t new taxa are added to the partial supertree (an optimum among all possible ways to add t leaves to the tree is picked). The second major step is to run the actual local search starting with the tree obtained in step one.

Clearly, the first step is rather slow, especially when it is costly to compute the distance measure for a supertree (as in our case). Even though many ideas could be suggested to accelerate the first step, we want to emphasize that it is not necessary to separate the two steps in the first place. That is, the local search heuristic, which is the main optimization engine, could be applied on every step of construction of the start tree. It could be argued that SPR-based local search brings in more flexibility than simply trying to add new taxa to a tree with a fixed structure. For estimation of PD median trees we implemented this novel hybrid heuristic using the introduced here local search algorithm.

Results and Discussion. Table 2 summarizes the results that we obtained from the conducted experiments with our heuristic PDM(SPR) in comparison with the published results for MMC, MRP, TH(SPR), and TH(TBR) [16]. As expected, all of the methods stand their ground. The MRP method proves to be

Table 2. Summary of the experimental evaluation. The best scores under each objective are highlighted in bold.

Data set	Method	PD score	Triplet-sim	MAST-sim	Pars. score
Marsuplial 158 **input trees 272** **taxa**	MMC	16,670.45	51.73 %	53.4 %	3901
	MRP	5,694.59	98.29 %	**71.6 %**	**2274**
	TH(SPR)	5,866.27	**98.99 %**	70.2 %	2317
	TH(TBR)	5,888.22	**98.99 %**	70.5 %	2317
	PDM(SPR)	**4,677.99**	68.43 %	63.4 %	3339
Cetartiodactyla 201 input trees 299 taxa	MMC	16,206.17	70.03 %	51.5 %	4929
	MRP	6,991.36	95.84 %	**64.7 %**	**2603**
	TH(SPR)	7,630.03	**97.28 %**	63.1 %	2754
	TH(TBR)	7,591.13	**97.28 %**	63.0 %	2754
	PDM(SPR)	**6,051.13**	59.49 %	52.2 %	4162

most effective according to the parsimony objective. In addition, MRP supertrees show the best fit over the input data in terms of our computed MAST-similarity scores – which could be seen as an "independent" objective in our evaluation. At the same time, both triplet heuristics, TH(SPR) and TH(TBR), produced the best supertrees under the triplet similarity objective. As for our method – PDM(SPR) – it was able to produce best supertrees with regards to the PD distance.

In order to rigorously assess the results of our heuristic, we should contemplate them with the distribution of the path-difference distance for the two datasets. However, such distributions, even for a single input tree, remain unknown [19]. Thus, following the approach from Steel and Penny [30], we estimate PD distance distributions based on sample data. For each dataset we generated two collections with 20,000 random supertrees using PAUP*. One collection was generated under the uniform binary tree distribution, and the other one was generated using the Markovian branching process [3]. Then, each collection was processed to obtain sample datasets with PD distance scores for every generated supertree. The obtained results are outlined in Fig. 2.

The figure makes it clear that even though our heuristic was able to obtain the best results for the two datasets, MRP and Triplet heuristics still produce trees that are significantly better than any of the randomly generated trees. As for the MMC algorithm, it performs much worse under the PD objective than simply constructing Markovian Binary trees; and, what is more, worse than drawing random trees from the uniform distribution.

Figure 2 suggests that there exists a positive correlation between the Parsimony, Triplet-similarity, and PD distance supertree objectives. On the other hand, according to Table 2, better PD supertrees do not necessarily score well in terms of parsimony and triplet measures. Thus, the PD heuristic might produce structurally new phylogenetic trees that have not been analyzed previously.

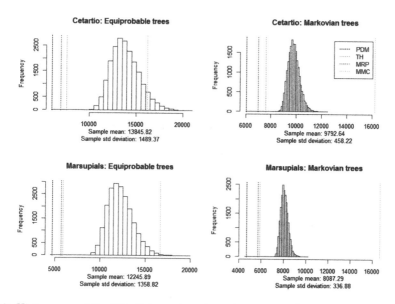

Fig. 2. Histograms of the PD distance based on the generated tree samples. All methods used for the evaluation are marked on each histogram with dotted lines.

6 Conclusion

We synthesized the first large-scale median trees under one of the oldest and widely popular tree distance metrics — the path-difference distance. While we show that the corresponding PD median tree problem is NP-hard, we demonstrated that it can be successfully approached by using our new SPR based local search heuristic. To make the heuristic applicable to real-world phylogenetic datasets, we have significantly improved its time complexity in comparison to the best known naïve approach.

Currently, no mainstream supertree method can construct edge-weighted supertrees. However, there has been an increased interest in such tools due to fast developing databases of time-annotated evolutionary trees (e.g., Time-Tree [15]). The path-difference distance on the other hand, is naturally extendable to account for edge-weights in phylogenetic trees. The introduced PD heuristic, in turn, could also be adapted to deal with edge-weighted supertrees. This property makes the PD distance even more appealing as a median tree objective and suggests further investigation in its theoretical and algorithmic means.

Acknowledgments. The authors would like to thank the two anonymous reviewers for their constructive comments that helped to improve the quality of this work.

References

1. Bansal, M.S., Burleigh, J.G., Eulenstein, O.: Efficient genome-scale phylogenetic analysis under the duplication-loss and deep coalescence cost models. BMC Bioinform. **11**(Suppl 1), S42 (2010)
2. Bansal, M.S., Burleigh, J.G., Eulenstein, O., Fernández-Baca, D.: Robinson-foulds supertrees. Algorithms Mol. Biol. **5**(1), 1–12 (2010)
3. Bean, N.G., Kontoleon, N., Taylor, P.G.: Markovian trees: properties and algorithms. Ann. Oper. Res. **160**(1), 31–50 (2007)
4. Bininda-Emonds, O.R. (ed.): Phylogenetic Supertrees: Combining Information to Reveal the Tree of Life. Computational Biology, vol. 4. Springer, The Netherlands (2004)
5. Bluis, J., Shin, D.: Nodal distance algorithm: calculating a phylogenetic tree comparison metric. In: 3rd IEEE International Symposium on BioInformatics and BioEngineering (BIBE 2003), 10–12 March 2003 Bethesda, pp. 87–94. IEEE Computer Society (2003)
6. Bryant, D.: Hunting for trees in binary character sets: efficient algorithms for extraction, enumeration, and optimization. J. Comput. Biol. **3**(2), 275–288 (1996)
7. Cardillo, M., Bininda-Emonds, O.R.P., Boakes, E., Purvis, A.: A species-level phylogenetic supertree of marsupials. J. Zool. **264**, 11–31 (2004)
8. Chaudhari, R., Burleigh, G.J., Eulenstein, O.: Efficient algorithms for rapid error correction for gene tree reconciliation using gene duplications, gene duplication and loss, and deep coalescence. BMC Bioinform. **13**(Suppl 10), S11 (2012)
9. Chaudhary, R., Bansal, M.S., Wehe, A., Fernández-Baca, D., Eulenstein, O.: iGTP: a software package for large-scale gene tree parsimony analysis. BMC Bioinform. **11**, 574 (2010)
10. Chen, D., Eulenstein, O., Fernández-Baca, D., Burleigh, J.: Improved heuristics for minimum-flip supertree construction. Evol. Bioinform. **2**, 347 (2006)
11. Cotton, J.A., Wilkinson, M.: Majority-rule supertrees. Syst. Biol. **56**(3), 445–452 (2007)
12. Farris, J.: A successive approximations approach to character weighting. Syst. Zool. **18**, 374–385 (1969)
13. Harris, S.R., Cartwright, E.J., Török, M.E., Holden, M.T., Brown, N.M., Ogilvy-Stuart, A.L., Ellington, M.J., Quail, M.A., Bentley, S.D., Parkhill, J., Peacock, S.J.: Whole-genome sequencing for analysis of an outbreak of meticillin-resistant Staphylococcus aureus: a descriptive study. Lancet Infect. Dis. **13**(2), 130–136 (2013)
14. Hufbauer, R.A., Marrs, R.A., Jackson, A.K., Sforza, R., Bais, H.P., Vivanco, J.M., Carney, S.E.: Population structure, ploidy levels and allelopathy of Centaurea maculosa (spotted knapweed) and C. diffusa (diffuse knapweed) in North America and Eurasia. In: Proceedings of the XI International Symposium on Biological Control of Weeds, Canberra Australia, pp. 121–126. USDA Forest Service, Forest Health Technology Enterprise Team, Morgantown (2003)
15. Leaché, A.D.: Integrative and Comparative Biology. In: Hedges, S.B., Kumar, S. (eds.) The Timetree of Life, vol. 50(1), pp. 141–142. Oxford University Press, New York (2010)
16. Lin, H.T., Burleigh, J.G., Eulenstein, O.: Triplet supertree heuristics for the tree of life. BMC Bioinform. **10**(Suppl 1), S8 (2009)
17. Lin, H.T., Burleigh, J.G., Eulenstein, O.: Consensus properties for the deep coalescence problem and their application for scalable tree search. BMC Bioinform. **13**(Suppl 10), S12 (2012)

18. Maddison, W.P., Knowles, L.L.: Inferring phylogeny despite incomplete lineage sorting. Syst. Biol. **55**(1), 21–30 (2006)
19. Mir, A., Rosselló, F.: The mean value of the squared path-difference distance for rooted phylogenetic trees. CoRR abs/0906.2470 (2009)
20. Moran, S., Rao, S., Snir, S.: Using semi-definite programming to enhance supertree resolvability. In: Casadio, R., Myers, G. (eds.) WABI 2005. LNCS (LNBI), vol. 3692, pp. 89–103. Springer, Heidelberg (2005)
21. Nik-Zainal, S., et al.: The life history of 21 breast cancers. Cell **149**(5), 994–1007 (2012)
22. Page, R.D.M.: Modified mincut supertrees. In: Guigó, R., Gusfield, D. (eds.) WABI 2002. LNCS, vol. 2452, p. 537. Springer, Heidelberg (2002)
23. Page, R.D., Holmes, E.: Molecular Evolution: A Phylogenetic Approach. Blackwell Science, Boston (1998)
24. Phipps, J.B.: Dendogram topology. Syst. Zool. **20**, 306–308 (1971)
25. Price, S.A., Bininda-Emonds, O.R.P., Gittleman, J.L.: A complete phylogeny of the Whales, Dolphins and even-toed hoofed mammals (cetartiodactyla). Biol. Rev. **80**(3), 445–473 (2005)
26. Puigbò, P., Garcia-Vallvé, S., McInerney, J.O.: TOPD/FMTS: a new software to compare phylogenetic trees. Bioinformatics **23**(12), 1556–1558 (2007)
27. Semple, C., Steel, M.A.: Phylogenetics. Oxford University Press, Oxford (2003)
28. Snir, S., Rao, S.: Quartets maxcut: a divide and conquer quartets algorithm. IEEE/ACM TCBB **7**(4), 704–718 (2010)
29. Steel, M.: The complexity of reconstructing trees from qualitative characters and subtrees. J. Classif. **9**(1), 91–116 (1992)
30. Steel, M.A., Penny, D.: Distributions of tree comparison metrics - some new results. Syst. Biol. **42**(2), 126–141 (1993)
31. Swofford, D.L.: PAUP*. Phylogenetic analysis using parsimony (and other methods), Version 4. Sinauer Associates, Sunderland, Massachusetts (2002)
32. Than, C., Nakhleh, L.: Species tree inference by minimizing deep coalescences. PLoS Comput. Biol. **5**(9), e1000501 (2009)
33. Williams, W., Clifford, H.: On the comparison of two classifications of the same set of elements. Taxon **20**(4), 519–522 (1971)

NEMo: An Evolutionary Model with Modularity for PPI Networks

Min Ye[1], Gabriela C. Racz[2], Qijia Jiang[3], Xiuwei Zhang[4],
and Bernard M.E. Moret[1(✉)]

[1] School of Computer and Communication Sciences, EPFL, Lausanne, Switzerland
`{min.ye,bernard.moret}@epfl.ch`
[2] Department of Mathematics, University of Zagreb, Zagreb, Croatia
[3] Department of Electrical and Computer Engineering, Stanford University,
Palo Alto, USA
[4] European Bioinformatics Institute (EMBL-EBI), Cambridge, UK

Abstract. Modelling the evolution of biological networks is a major challenge. Biological networks are usually represented as graphs; evolutionary events include addition and removal of vertices and edges, but also duplication of vertices and their associated edges. Since duplication is viewed as a primary driver of genomic evolution, recent work has focused on duplication-based models. Missing from these models is any embodiment of modularity, a widely accepted attribute of biological networks. Some models spontaneously generate modular structures, but none is known to maintain and evolve them.

We describe NEMo (Network Evolution with Modularity), a new model that embodies modularity. NEMo allows modules to emerge and vanish, to fission and merge, all driven by the underlying edge-level events using a duplication-based process. We introduce measures to compare biological networks in terms of their modular structure and use them to compare NEMo and existing duplication-based models and to compare both generated and published networks.

Keywords: Generative model · Evolutionary model · PPI network · Evolutionary event · Modularity · Network topology

1 Introduction

The rapid growth of experimentally measured data in biology requires effective computational models to uncover biological mechanisms in the data. Networks are commonly used to represent key processes in biology; examples include transcriptional regulatory networks, protein-protein interaction (PPI) networks, metabolic networks, etc. The model is typically a graph, directed or undirected, where edges or arcs represent interactions and vertices represent actors (genes, proteins, etc.). Establishing experimentally the existence of a particular interaction is expensive and time-consuming, so most published networks have been inferred through computational methods ranging from datamining the literature

© Springer International Publishing Switzerland 2016
A. Bourgeois et al. (Eds.): ISBRA 2016, LNBI 9683, pp. 224–236, 2016.
DOI: 10.1007/978-3-319-38782-6_19

(see, e.g., [1,10,15]) to inferring the evolutionary history of the networks from present observations [8,26,35,36]. (Makino and McLaughlin [14] present a thorough discussion of evolutionary approaches to PPI networks.) Often these networks are built through a process of accretion, by adding new actors and new interactions as they are observed, published, or inferred, with the result that errors in many current biological networks tend to be false positives (errors of commission) rather than false negatives (errors of omission). A variety of databases store inferred networks and range from large graphs, such as the human PPI network in the STRING database (ca. 3'000'000 interactions) [9,31], down to quite small ones, such as the manually curated Human Protein Reference Database (ca. 40'000 interactions) [23]. Even a cursory reading of the literature shows that agreement among findings is rather limited, due in part to the variety of samples used and the dynamic nature of the networks, but also in good part because of the difficulty of inference.

This intrinsic difficulty has led researchers to go beyond the inference of a single network from data about one organism and to use comparative methods. Pairwise comparative methods, while more powerful, offer only limited protection against noise and high variability. This weakness in turn has led to the use of evolutionary methods that use several organisms and carry out simultaneous inference on all of them [8,14,36]—a type of inference that falls within the category of transfer learning [20]. A unique feature in these approaches is their use of evolutionary models (not commonly associated with transfer learning). These approaches posit a model of evolution for the networks, typically based on inserting and deleting edges and duplicating or losing vertices, and then seek to infer present-day networks as well as ancestral networks that, under the chosen evolutionary model, would best explain the data collected. The evolutionary model is thus the crucial component of the inference procedure.

An early finding about biological networks such as regulatory networks and PPI networks was the clear presence of modularity [11]: these networks are not homogeneous, with comparable connectivity patterns at every vertex, but present a higher-level structure consisting of well connected subgraphs with less substantial connectivity to other such subgraphs. Modularity is now widely viewed as one of the main characteristics of living systems [28]. While some of the models devised for networks lead automatically to the emergence of modules within the network [30], these models are purely generative—increasing the size of the network at each step—and thus do not match biological reality. There is thus a need for an evolutionary model for PPI networks that, while still based on the gain and loss of vertices and edges, takes into account modularity.

In this paper, we introduce NEMo, a network evolutionary model with modularity for PPI networks that includes both growth and reduction operators, and that explicitly models the influence of modularity on network evolution. While modules remain the product of purely local events (at the level of single vertices or edges), they are subject to slightly different selection constraints once they have emerged, so that our model allows modules to emerge, to disappear, to merge, and to split. We present the results of simulations and compare the networks thus produced to the consensus networks currently stored in a variety of databases for model organisms. Our comparisons are based on both network alignment ideas and new measures aimed at quantifying modularity, so we also discuss the usefulness of these measures and evaluate published PPI networks with respect to these measures. Our measures of modularity can be used to analyze the general characteristics of PPI networks and clearly distinguish the various models organisms. Our findings support the accepted bias of published networks towards false positives and the often reported distribution of modules into a few large subgraphs and a collection of much smaller subgraphs; NEMo produces networks with the latter characteristic and maintains it even when it has reached a target range of sizes and simply makes small changes to the structure of the network.

2   Current Generative Models for PPI Networks

All evolutionary models to date are based on the addition or removal of the basic constituent elements of the network: vertices (proteins) and edges (pairwise interactions). In terms of complexity and verisimilitude, however, models proposed to date vary widely. Most of the recent models are based on duplication followed by divergence, denoted D&D [4,24], in which a vertex is duplicated (think of a gene duplication) and inherits some randomly chosen subset of the connections of the original vertex (the copy of the gene initially produces much the same protein as the original and so enters into much the same interactions). Most evolutionary biologists view gene duplication (single gene, a segment of genes, or even the entire genome) as the most important source of diversification in genomic evolution [13,19], so models based on D&D have become widely used for PPI networks.

The full D&D model considers both specialization and gene duplication events. Following a specialization event, interactions (edges) can be gained or lost with specified probabilities. A duplication event duplicates all interactions of the original copy, but some interactions for both the original and the duplicated copies are immediately lost with some probability. A recent variation on the D&D model is the duplication-mutation-complementarity (DC) model [16,17,32], in which the same interaction cannot be lost simultaneously in the original and in the copy and in which the duplicated gene itself may gain a direct interaction with the original gene. The DMR (random mutation) model [29] is another variation, in which new interactions (not among those involving the original vertex) can be introduced between the duplicate vertex and some random vertex in the network.

3   NEMo

While, as noted earlier, the D&D model (and, by extension, its various derivatives) will automatically give rise to modular structures, it does so in scenarios of unrestricted growth: no edge deletions are allowed other than those that occur as part of a vertex duplication and a vertex gets deleted only indirectly, if and when its degree is reduced to zero. In that sense, the D&D, while a generative model, is not an evolutionary model: it can only grow networks, not evolve them while keeping their size within some fixed range. The same is true of its several variants.

Our aim is to produce a generative model that is also an evolutionary model and that we can later use for reconstructing the evolutionary history of PPI networks. Under such a model, a network may grow, shrink, or, most commonly, vary in size within some bounded range. Since the dominant growth operator is duplication and since this operator typically adds multiple edges to the network, random (i.e., unrelated to other events) loss of edges must be common. We conjecture that, under such a model, modularity might not be preserved—because, under such a model, the selection of which interactions to lose is independent of the modular structure. Since modules appear both necessary to life and quite robust against mutations, a model of evolution of PPI networks that is biased (as nature appears to be) in favor of the survival of modules would need to "know" about the module structure.

We therefore decided to design a two-level model. The lower level is just a variant of the DC model, except that it allows random mutations for each vertex—a vertex can be lost at any step rather than just when its degree is reduced to zero—and that, due to the same random mutations, arbitrary edges can be added to or removed from the network. The higher level, however, is "module-aware" so that interactions can be classified as within a module, between modules, or unrelated to modules. This classification allows us to treat these three types differently in the evolutionary model, with interactions within modules less likely to be lost. Our model represents a PPI network as a graph, with the set of vertices representing proteins and the set of undirected edges representing undirected interactions between the proteins. In addition, the graph is at all times subdivided into subgraphs, which correspond to modules.

The events directly affecting vertices and edges are similar to those of the D&D model and its relatives and can be classified into four categories: protein gain, protein loss, interaction gain, and interaction loss. Protein gain is exclusively through duplication and thus also includes interaction gains for the newly added protein. Protein loss removes a randomly chosen vertex; it can be a consequence of, e.g., pseudogene formation. (As in the DMC model, it is also possible to lose a vertex through progressive loss of interactions until the vertex has degree zero.) Interaction loss removes a randomly chosen edge; it can come about through domain mutations, structural mutations, subfunctionalization, and the like. Interaction gains come in two varieties: those caused by vertex duplication and those arising purely at random, by connecting a previously unconnected pair of vertices, which could arise, like loss, through domain or structural mutations, or through progressive neofunctionalization (Fig. 1).

We use the module level to influence the event chain as follows. First, we allow events to arise within the same time frame in different modules; whereas existing models treat the network as one unit and allow a single event at a time, our model treats the network as a collection of subgraphs and allows up to one event in each subgraph. Multiple events within the same time frame can

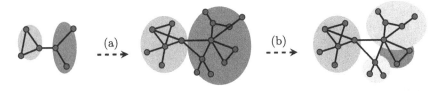

Fig. 1. NEMo: schema of the evolutionary process: (a) after multiple timesteps; (b) after reclustering

more closely model events such as segmental duplication (in which many genes are duplicated together). Second, we distinguish intramodular events (all four events can be intramodular) from intermodular events (only edge gains and losses can be intermodular), allowing us to use different parameters for the two types. While we automatically place a duplicate vertex within the same module as the original vertex, we also periodically recompute the subgraph decomposition, thereby inducing changes in the module structure, including both fission and fusion of modules. Thus there is no specific evolutionary event associated with changes in the module structure: rather, it is a recognition that underlying events have, over some number of steps, sufficiently altered the network as to have altered, destroyed, or created some modules.

Such a model as this requires the identification of modules within a network and the extraction and quantification of some high-level attributes that can be used to measure similarity. Methodologies used in much of the work on the identification of functional modules [2,6,7] are not applicable here, as we deal with an anonymous graph, not with annotated proteins. We rely in part on clustering algorithms (to detect clusters, which we regard as potential modules, within the graph) and in part on matching high-level attributes of actual PPI networks and using these attributes to measure drift in the course of evolution. There are several families of clustering algorithms used in the biological domain. In this study, we use ClusterOne [18], a graph clustering algorithm that allows overlapping clusters. It has been useful for detecting protein complexes in PPI networks tolerating nodes to have multiple-module membership.

4   Assessing Modularity

In order to evaluate the output of NEMo, we must first quantify significant attributes of PPI networks. The resulting features can then be used to measure the similarity of our generated networks to real networks, as well as the differences between networks generated by our model and networks generated under existing models. Similarity here refers to structural and topological features such as modularity and connectivity: we need to compare networks very different in size and composition and so cannot use tools such as network alignment methods. We thus propose a set of features applicable to hall networks, features chosen to measure global properties of networks and to quantify aspects of modularity.

Most of the features proposed here are commonly used in the analysis of networks [2,3]; several are modified so as to provide a level of independence from size—bacterial PPI networks are necessarily smaller than mammalian PPI networks, while simulations can be run at all sizes. For each network, we compute the number of nodes, the number of edges, and the degree distribution; we also run the ClusterOne cluster algorithm (always with the same parameters) and store the number of clusters as well as the size and composition of each cluster. We then compute the following five global measures.

Cluster Coefficient (CC): The CC is based on triplets of vertices. A triplet is open if connected with two edges, closed if connected with all three edges. The CC is just the ratio of the number of closed triplets divided by the total number of (open or closed) triplets [34].

Graph Density (GD): The density of a graph is the ratio of the actual number of edges to the number of possible edges.

Diameter (⊘): The diameter of a graph is the length of the longest simple path in the graph.

Fraction of Edges Inside (FEI): FEI is the fraction of edges contained within modules. We expect it to be high since PPI networks contain highly connected substructures (modules) that have only few connections to vertices outside the substructure [3,12,33].

Tail Size (TS): A simple representation of the tail of the degree distribution, TS is fraction of the number of nodes with degree higher than one-third of that maximum node degree.

5   Results on Natural PPI Networks

For the data, we chose to work with model organisms, as they have large numbers of high-confidence interactions. We chose to download the following species since they have the largest number of well documented interactions: *E. Coli*, *S. Cerevisiae*, and *H. Sapiens*. Different sources were considered to emphasize the discrepancies of the networks stored and provided in existing datasets of real world PPI networks.

One source is the STRING database [9] that aims to provide a global perspective for as many organisms as feasible, tolerating lower-quality data and computational predictions. With this purpose the database holds a large part of false positive interactions. Although the STRING database stores evidence scores for each protein-protein interaction to allow elimination of as many false positive entries as possible by the user, it is still very much biased. For other sources, we consulted the manually curated *H. sapiens* PPI network HPRD [22] database and the experimental setup of the MAGNA++ algorithm [27] that aims at maximizing accuracy in global network alignment: an *H. sapiens* PPI network of 9141 proteins and 41456 interactions (Radivojac et al., 2008 [25]), an *E. coli* PPI network [21] of high-confidence of 1941 proteins with 3989 interactions, and a yeast *S. cerevisiae* PPI network with 2390 proteins and 161277 PPIs (Collins et al., 2007 [5]).

Table 1. General characteristics of the three PPI networks in various databases

Species	# nodes	# edges	# clusters
H. sapiens STRING	19247	4274001	2077
E. coli STRING	4145	568789	16
S. cerevisiae STRING	6418	939998	159
H. sapiens HPRD	9673	39198	2886
E. coli MAGNA++exp	1941	3989	393
S. cerevisiae MAGNA++exp	2390	16127	360
H. sapiens MAGNA++exp	9141	41456	2306

Table 2. Values of our features for the three PPI networks in various versions

Species	CC	GD	⊘	FEI	TS
H. sapiens STRING	0.23058	0.02308	18	0.94506	0.99777
E. coli STRING	0.21368	0.06623	9	1.00942*	0.80555
S. cerevisiae STRING	0.27757	0.04565	20	1.08949*	0.99564
H. sapiens HPRD	0.19602	0.00084	30	0.53896	0.99369
E. coli MAGNA++exp	0.3394	0.00212	33	0.92454	0.98454
S. cerevisiae MAGNA++exp	0.43854	0.00565	34	0.97055	0.95105
H. sapiens MAGNA++exp	0.16377	0.00099	30	0.56549	0.99103

* (FEI > 1) comes from the multiple membership of nodes. Edges shared by two nodes that belong to more than one same module are counted more than once.

We downloaded PPI networks from the STRING database [31] and used a high threshold (900) on the supplied confidence scores to retain only high-confidence interactions. Table 1 provides a brief description of these three PPI networks in the various databases and versions.

We then computed our network features for each of these networks, as shown in Table 2.

6   Results on Simulations

6.1   Simulation Goals and Setup

The goal of our simulations was to verify the ability of NEMo to produce networks with characteristics similar to those of the natural PPI networks and also to compare the networks it produces with those produced without the module-aware level and with those produced by D&D models. In particular, we wanted to test the ability of NEMo to sustain modules in networks not undergoing growth, but subject only to change—where gain of proteins and interactions is balanced by loss of same. Therefore we ran two distinct series of simulations, one for generation and one for evolution.

The first series uses both the DMC model [32], perhaps the most commonly used model in the D&D family today, and NEMo to grow networks to fixed sizes. We then compute our features on these networks and compare both types of generated networks with the PPI networks of the model organisms. Since DC is not module-aware, but claimed to generate modular networks, whereas NEMo is explicitly module-aware, we want to see how well the characteristics of each type of generated network compare to the PPI networks of the model organisms.

In the second series of simulations, we use NEMo in steady-state mode (balanced gains and losses) over many steps to evolve networks produced during the first simulation series. Our main intent here is to observe the evolution (mostly in terms of size, edge density, and modules) of the networks. We use parameters for NEMo that give it a slight bias towards growth, mostly to avoid the natural variance of the process from "starving" too many of the networks.

6.2   Results for Network Generation

We set parameters of our model for simulating growth of the network and compared the resulting networks with those built with the standard DMC model for similar sizes, as well as with the PPI networks from the three model organisms.

We then computed our network features for each of these networks, as shown in Table 3, where they can be compared to the same features shown for PPI networks (from Table 2). Both DC and NEMo generated networks with features comparable to those observed in the PPI networks collected from MAGNA and HPRD, although the significantly lower clustering coefficient of the DC-generated network (0.04 as compared to 0.14 for the NEMo-generated network) indicates a less resolved modular structure. Note that all PPI networks from databases have larger clustering coefficients than the generated networks, a difference attributable in good part to the generation mode.

Table 3. Values of our features for the generated networks and the three PPI networks in various versions

Species	CC	GD	⊘	FEI	TS
H. sapiens HPRD	0.196021	0.000837947	30	0.538956	0.993694
E. coli MAGNA++exp	0.3394	0.00211869	33	0.924542	0.984544
S. cerevisiae MAGNA++exp	0.438538	0.00564897	34	0.970546	0.951046
H. sapiens MAGNA++exp	0.163768	0.000992379	30	0.565491	0.991029
DC-generated net500	0.0478	0.0040	12	0.9520	0.9880
NEMO-generated net500	0.1417	0.0078	31	0.9559	0.9519

6.3   Results for Network Evolution

In the second step of our experiments we test the ability of NEMo to simulate the evolution of a PPI network (with roughly balanced ngain and loss rates)

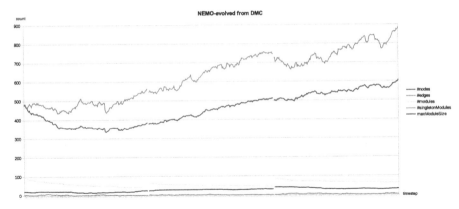

(a) evolution from a DMC-generated initial network

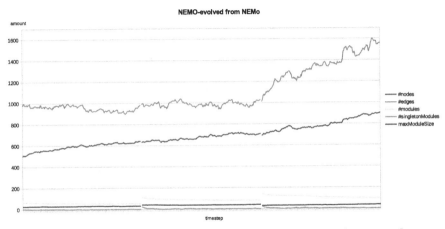

(b) evolution from a NEMo-generated initial network

Fig. 2. Evolution of network characteristics under the NEMo model over 600 steps, with reclustering into modules at 200 and 400 steps. Top line shows the total number of edges, second line the number of vertices, third line the number of modules, fourth line the size of the largest module, and bottom line the number of singleton modules.

while preserving modularity and also test how NEMo's behavior is affected by its initial condition by using for DMC- and NEMo-generated networks at time zero. Figure 2 shows the changes in network size (numbers of edges and vertices) and structure (numbers of modules) as an initial n-etwork is evolved through 600 steps, with reclustering into modules taking place after 200 and 400 steps.

The main observation here is that NEMo, when started with a DMC-generated network (part (a) of the figure), begins by reconfiguring the network, reducing its number of vertices by about one third over the first hundred steps and replacing edges. It then moves into much the same mode as depicted in part (b) of the figure, which shows a steady evolutionary behavior mixed with a small bias towards growth. The implication is that, while the DC-generated network may have a modular structure, that structure is less well structured (as observed above) as well as not well supported under the evolutionary model. We can observe that the graph density of the DMC-generated network is low and gets swiftly increased by NEMo, while the initial number of modules is high and gets switfly decreased by NEMo as a consequence of the removal of many nodes. After the first 200 steps and the first reclustering of modules, the evolution follows the same path as that followed immediately when working from a NEMo-generated intial graph, as seen in part (b) of the figure. Part (b) shows variance in the rate of increase in the number of edges, partly a consequence of the node duplication process—duplicating a few high-degree nodes in rapid succession quickly increases the overall degree of the network, while also increasing the number of high-degree nodes.

The mild generative bias we deliberately introduced into the evolutionary simulations can be harmlessly removed for evolving NEMo-generated networks and, through larger numbers of steps, evolving a modular structure closer to that of the PPI networks from the databases.

It is worth noting that the module-aware level of NEMo is very limited in its effects: its power derives from its distinguishing intermodular from intramodular events, but NEMo uses this power in quite a minimal way, by assigning slightly different probabilities to the two classes of events—in evolutionary terms, it simulates a slightly stronger negative selection for intermodular events than for intramodular events. The distinction between the two classes of events could be used to a much larger extent, but our results show that even this minimal intervention, consistent with a selective pressure to preserve modularity while allowing modules themselves to adapt, suffices to create a significant difference in the type of networks generated.

7   Discussion and Future Work

We presented NEMo, a module-aware evolutionary model for PPI networks. The emphasis of NEMo, as compared to existing models for PPI networks, is on evolution rather than generation: whereas existing models (and the first layer of NEMo, which is a variant of existing models) are know to generate modularity when growing networks, we were interested in a model that would evolve existing networks, using the same basic set of evolutionary events.

The salient feature of NEMo is a module-aware layer that sits above the event layer and distinguishes between intermodular and intramodular events. The awareness is achieved through periodic recomputation (triggered by sampling and analysis for drift) of the modular structure. The uses to which this awareness are put are minimal: NEMo simply gives a slightly higher probability to intramodular events than to intermodular events, thereby slightly favoring conservation of modules and evolution of internal module structure. The details of the model are broadly adjustable: the algorithm used to detect modules, the number and nature of parameters used to control intra- vs. intermodular events, the features chosen to characterize the network, and the distance measure used to measure drift in order to decide when to re-evaluate the composition of modules, are all flexible.

Our simulation results show that this second layer enables NEMo to run through large numbers (as compared to the size of the network) of evolutionary events, balanced so as not to affect the expected size of the network, while preserving the characteristics of its original (growth-derived) modular structure. To the best of our knowledge, this is the first such result and it paves the way for phylogenetic analyses as well as population studies of PPI networks.

As discussed by Makino and McLaughlin [14], however, the number of factors that could affect the evolution of PPI networks is very large. NEMo captures only a small subset of these factors, since it works just on the graph structure and, at the level of individual events, makes the same independence assumptions as current models. Interdependent events or hidden underlying events present serious challenges. Incorporating externally supplied data (in addition to the network itself) makes sense in a data-rich era, but will require, for each type of data, further development of the model.

Acknowledgments. MY wishes to thank Mingfu Shao for many helpful discussions.

References

1. Abi-Haidar, A., et al.: Uncovering protein interaction in abstracts and text using a novel linear model and word proximity networks. Genome Biol. **9**(Suppl 2), S11 (2008)
2. Aittokallio, T.: Module finding approaches for protein interaction networks. In: Li, X.L., Ng, S.K. (eds.) Biological Data Mining in Protein Interaction Networks, pp. 335–353. IGI Publishing, Hershey (2009)
3. Barabási, A.L., Oltvai, Z.: Network biology: understanding the cell's functional organization. Nat. Rev. Genet. **5**, 101–113 (2004)
4. Bhan, A., Galas, D., Dewey, T.: A duplication growth model of gene expression networks. Bioinformatics **18**(11), 1486–1493 (2002)
5. Collins, S., Kemmeren, P., Zhao, X., et al.: Toward a comprehensive atlas of the physical interactive of saccharomyces cerevisiae. Mol. Cell. Proteomics **6**(3), 439–450 (2007)
6. Dittrich, M., et al.: Identifying functional modules in protein-protein interaction networks: an integrated exact approach. In: Proceedings of 16th International Conference on Intelligent Systems for Molecular Biology (ISMB 2008), in Bioinformatics, vol. 24, pp. i223–i231 (2008)
7. Dutkowski, J., Tiuryn, J.: Identification of functional modules from conserved ancestral protein interactions. Bioinformatics **23**(13), i149–i158 (2007)
8. Dutkowski, J., Tiuryn, J.: Phylogeny-guided interaction mapping in seven eukaryotes. BMC Bioinform. **10**(1), 393 (2009)
9. Franceschini, A., et al.: String v9.1: protein-protein interaction networks, with increased coverage and integration. Nucleic Acids Res. **41**, D808–D815 (2013)

10. Hao, Y., Zhu, X., Huang, M., Li, M.: Discovering patterns to extract protein-protein interactions from the literature. Bioinformatics **21**(15), 3294–3300 (2005)
11. Hartwell, L., Hopfield, J., Leibler, S., Murray, A.: From molecular to modular cell biology. Nature **402**(6761), C47–C52 (1999)
12. Jin, Y., Turaev, D., Weinmaier, T., Rattei, T., Makse, H.: The evolutionary dynamics of protein-protein interaction networks inferred from the reconstruction of ancient networks. PLoS ONE **8**(3), e58134 (2013)
13. Lynch, M., et al.: The evolutionary fate and consequences of duplicate genes. Science **290**(5494), 1151–1254 (2000)
14. Makino, T., McLaughlin, A.: Evolutionary analyses of protein interaction networks. In: Li, X.L., Ng, S.K. (eds.) Biological Data Mining in Protein Interaction Networks, pp. 169–181. IGI Publishing, Hershey (2009)
15. Marcotte, E., Xenarios, I., Eisenberg, D.: Mining literature for protein protein interactions. Bioinformatics **17**, 359–363 (2001)
16. Middendorf, M., Ziv, E., Wiggins, C.: Inferring network mechanisms: the drosophila melanogaster protein interaction network. Proc. Nat. Acad. Sci. USA **102**(9), 3192–3197 (2005)
17. Navlakha, S., Kingsford, C.: Network archaeology: uncovering ancient networks from present-day interactions. PLoS Comput. Biol. **7**(4), e1001119 (2011)
18. Nepusz, T., Yu, H., Paccanaro, A.: Detecting overlapping protein complexes in protein-protein interaction networks. Nat. Methods **9**, 471–472 (2012)
19. Ohno, S.: Evolution by Gene Duplication. Springer, Berlin (1970)
20. Pan, S., Yang, Q.: A survey on transfer learning. IEEE Trans. Knowl. Data Eng. **22**, 1345–1359 (2010)
21. Peregrin-Alvarez, J., et al.: The modular organisation of protein interactions in escherichia coli. PLoS Comput. Biol. **5**(10), e1000523 (2009)
22. Prasad, T.S.K., et al.: Human protein reference database-2009 update. Nucleic Acids Res. **37**, D767–D772 (2009)
23. Prasad, T., et al.: The human protein reference database-2009 update. Nucleic Acids Res. **37**, D767–D772 (2009)
24. Qian, J., Luscombe, N., Gerstein, M.: Protein family and fold occurrence in genomes: powerlaw behaviour and evolutionary model. J. Mol. Biol. **313**, 673–689 (2001)
25. Radivojac, P., Peng, K., Clark, W., et al.: An integrated approach to inferring gene-disease associations in humans. Proteins **72**(3), 1030–1037 (2008)
26. Sahraeian, S., Yoon, B.J.: A network synthesis model for generating protein interaction network families. PLoS ONE **7**(8), e41474 (2012)
27. Saraph, V., Milenkovi, T.: Magna: maximizing accuracy in global network alignment. Bioinformatics **30**(20), 2931–2940 (2014). http://bioinformatics.oxfordjournals.org/content/30/20/2931.abstract
28. Schlosser, G., Wagner, G.: Modularity in Development and Evolution. University of Chicago Press, Chicago (2004)
29. Sole, R., Pastor-Satorras, R., Smith, E., Kepler, T.: A model of large-scale proteome evolution. Adv. Complex Syst. **5**, 43–54 (2002)
30. Sole, R., Valverde, S.: Spontaneous emergence of modularity in cellular networks. J. R. Soc. Interface **5**(18), 129–133 (2008)
31. Szklarczyk, D., et al.: String v10: protein-protein interaction networks, integrated over the tree of life. Nucleic Acids Res. **43**, D447–D452 (2015)
32. Vazquez, A., Flammini, A., Maritan, A., Vespignani, A.: Global protein function prediction from protein-protein interaction networks. Nat. Biotech. **21**(6), 697–700 (2003)

33. Wagner, A.: The yeast protein interaction network evolves rapidly and contains few redundant duplicate genes. Mol. Biol. Evol. **18**, 1283–1292 (2001)
34. Wasserman, S., Faust, K.: Social Network Analysis: Methods and Applications. Cambridge University Press, UK (1994)
35. Zhang, X., Moret, B.: Refining transcriptional regulatory networks using network evolutionary models and gene histories. Algorithms Mol. Biol. **5**(1), 1 (2010)
36. Zhang, X., Moret, B.: Refining regulatory networks through phylogenetic transfer of information. ACM/IEEE Trans. Comput. Biol. Bioinf. **9**(4), 1032–1045 (2012)

Multi-genome Scaffold Co-assembly Based on the Analysis of Gene Orders and Genomic Repeats

Sergey Aganezov[1,2] and Max A. Alekseyev[1(✉)]

[1] The George Washington University, Washington, DC, USA
{maxal,aganezov}@gwu.edu
[2] ITMO University, St. Petersburg, Russia

Abstract. Advances in the DNA sequencing technology over the past decades have increased the volume of raw sequenced genomic data available for further assembly and analysis. While there exist many software tools for assembly of sequenced genomic material, they often experience difficulties with reconstructing complete chromosomes. Major obstacles include uneven read coverage and long similar subsequences (repeats) in genomes. Assemblers therefore often are able to reliably reconstruct only long subsequences, called scaffolds.

We present a method for simultaneous co-assembly of all fragmented genomes (represented as collections of scaffolds rather than chromosomes) in a given set of annotated genomes. The method is based on the analysis of gene orders and relies on the evolutionary model, which includes genome rearrangements as well as gene insertions and deletions. It can also utilize information about genomic repeats and the phylogenetic tree of the given genomes, further improving their assembly quality.

Keywords: Genome assembly · Scaffolding · Gene order

1 Introduction

Genome sequencing technology has evolved over time, increasing availability of sequenced genomic data. Modern sequencers are able to identify only short subsequences (*reads*) in the supplied genomic material, which then become an input to genome assembly algorithms aimed at reconstruction of the complete genome. Such reconstruction is possible (but not guaranteed) only if each genomic region is covered by sufficiently many reads. Lack of comprehensive coverage (particularly severe in single-cell sequencing), presence of long similar subsequences (*repeats*) and polymorphic polyploid genomic structure pose major obstacles for existing assembly algorithms. They often are able to reliably reconstruct only

---

The work is supported by the National Science Foundation under Grant No. IIS-1462107.

A. Bourgeois et al. (Eds.): ISBRA 2016, LNBI 9683, pp. 237–249, 2016.
DOI: 10.1007/978-3-319-38782-6_20

long subsequences of the genome (interspersed with low-coverage, highly polymorphic, or repetitive regions), called *scaffolds*.

The challenge of reconstructing a complete genomic sequence from scaffolds is known as the *scaffolds assembly (scaffolding)* problem. It is often addressed technologically by generating so-called long-jump libraries or by using a related complete genome as a reference. Unfortunately, the technological solution may be expensive and inaccurate [9], while the reference-based approach is obfuscated with structural variations across the genomes [7].

In the current study, we assume that constructed scaffolds are accurate and long enough to allow identification of homologous genes. The scaffolds then can be represented as ordered sequences of genes and we pose the scaffolds assembly problem as the reconstruction of the global gene order (along genome chromosomes) by identifying pairs of scaffolds extremities (*assembly points*) to be glued together. We view such gene sub-orders as the result of both evolutionary events and technological fragmentation in the genome. In the course of evolution, gene orders are changed by *genome rearrangements* (including *reversals*, *fusions*, *fissions*, and *translocations*) as well as by gene *insertions* and *deletions* commonly called *indels*. Technological fragmentation can be modeled by artificial "fissions" that break genomic chromosomes into scaffolds. This observation inspires us to employ the genome rearrangement analysis techniques for scaffolding purposes.

Earlier we proposed to address the scaffolding problem with a novel method [1] based on the comparative analysis of multiple related genomes, some or even all of which may be fragmented. The core of the proposed approach can be viewed as a generalization of the reference-based assembly to the case of multiple reference genomes (possibly of different species), which themselves may be fragmented. In the current work we extend our initial method in a number of ways. First, we lift the previously imposed restriction of having a *uniform* gene content across the given genomes (i.e., the requirement for each gene to present exactly once in each of the genomes), thus enabling analysis of gene orders at a higher resolution. Second, our new method can utilize and greatly benefit from the information (when available) about (i) a phylogenetic tree of the given genomes, and (ii) flanking genomic repeats at scaffolds ends. Furthermore, our new method can be integrated with the ancestral genome reconstruction software MGRA2 [5], effectively incorporating its multi-genome rearrangement and gene indel analysis into the framework.

We evaluate performance of our method on both simulated and real genomic datasets. First, we evaluate it on randomly fragmented mammalian genomes, representing the case of no prior knowledge about the fragmentation nature. These experiments demonstrate that the method is robust with respect to the number of fragments and produces assembly of high quality. Second, we run our method on mammalian genomes artificially fragmented at the positions of long repeats, thus simulating genomes obtained from a conventional genome assembler. Most of the reported assembly points in this case are correct, implying that our method can be used as a reliable step in improving existing incomplete assemblies. We also compare the performance of our new method with its initial

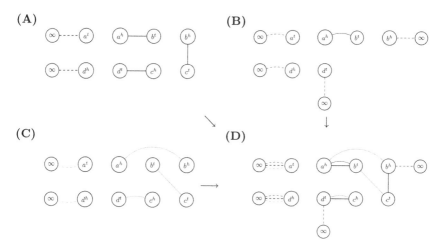

Fig. 1. (A) Genome graph GG(A) of genome $A = [a, b, c, d]$ colored black. **(B)** Genome graph GG(B) of genome $B = [a, b][d]$ colored purple. **(C)** Genome graph GG(C) of genome $C = [a, -b, c, d]$ colored green. **(D)** The breakpoint graph BG(A, B, C). Regular and irregular (multi)edges are shown as solid and dashed, respectively. Irregular vertex is shown in multiple copies representing endpoints of different irregular (multi)edges. (Color figure online)

version [1] on fragments with uniform gene content. These experiments demonstrated that while the new method outperforms its initial version on genomes with a uniform gene content, the assembly quality further increases when nonuniform genes are taken into account. Third, we evaluate our method on real incomplete *Anopheles* mosquito genomes without a complete reference genome. Comparison of the resulting assembly of *A. funestus* genome to the one obtained by [4] reveals that our method is able to assemble a larger number of scaffolds.

2   Background

Our method employs the breakpoint graph of multiple genomes [2] to represent their gene orders. We start with describing the graph representation for a single genome, which may consist one or more linear/circular chromosomes and/or scaffolds, which we commonly refer to as *fragments*. A circular fragment with n genes is represented as an undirected graph on $2 \cdot n$ *regular* vertices that encode gene extremities (namely, each gene x corresponds to a pair of vertices x^t and x^h representing its "tail" and "head" extremities). For linear fragments, we introduce one additional *irregular* vertex labeled ∞, which encodes the fragment ends (i.e., telomeres if the fragment represents a chromosome). Undirected edges in this graph connect pairs of vertices encoding genes/fragment extremities that are adjacent on the fragment. An edge is called *irregular* if it is incident to an irregular vertex; otherwise it is called *regular*. The *genome graph* GG(P) of a genome P is formed by the union of the graphs representing all fragments of P.

For a set of k genomes $\{G_1, G_2, \ldots, G_k\}$, we construct their individual genome graphs $GG(G_1)$, $GG(G_2)$, ..., $GG(G_k)$ and assume that edges in each graph $GG(G_i)$ are colored in a unique color, which we refer to as the color G_i. Edges colored into the color G_i are called G_i-edges. The *breakpoint graph* $BG(G_1, G_2, \ldots, G_k)$ is defined as the superposition of the colored genome graphs $GG(G_1)$, $GG(G_2)$, ..., $GG(G_k)$ (Fig. 1). Alternatively, it can be obtained by "gluing" the identically labeled vertices in these genome graphs.

All edges that connect the same pair of vertices u and v in the breakpoint graph $BG(G_1, G_2, \ldots, G_k)$ form a *multiedge* $[u, v]$, whose *multicolor* is defined as the set of individual colors of corresponding edges (e.g., in Fig. 1D vertices a^h and b^t are connected by a multiedge of the purple–black multicolor). A multicolor can be viewed as a subset of the set of all colors $\mathbb{G} = \{G_1, G_2, \ldots, G_k\}$, which we refer to as the *complete multicolor*. Multiedges of multicolor $M \subseteq \mathbb{G}$ are called M-*multiedges*.

We find it convenient to view irregular multiedges of the breakpoint graph as each having a separate irregular vertex (rather than sharing a single irregular vertex) as an endpoint. Relatedly, we consider connected components in the breakpoint graph $BG(G_1, G_2, \ldots, G_k)$ with respect to these disjoint representation of irregular vertices (i.e., no connection between regular vertices happen through irregular ones).

Let T be the *evolutionary tree* of a set of genomes $\mathbb{G} = \{G_1, G_2, \ldots, G_k\}$, i.e., the leaves of T represent individual genomes from $\mathbb{G}$, while its internal nodes represent their ancestral genomes. We label every internal node in T with the multicolor formed by all leaves that are descendants of this vertex and call such multicolor $\vec{T}$-*consistent* (Fig. 2). Evolutionary events that happen along the branch (U, V) in T, where $V \subset U$, affect all genomes in the set V and can be modeled as operating on V-multiedges in $BG(G_1, G_2, \ldots, G_k)$ [2].

3   Scaffolding Algorithm

Our scaffolding method, initially described in [1], takes as an input a set of genomes $\mathbb{G} = \{G_1, G_2, \ldots, G_k\}$, some or even all of which may be fragmented, where genome chromosomes or fragments are represented as ordered sequences of homologous genes. We allow specification of a subset $\mathbb{G}_T \subseteq \mathbb{G}$ of genomes targeted for assembly; by default $\mathbb{G}_T = \mathbb{G}$. The method assembles genomes from $\mathbb{G}_T$ based on the orders of homologous genes along chromosomes and/or fragments in the input set of genomes $\mathbb{G}$.

In the current work, we present an extension of the initial method by not only utilizing additional information (when available), but also by introducing new processing stages, which improve the assembly quality. In particular, we integrate the rearrangement and gene indel analysis from MGRA2 [5] into in our method framework and lift the previously imposed limitation on the input data and allow non-uniform gene content, i.e., all homologous genes may now be present *at most* once in each of the input genomes.

The method operates on the breakpoint graph $BG(G_1, G_2, \ldots, G_k)$ of the input genomes. We remark that a missing link between two fragments in a

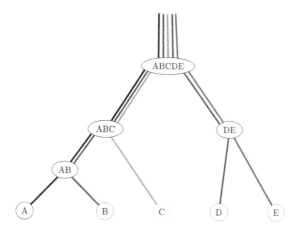

Fig. 2. A phylogenetic tree T of genomes A, B, C, D, and E represented as the leaves of T and colored in black, purple, green, blue, and red colors, respectively. The internal nodes of T represent various common ancestors of these genomes. The branches of T are represented by parallel edges of the colors of descendant genomes. (Color figure online)

fragmented genome G_i can be viewed as the absence of a G_i-edge (u, v) in $BG(G_1, G_2, \ldots, G_k)$, where u and v represent the extremities of the outermost genes at the fragments ends. Namely, instead of the edge (u, v) we observe a pair of irregular G_i-edges (u, ∞) and (v, ∞). Thus, assembling a pair of fragments in the genome G_i can be posed as finding a suitable pair of irregular G_i-edges (u, ∞) and (v, ∞) and replacing them with a G_i-edge (u, v).

Since genomic regions (e.g., repeats) that are problematic for existing genome assemblers may be present in multiple genomes in the $\mathbb{G}_T$ due to the inheritance from their common ancestor, we find it beneficial for our new method to perform simultaneous assembly of multiple given genomes. Namely, we identify all T-consistent multicolors $MC = \{mc_1, mc_2, \ldots, mc_n\}$ $(mc_i \subseteq \mathbb{G}_T)$ and attempt to co-assemble genomes in each mc_i $(i = 1, 2, \ldots, n)$.[1] To accomplish this task, we generalize the search for pairs of irregular edges of the same color to the search for pairs of irregular multiedges both containing some $mc_i \in MC$ as a subset (called mc_i^+-*multiedges*).

We iterate over the elements $tm \in MC$ in the evolutionary order, starting with most ancient multicolors (ancestral genomes) and ending with singleton multicolors (genomes) from $\mathbb{G}_T$.

3.1   Connected Components

Earlier we showed [1] that connected components in the breakpoint graph of multiple genomes are robust with respect to the fragmentation in the genomes

---

[1] Since each singleton multicolor mc (i.e., $|mc| = 1$) is T-consistent, our new method involves independent assembly of single genomes as a particular case.

in the case of uniform gene content and genome rearrangements as the only evolutionary events. The case of non-uniform gene content and evolutionary events including gene indels in addition to genome rearrangements can be reduced to the case of uniform gene content, where gene indels are modeled as genome rearrangements, as it was illustrated [5]. Namely, if gene g is missing in genome $G_i \in \mathbb{G}$, then there are no G_i-edges incident to vertices g^t and g^h in $BG(G_1, G_2, \ldots, G_k)$. Following the MGRA2 approach, we add such *prosthetic* G_i-edges (g^t, g^h) for each gene g and genome G_i, making the breakpoint graph $BG(G_1, G_2, \ldots, G_k)$ *balanced* (i.e., each vertex becomes incident to edges of all k colors). Gene indels are then modeled as genome rearrangements operating on prosthetic (multi)edges, which allow us to apply heuristics we developed earlier [1].

A particularly powerful heuristics relies on the robustness of connected components and performs assembly independently in each connected component cc of $BG(G_1, G_2, \ldots, G_k)$. In particular, if cc contains only a single pair of irregular tm^+-multiedges, we perform assembly on these multiedges. Below we describe new approaches that can process connected components with more than two irregular tm^+-multiedges.

3.2   Integration with MGRA2

For given genomes $G_1, G_2, \ldots, G_k$ and their evolutionary tree T, MGRA2 [5] reconstructs the gene order in ancestral genomes at the internal nodes of T. Namely, MGRA2 recovers genome rearrangements and gene indels along the branches of T and transforms the breakpoint graph $BG(G_1, G_2, \ldots, G_k)$ into $BG(X, X, \ldots, X)$ of some single genome X. We remark that in the breakpoint graph $BG(X, X, \ldots, X)$ all multiedges have multicolor $\mathbb{G}$ and all connected components consist of either 1 or 2 regular vertices.

Our method benefits from integration with MGRA2, which we use to reduce the size of connected components in the breakpoint graph. Namely, we run most reliable (non-heuristic) stages of MGRA2 to identify evolutionary genome rearrangements and gene indels and transform the breakpoint graph of input genomes closer to the breakpoint graph of single ancestral genome. In the course of this transformation, large connected components can only be broken into smaller ones, thus narrowing the search for assembly points.

3.3   Evolutionary Scoring

In this section, we describe a new stage in our algorithm that can identify pairs of tm^+-multiedges resulted from fragmentation in a (large) connected component of the breakpoint graph. We start with the description of the relationship of the multicolors structure and the topology of T. Each multicolor mc can be uniquely partitioned into the smallest number $\vec{T}$-consistent multicolors [5]. We denote the cardinality of such minimal partition of mc as $ps(mc)$.

For a mc-multiedge $e = [u, v]$ in $BG(G_1, G_2, \ldots, G_k)$, we consider two cases depending on whether $\mathbb{G}$-multiedge $[u, v]$ is present in $BG(X, X, \ldots, X)$:

$+e$: $\mathbb{G}$-multiedge $[u, v]$ is present in $BG(X, X, \ldots, X)$, which then in the course of evolution loses colors from the set $\mathbb{G}\backslash mc$;

$-e$: $\mathbb{G}$-multiedge $[u, v]$ is absent in $BG(X, X, \ldots, X)$ and the mc-multiedge $[u, v]$ is then created in the course of evolution by gaining colors from mc.

We define the *evolutionary scores* $es(+e)$ and $es(-e)$ as the smallest number of rearrangements required to create an mc-multiedge in $BG(G_1, G_2, \ldots, G_k)$ in the course of evolution under the assumption that the $\mathbb{G}$-multiedge $[u, v]$ is present or absent in $BG(X, X, \ldots, X)$, respectively. We claim that they can be computed as $es(+e) = d(\mathbb{G}\backslash mc)$ and $es(-e) = d(mc)$, where

$$d(mc) = \min_{c \subsetneq (\mathbb{G}\backslash mc)} ps(mc \cup c) + ps(c).$$

Candidate Assembly Points Identification. The new stage of our algorithm searches for regular multiedges $sm = [u, v]$ (*support multiedges*) that are incident to a pair of irregular tm^+-multiedges $im_1 = [u, \infty]$ and $im_2 = [v, \infty]$ of multicolors ic_1 and ic_2, respectively. For each such triple (sm, im_1, im_2), we estimate the likeliness for the irregular multiedges to arise from fragmentation rather than in the course of evolution. Namely, we compare two potential outcomes:

(O1) im_1 and im_2 are created by evolution (Fig. 3A), in which case the multi-edges remain intact;

(O2) im_1 and im_2 arose from fragmentation, in which case we perform their assembly in tm genomes (Fig. 3B). As a result of this assembly the multiedges sm, im_1, im_2 change their multicolors to $sm' = sm \cup tm$, $ic_1' = ic_1 \backslash tm$, and $ic_2' = ic_2 \backslash tm$, respectively.

In both these outcomes, we consider the following five mutually exclusive evolutionary scenarios (depending on presence or absence of the multiedges sm, im_1, im_2 in $BG(X, X, \ldots, X)$): $s_1 = \{-sm, -im_1, -im_2\}$; $s_2 = \{+sm, -im_1, -im_2\}$; $s_3 = \{-sm, -im_1, +im_2\}$; $s_4 = \{-sm, +im_1, -im_2\}$; $s_5 = \{-sm, +im_1, +im_2\}$. For each evolutionary scenario s_i, we compute the evolutionary score $es(s_i) = \sum_{m \in s_i} es(m)$ and let Δ_i be the difference between $es(s_i)$ in (O2) and $es(s_i)$ in (O1). We compute $\Delta = \min_i \Delta_i$ and if $\Delta > 1$,[2] we consider $\{im_1, im_2\}$ as a *candidate assembly point*.

Selection of Assembly Points. Since candidate assembly points may share an irregular multiedge, we need to choose a pairwise disjoint subset of them. To do so, we construct a candidate assembly graph CAG as the superposition of the corresponding support multiedges sm with the weight $w(sm) = \Delta$. We obtain a pairwise disjoint set of candidate assembly points by computing a maximum matching in CAG, and perform the actual scaffold assembly in this set.

---

[2] The value of $\Delta = 1$ corresponds to a typical fusion. We consider potential assembly only if it could achieve a better gain in the evolutionary score than a fusion.

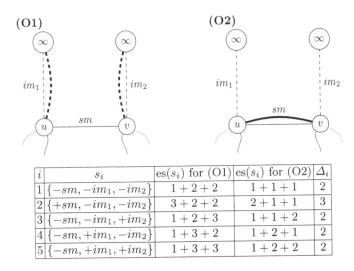

i	s_i	es(s_i) for (O1)	es(s_i) for (O2)	Δ_i
1	$\{-sm, -im_1, -im_2\}$	$1 + 2 + 2$	$1 + 1 + 1$	2
2	$\{+sm, -im_1, -im_2\}$	$3 + 2 + 2$	$2 + 1 + 1$	3
3	$\{-sm, -im_1, +im_2\}$	$1 + 2 + 3$	$1 + 1 + 2$	2
4	$\{-sm, +im_1, -im_2\}$	$1 + 3 + 2$	$1 + 2 + 1$	2
5	$\{-sm, +im_1, +im_2\}$	$1 + 3 + 3$	$1 + 2 + 2$	2

Fig. 3. Examples of potential outcomes (O1) and (O2), where the target multicolor is $tm = \{black\}$ (shown in bold). Outcome (O1) corresponds to evolutionary origin of im_1 and im_2, while outcome (O2) corresponds to assembly of irregular tm-multiedges. The table gives evolutionary scores for different evolutionary scenarios for (O1) and (O2) and corresponding values Δ_i, implying that $\Delta = \min_i \Delta_i = 2$.

3.4 Flanking DNA Repeats

Long DNA repeats represent one of the major obstacles for typical genome assemblers, which often cannot reconstruct the order of fragments interspersed with repeats. As a result they produce scaffolds ending at positions of (i.e., flanked by) repeats.

We introduce a new stage in our algorithm, which incorporates the information about flanking DNA repeats into the processing pipeline. Namely, we use flanking repeats (and their orientation) to label corresponding irregular edges in $BG(G_1, G_2, \ldots, G_k)$ as follows. If a repeat p appears at the end of a fragment after a gene c, we label the edge (c^h, ∞) with p^t; similarly, if repeat $-p$ (i.e., reverse complement of p) appears after a gene z, we label the edge (z^h, ∞) with p^h; etc. (Fig. 4A, D).

If two consecutive fragments f_1 and f_2 in a complete genome are interspersed with a copy of some repeat p, the corresponding irregular edges in these fragments should have labels p^t and p^h. The new stage of our algorithm uses such matching labels to allow or disallow assembly of irregular edges (Fig. 4B, E).

We also address the case when two fragments with annotated genes are interspersed with repeats and fragments without annotated genes. We view a fragment with no genes and flanked by repeats p and q as a *bridge* allowing assembly of irregular edges labeled p^t and q^h (Fig. 4C, F). This convention naturally extends to the case when there exist multiple bridges in between of two fragments with annotated genes.

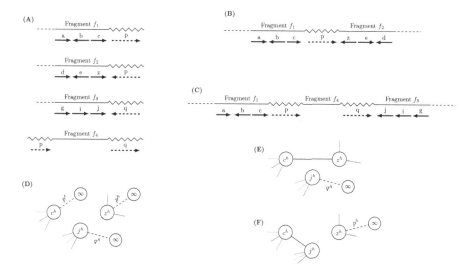

Fig. 4. (A): Extremities of several fragments in genome $G_1 \in \mathbb{G}_T$ flanked by copies of repeats p and q, where annotated genes are represented by labeled solid arrows. **(B):** Assembly of fragments f_1 and f_2, which are interspersed with a copy of the repeat p. **(C):** Assembly of fragments f_2 and f_3, which is permitted by bridge f_4. **(D):** The subgraph of $BG(G_1, G_2, \ldots, G_k)$ containing irregular G_1-edges that correspond to the extremities of fragments f_1, f_2, and f_3. Irregular edges are labeled by corresponding flanking repeats. **(E):** Assembly of irregular G_1-edges corresponding to the assembly in (B). **(F):** Assembly of irregular G_1-edges corresponding to the assembly in (C).

4   Evaluation

4.1   Artificially Fragmented Genomes

We start evaluation of our method by running it on artificially fragmented mammalian genomes. We used a set of seven mammalian genomes: (Hs) *Homo sapiens* (hg38), (Pt) *Pan troglodytes* (panTor2.1), (Mm) *Mus musculus* (mm38), (Rn) *Rattus norvegicus* (rn6.0), (Cf) *Canis familiaris* (canFam3.1), (Md) *Monodelphis demestica* (monDom5), and (Fc) *Felis catus* (felCat6.2). We obtained the information about homologous genes between each pair of these genomes from Ensembl BioMart [10] and constructed multi-genome homologous genes families from this data. We filtered out all copies of the genes that are present more than once in at least one genome. A phylogenetic tree for these genomes is given in [13].

We use two different approaches for artificial fragmentation of complete genomes: random and repeat-based fragmentation. Random fragmentation allows us to evaluate our method in the case when we have no insight about the nature of fragmentation, while repeat-based fragmentation allows us to simulate the case when fragmented genomes are constructed by a conventional genome assembler having difficulties with long repeats. Since our method

Table 1. Averaged assembly quality for genomes in $\mathbb{G} = \{Hs, Pt, Fc, Cf, Mm, Rn, Md\}$ all of which are randomly fragmented at k positions.

k	100	200	300	400	500	600	700	800	900	1000
TP	89.3%	90.6%	90%	88.6%	88.1%	87.3%	86.8%	86%	85.4%	84.7%
SP	1.4%	1.2%	1.1%	1.5%	1.6%	2.3%	2.9%	3.4%	4.7%	5.5%
FP	2.4%	2.2%	2.6%	2.8%	3%	3.3%	3.7%	3.8%	3.8%	4%

identifies assemblies only between fragments with annotated genes, we compare the reported assemblies to the complete genomes where the fragments without genes are omitted.

We report the following metrics for the resulting assemblies:

1. **True Positive** (*TP*), the percentage of correctly identified assembly points, i.e., pairs of assembled fragments that have correct orientation and are adjacent in the complete genome;
2. **Semi Positive** (*SP*), the percentage of semi-correctly identified assembly points, i.e., pairs of assembled fragments that appear in the same order and orientation on some chromosome in the complete genome but are interspersed with other fragments;
3. **False Positive** (*FP*), the percentage of incorrectly identified assembly points, i.e., pairs of assembled fragments that have wrong relative orientation or come from different chromosomes in the complete genome.

Random Fragmentation. We randomly fragment each mammalian genome in $k \in \{100, 200, \ldots, 1000\}$ locations. For each value k, we create 20 different sets of fragmented genomes, execute our method on each of the sets, and assess the quality of the produced assembly (averaged over the 20 sets).

The assembly results in Table 1 demonstrate the high quality, which is slightly better than the assembly quality of the initial version of our method and similar to that reported in [3].

Repeat-Based Fragmentation. Locations of DNA repeats in the observed mammalian genomes were obtained from RepeatMasker [13] database. We performed a number of experiments with different values of the repeat threshold RT, where repeats longer than RT specify breakpoints for artificial fragmentation. The resulting fragments and the information about their flanking repeats were provided as an input to our method.

First we compared the performance of our new method to its initial version on genomes with uniform gene content and then evaluated how the assembly quality changes when non-uniform genes are taken into account. For this experiment we fragmented genome *Cf* with repeats longer than $RT = 400$ and then executed both versions of our method on the set of genomes $\mathbb{G} = \{Cf, Hs, Pt, Mm, Rn, Md\}^3$ targeting $\mathbb{G}_T = \{Cf\}$ for assembly. The results in

---

[3] Genome *Fc* is omitted to simulate the case when no closely related reference genome is available.

Table 2. Assembly quality for genome Cf fragmented at repeats longer than $RT = 400$; $\mathbb{G} = \{Cf, Hs, Pt, Mm, Rn, Md\}$.

	Initial version			New version		
	TP	SP	FP	TP	SP	FP
uniform gene content	18.23 %	6.02 %	5.2 %	26.9 %	13.5 %	8.3 %
non-inform gene content	-	-	-	30.8 %	9.8 %	5.1 %

Table 3. Assembly quality for genome Cf fragmented at repeats longer than $RT = 2000$ without and with the information about flanking repeats; $\mathbb{G} = \{Cf, Hs, Pt, Mm, Rn, Md\}$.

	TP	SP	FP
without repeats info	21 %	5.1 %	5 %
with repeats info	22.7 %	5.9 %	3 %

Table 2 show that while the new method outperforms its initial version on genomes with uniform gene content, the assembly quality further increases when non-uniform genes are taken into account.

Then we evaluated how the assembly quality changes with respect to utilization of the information about flanking repeats. For this task we fragmented genome Cf at repeats longer than $RT = 2000$ and provided the same set of genomes $\mathbb{G}$ as an input for our method. Table 3 shows that utilization of flanking repeats increases both TP and SP, while decreasing the FP values for the produced assembly.

We also simulated a scenario when all input genomes are fragmented, but only one of them is targeted for assembly. To do so we took $\mathbb{G} = \{Fc, Cf, Pt, Mm, Rn, Md\}$, fragmented all these genomes at repeats longer than $RT = 400$, and targeted $\mathbb{G}_T = \{Pt\}$ for assembly. Table 4(A) demonstrates that even when there are no complete genomes available our method is still able to identify a large number of correct assembly points and make small number of errors.

Last but not least, we evaluated our method in the case of multi-genome co-assembly, i.e., when multiple input genomes are targeted for assembly. Namely, we used all seven mammalian genomes as an input, among which we fragmented genomes Fc and Cf (i.e., the carnivore representatives) at repeats longer than $RT = 400$ and targeted both of these genomes for assembly. As shown in Table 4(B), multi-genome co-assembly yields additional assembly points as compared to single-genome assembly of Cf shown in Table 2.

4.2   Incomplete Genomes

To evaluate our method on real incomplete genomes (represented as collections of scaffolds), we have obtained from VectorBase [11] the current (incomplete) assembly and homologous gene maps of the following seven Anopheles

Table 4. (A): Assembly quality for genome Pt from $\mathbb{G} = \{Fc, Cf, Pt, Mm, Rn, Md\}$, where all genomes are fragmented at repeats longer than $RT = 400$. **(B)**: Assembly quality for genomes Fc and Cf both fragmented at repeats longer than $RT = 400$; $\mathbb{G} = \{Cf, Fc, Hs, Pt, Mm, Rn, Md\}$.

(A)

Genome	TP	SP	FP
Pt	21.2%	3.3%	1.3%

(B)

Genome	TP	SP	FP
Fc	31%	5.3%	5%
Cf	32.3%	10.4%	6.5%

Table 5. Assembly results for *A. funestus*, *A. arabiensis*, *A. minimus* genomes. For each genome, AF and AAF denote the percentage of the fragments involved in the produced assemblies among all of the input fragments and the input fragments with annotated genes, respectively.

Genome	# Assembly points	AF	AAF
A. arabiensis	51	14.3 %	24.8 %
A. funestus	250	54.34 %	64.5 %
A. minimus	15	10.3 %	19.7 %

mosquito genomes: *A. funestus*, *A. gambiae* PEST, *A. arabiensis*, *A. minimus*, *A. albimanus*, *A. dirus*, and *A. atropavarus*. Their evolutionary tree is given in [12]. We targeted our new method for assembly of genomes $\mathbb{G}_T = \{A.\ funestus, A.\ arabiensis, A.\ minimus\}$.

Table 5 shows the number of assembly points identified by our method in each of the genomes in $\mathbb{G}_T$. We remark that it was able to identify 250 assembly points (involving 338 scaffolds, which represent 64.5 % of all scaffolds containing at least one gene) in *A. funestus* (for comparison, the assembly of *A. funestus* reported in [4] uses 51.8 % of scaffolds).

In order to evaluate our method on incomplete genomes with flanking repeats, we inquired about obtaining this information from genome assemblers ALLPATHS-LG [8] (used for assembly of Anopheles genomes) and SPADES [6]. While the developers generally admit a possibility of providing such information along with the assembled scaffolds, this feature is not readily available yet.

We find it promising that the information about flanking repeats is supported by the recently emerged *Graphical Fragment Assembly* (GFA) format [14]. While none of the major genome assemblers fully supports this format yet, it becomes more and more popular within the genome assembly community and we anticipate that the flanking repeat data may become available in near future.

References

1. Aganezov, S., Sydtnikova, N., AGC Consortium, Alekseyev, M.A.: Scaffold assembly based on genome rearrangement analysis. Comput. Biol. Chem. **57**, pp. 46–53 (2015)
2. Alekseyev, M.A., Pevzner, P.A.: Breakpoint graphs and ancestral genome reconstructions. Genome Res. **19**(5), 943–957 (2009)

3. Anselmetti, Y., Berry, V., Chauve, C., Chateau, A., Tannier, E., Bérard, S.: Ancestral gene synteny reconstruction improves extant species scaffolding. BMC Genomics **16**(Suppl. 10), S11 (2015)

4. Assour, L., Emrich, S.: Multi-genome synteny for assembly improvement. In: Proceedings of 7th International Conference on Bioinformatics and Computational Biology, pp. 193–199 (2015)

5. Avdeyev, P., Jiang, S., Aganezov, S., Hu, F., Alekseyev, M.A.: Reconstruction of ancestral genomes in presence of gene gain and loss. J. Comput. Biol. **23**(3), 1–15 (2016)

6. Bankevich, A., Nurk, S., Antipov, D., Gurevich, A.A., Dvorkin, M., Kulikov, A.S., Lesin, V.M., Nikolenko, S.I., Pham, S., Prjibelski, A.D., et al.: SPAdes: a new genome assembly algorithm and its applications to single-cell sequencing. J. Comput. Biol. **19**(5), 455–477 (2012)

7. Feuk, L., Carson, A.R., Scherer, S.W.: Structural variation in the human genome. Nat. Rev. Genet. **7**(2), 85–97 (2006)

8. Gnerre, S., MacCallum, I., Przybylski, D., Ribeiro, F.J., Burton, J.N., Walker, B.J., Sharpe, T., Hall, G., Shea, T.P., Sykes, S., et al.: High-quality draft assemblies of mammalian genomes from massively parallel sequence data. Proc. Natl. Acad. Sci. **108**(4), 1513–1518 (2011)

9. Hunt, M., Newbold, C., Berriman, M., Otto, T.D.: A comprehensive evaluation of assembly scaffolding tools. Genome Biol. **15**(3), R42 (2014)

10. Kasprzyk, A.: BioMart: driving a paradigm change in biological data management. Database **2011**, bar049 (2011)

11. Megy, K., Emrich, S.J., Lawson, D., Campbell, D., Dialynas, E., Hughes, D.S., Koscielny, G., Louis, C., MacCallum, R.M., Redmond, S.N., et al.: Vector-Base: improvements to a bioinformatics resource for invertebrate vector genomics. Nucleic Acids Res. **40**(D1), D729–D734 (2012)

12. Neafsey, D.E., Waterhouse, R.M., Abai, M.R., Aganezov, S.S., Alekseyev, M.A., et al.: Highly evolvable malaria vectors: the genomes of 16 Anopheles mosquitoes. Science **347**(6217), 1258522 (2015)

13. Smit, A., Hubley, R., Green, P.: RepeatMasker Open-3.0 (1996–2010). http://www.repeatmasker.org

14. The GFA Format Specification Working Group: Graphical Fragment Assembly (GFA) Format Specification. https://github.com/pmelsted/GFA-spec

Sequence and Image Analysis

Selectoscope: A Modern Web-App for Positive Selection Analysis of Genomic Data

Andrey V. Zaika[1]([⊠]), Iakov I. Davydov[3,4], and Mikhail S. Gelfand[1,2]

[1] A.A. Kharkevich Institute for Information Transmission Problems,
Russian Academy of Sciences, Moscow, Russia
anzaika+publications@gmail.com
[2] Faculty of Bioengineering and Bioinformatics,
Lomonosov Moscow State University, Moscow, Russia
[3] Department of Ecology and Evolution, University of Lausanne,
Lausanne, Switzerland
[4] Swiss Institute of Bioinformatics, Lausanne, Switzerland

Abstract. Selectoscope is a web application which combines a number of popular tools used to infer positive selection in an easy to use pipeline. A set of homologous DNA sequences to be analyzed and evaluated are submitted to the server by uploading protein-coding gene sequences in the FASTA format. The sequences are aligned and a phylogenetic tree is constructed. The `codeml` procedure from the `PAML` package is used first to adjust branch lengths and to find a starting point for the likelihood maximization, then `FastCodeML` is executed. Upon completion, branches and positions under positive selection are visualized simultaneously on the tree and alignment viewers. Run logs are accessible through the web interface. Selectoscope is based on the `Docker` virtualization technology. This makes the application easy to install with a negligible performance overhead. The application is highly scalable and can be used on a single PC or on a large high performance clusters. The source code is freely available at https://github.com/anzaika/selectoscope.

Keywords: Positive selection · Codeml · Fastcodeml

1 Introduction

Positive selection is the major force standing behind the innovation during the evolution. Methods based on the ratio of non-synonymous (dN) to synonymous (dS) mutations allow the detection of ancient positive selection in protein-coding genes. A number of methods have been developed over the years, they differ in their data partitioning approach, power, and computational performance [3–6] and many others.

Not only the method choice has a strong impact on the results, but also the intermediate stages of data processing. It has been shown that positive selection codon models are sensitive to the quality of multiple sequence alignments [7,8]

A. Bourgeois et al. (Eds.): ISBRA 2016, LNBI 9683, pp. 253–257, 2016.
DOI: 10.1007/978-3-319-38782-6_21

and the gene tree [9]. All this makes detection of positive selection on a genomic scale not only computationally intensive, but also difficult to perform properly.

We created a web application that does not require prior experience with tools for positive selection analysis, while allowing more experienced users to reconfigure parts of the pipeline without writing any code and to share easily their successful designs with colleagues.

The pipeline is based on the branch-site model of positive selection, as this model is widely used, highly sensitive, and straightforward to interpret [4]. We use a fast implementation of the method in `FastCodeML` [1].

2   Implementation

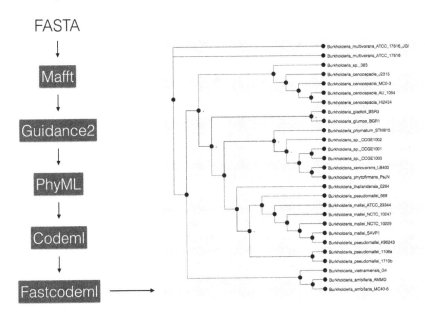

One starts with uploading FASTA files containing nucleotide sequences of orthologous protein-coding genes. The files can be uploaded one by one or in batches of up to hundreds.

The sequences are aligned with `mafft` [12], and low-quality regions are filtered out using `guidance2` [10].

The pipeline includes `phyml` [11] to construct multiple phylogenetic trees. By default a tree is constructed for every orthologous group independently, but it is also possible to create a single tree based on the concatenated alignment. This approach should provide a better tree in the case of homogeneous substitution rates over the genes and in the absense of horizontal gene transfers.

We use `codeml` with model M1a [13] to refine the branch lengths and estimate the transition to transversion ratio (κ).

`FastCodeML` uses the branch-site model [4] to infer branches and sites under positive selection. The phylogenetic tree estimated during the `codeml` run is used.

By default the branch lengths are not optimized during the `FastCodeML` for the sake of computational performance.

After the computations are finished, the phylogenetic tree is displayed. All branches, for which positive selection has been detected, are highlighted in the tree, and positions under positive selection are highlighted in the alignment viewer.

Our application uses pure JavaScript libraries provided by BioJS [14] project for displaying the alignments and phylogenetic trees. Thus it does not require any installation procedures on the user's computer.

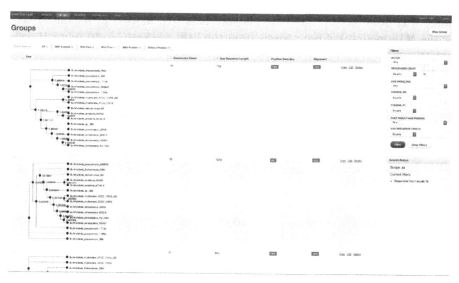

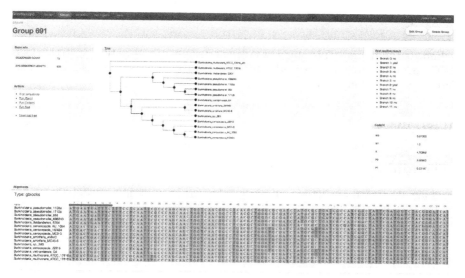

3   Tools

The core of the pipeline is `FastCodeML`, an extremely optimized software for detecting positive selection using Yang's branch-site model of positive selection. `FastCodeML` uses highly optimized matrix computation libraries (BLAS, LAPACK), supports multicore (OpenMP) and multihost (MPI) parallelization.

The sequence alignment is performed using `mafft`. Low-quality regions of the multiple sequence alignment are detected using `guidance2` and filtered out according to the threshold (0.93 by default).

To construct the phylogenetic tree we first use `phyml` with the `GTR+Gamma+I` model with four gamma rate site classes. This tree is used as an initial tree for the M1a model. We use `codeml` from the `PAML` package [2], to refine transition/transversion ratio (*kappa*) and branch lengths.

All of the application code runs in the Docker containers. Docker is a virtualisation technology that allows one to run a complete operating system within a container without sacrifices in the computational power as opposed to other virtualization technologies such as Virtualbox. This is achieved using resource isolation features of the Linux kernel - cgroups and kernel namespaces. This allows the processes running in the container to run on the host machine without being able to interact with host processes in any way. Docker provides the following advantages:

- Easy dependency management. No need to install or configure any part of the pipeline. Docker is available for all major operating system and its installation process is described in detail on its website [15].
- The container can be launched both as a web frontend server or as a worker that runs parts of the pipeline. This provides a considerable flexibility in launching the application on a single workstation or on a fleet of machines in a computational cluster.

Acknowledgements. This study was supported by the Scientific & Technological Cooperation Program Switzerland-Russia (RFBR grant 16-54-21004 and Swiss National Science Foundation project ZLRZ3_163872).

References

1. Valle, M., Schabauer, H., Pacher, C., Stockinger, H., Stamatakis, A., Robinson-Rechavi, M., Salamin, N.: Optimisation strategies for fast detection of positive selection on phylogenetic trees. Bioinformatics **30**(8), 1129–1137 (2014)
2. Yang, Z.: PAML 4: phylogenetic analysis by maximum likelihood. Mol. Biol. Evol. **24**(8), 1586–1591 (2007)
3. Yang, Z., Nielsen, R., Goldman, N., Pedersen, A.M.: Codon-substitution models for heterogeneous selection pressure at amino acid sites. Genetics **155**(1), 431–449 (2000)
4. Zhang, J., Nielsen, R., Yang, Z.: Evaluation of an improved branch-site likelihood method for detecting positive selection at the molecular level. Mol. Biol. Evol. **22**(12), 2472–2479 (2005). Epub 2005 Aug 17

5. Kosakovsky Pond, S.L., Murrell, B., Fourment, M., Frost, S.D., Delport, W., Scheffler, K.A.: A random effects branch-site model for detecting episodic diversifying selection. Mol. Biol. Evol. **28**(11), 3033–3043 (2011). doi:10.1093/molbev/msr125. Epub 2011 Jun 13

6. Murrell, B., Wertheim, J.O., Moola, S., Weighill, T., Scheffler, K., Kosakovsky Pond, S.L.: Detecting individual sites subject to episodic diversifying selection. PLoS Genet. **8**(7), e1002764 (2012). doi:10.1371/journal.pgen.1002764. Epub 2012 Jul 12

7. Redelings, B.: Erasing errors due to alignment ambiguity when estimating positive selection. Mol. Biol. Evol. **31**(8), 1979–1993 (2014). doi:10.1093/molbev/msu174. Epub 2014 May 27

8. Fletcher, W., Yang, Z.: The effect of insertions, deletions, and alignment errors on the branch-site test of positive selection. Mol. Biol. Evol. **27**(10), 2257–2267 (2010). doi:10.1093/molbev/msq115. Epub 2010 May 5

9. Diekmann, Y., Pereira-Leal, J.B.: Gene tree affects inference of sites under selection by the branch-site test of positive selection. Evol. Bioinform. **11**(Suppl. 2), 11–17 (2016). doi:10.4137/EBO.S30902. eCollection 2015

10. Sela, I., Ashkenazy, H., Katoh, K., Pupko, T.: GUIDANCE2: accurate detection of unreliable alignment regions accounting for the uncertainty of multiple parameters. Nucleic Acids Res. **43**(W1), W7–14 (2015). doi:10.1093/nar/gkv318. Epub 2015 Apr 16

11. Guindon, S., Dufayard, J.F., Lefort, V., Anisimova, M., Hordijk, W., Gascuel, O.: New algorithms and methods to estimate maximum-likelihood phylogenies: assessing the performance of PhyML 3.0. Syst. Biol. **59**(3), 307–321 (2010)

12. Katoh, K., Standley, D.M.: MAFFT multiple sequence alignment software version 7: improvements in performance and usability. Mol. Biol. Evol. **30**(4), 772–780 (2013). doi:10.1093/molbev/mst010. Epub 2013 Jan 16

13. Yang, Z., Nielsen, R., Goldman, N., Pedersen, A.M.: Codon-substitution models for heterogeneous selection pressure at amino acid sites. Genetics **155**(1), 431–449 (2000)

14. BioJS, the leading, open-source JavaScript visualization library for life sciences. https://www.biojs.net/

15. Docker installation guide. https://docs.docker.com/engine/installation/

Methods for Genome-Wide Analysis of MDR and XDR Tuberculosis from Belarus

Roman Sergeev[1(✉)], Ivan Kavaliou[1,3], Andrei Gabrielian[2],
Alex Rosenthal[2], and Alexander Tuzikov[1]

[1] United Institute of Informatics Problems NASB,
6, Surganov Str., 220012 Minsk, Belarus
roma.sergeev@gmail.com
[2] Office of Cyber Infrastructure and Computational Biology, National Institute
of Allergy and Infectious Diseases, NIH, Bethesda, MD, USA
[3] EPAM Systems, 1/1 Academician Kuprevich Str., 220141 Minsk, Belarus

Abstract. Emergence of drug-resistant microorganisms has been recognized as a serious threat to public health since the era of chemotherapy began. This problem is extensively discussed in the context of tuberculosis treatment. Alterations in pathogen genomes are among the main mechanisms by which microorganisms exhibit drug resistance. Analysis of the reported cases and discovery of new resistance-associated mutations may contribute greatly to the development of new drugs and effective therapy management. The proposed methodology allows identifying genetic changes and assessing their contribution to resistance phenotypes.

Keywords: Genome-wide association study · Multi drug-resistant tuberculosis · Genotype · Single nucleotide polymorphisms

1 Introduction

Rapid development of high-throughput sequencing technologies has changed dramatically the nature of biological researches and activated establishment of personalized medicine. However, making sense of massively generated genomic data can be a challenging task. This study aims at developing an approach to analyze *Mycobacterium tuberculosis* whole-genome sequences and identifying genomic markers of drug resistance that may become important for early diagnosis of tuberculosis and better understanding of biological foundations behind drug resistance mechanisms.

Although there has been substantial progress in tuberculosis (TB) control, drug-resistant tuberculosis is an important public health problem in Belarus and worldwide [1]. The situation has been complicated with emergence and development of multi drug-resistant (MDR) and extensively drug-resistant (XDR) TB that require long-term treatment. According to a Belarus nationwide survey in 2010–2011 MDR-TB was found in 32.3 % of new patients and 75.6 % of those previously treated for TB. XDR-TB was reported in 11.9 % of MDR-TB patients [2]. Belarus is still among countries having the highest multi-resistant tuberculosis incidence.

© Springer International Publishing Switzerland 2016
A. Bourgeois et al. (Eds.): ISBRA 2016, LNBI 9683, pp. 258–268, 2016.
DOI: 10.1007/978-3-319-38782-6_22

There are several lines of drugs used in tuberculosis therapy. First-line drugs are used in standard course of treatment for newly diagnosed TB cases or fully sensitive organisms. Treatment of resistant TB requires second-line drugs, which is usually more toxic and much more expensive.

The ability of TB agent to resist treatment is strongly connected with mutations in specific coding genes and intergenic regions (IGRs) of the bacteria genome. There has been a number of studies to address genomic aspects of drug resistance in *M. tuberculosis*. Early research projects were aimed at capturing mutations in a limited number of coding genes and IGRs, while the full spectrum of genetic variations remained unclear. Most recent studies [3–5] were focused on comparative genome-wide analysis of TB strains and have identified previously undescribed mutations with a role that is not completely clear.

Despite the diversity of the mutations investigated so far, they do not fully explain all cases of drug resistance observed. For example, analysis of TBDreamDB database [6] and GenoType MTBDRplus/MTBDRsl assays [7, 8] has shown that only 85.7 % of ofloxacin and 51.9 % of pyrazinamide resistance could be explained by the presented high-confidence mutations. For the second-line drugs (ethionamide, kanamycin, amikacin, capreomycin, cycloserine) these values are even lower. This indicates that the genetic basis of drug resistance is more complex than previously anticipated which encourages honing investigation of unexplained resistance in the experiments. Comprehensive analysis of mutations in MDR/XDR TB sequences may become very valuable for choosing an adequate treatment regimen and preventing therapy failure.

2 Strain Selection and Sequencing

To discover genetic markers of drug resistance we performed whole-genome sequencing of *M. tuberculosis* isolates obtained from patients in Belarus. Strain selection was performed to have 17.65 % of drug-sensitive, 10.29 % of MDR, 22.06 % of preXDR, 27.21 % of XDR and 19.85 % of totally drug-resistant (TDR) tuberculosis.

Two sequencing libraries were created to capture genetic variations for each sample: fragment library with 180 bp insert size and jumping library with ~ 3–5 kb insert size. Both libraries were sequenced on Illumina HiSeq2000 instruments to generate 101 bp paired-end reads at 140 x coverage of the genome. Data were assembled and aligned against H37Rv reference genome (GeneBank accession NC_018143.2) to identify and annotate variants, since H37Rv is the most studied drug-susceptible laboratory strain that retains its virulence in different animal models. Pilon tool [9] was used for variant calling to capture SNPs and indels. Among 144 assemblies submitted to GeneBank, we selected genomes with consistent lab data for further analysis. Finally, we performed a genome-wide association study (GWAS) for 132 annotated nucleotide sequences (contaminated sequences and strains with ambiguous resistance status were excluded) to check for statistically significant differences between drug-resistant (cases) and drug-susceptible (control) samples.

3 Data Analysis

Data analysis procedure comprised several steps. Based on phenotypic resistance status we grouped microbial genomes into datasets. Then we performed comparative sequence analysis and investigated population structure of TB strains. Next steps were aimed to uncover genome variations highly associated with phenotypic resistance to known drugs. We tried a number of methods based on different principles to analyze SNP data, including single- and multi-marker association tests. Lists of known associations were used to validate predictions.

3.1 Data Preparation and Filtering

Let $S_i = s_{i1}s_{i2}\ldots s_{il}$, $s_{ij} \in \{A, T, C, G, -\}$ be a set of input genomic sequences, such that A, T, G, C designate four nucleotides and "$-$" means that the element of the sequence isn't defined. We will consider any variation in a single nucleotide from reference genome S_0, which may occur at some specific position, as a single nucleotide polymorphism (SNP). For more compact representation, we will introduce matrix X of genotypes with elements $x_{ij} = 1$ if mutation occurred for sequence $i = 1$ to n and position $j = 1$ to m, or $x_{ij} = 0$ otherwise. We will code information on drug susceptibility testing (DST) results to some medicine in a binary vector Y, where $y_i = 1$ in case of drug-resistance recognized for the i th organism.

To reduce the number of parameters in the models we applied a series of filters to variant calling results. We filtered out long insertions and deletions to create pure SNP lists with acceptable quality scores. We then removed positions that mutated in less than 5 % of the organisms. Within the resulting alignment, we recorded SNP positions and merged identical columns in the corresponding matrix of genotypes. Consequently, we reduced the search space from approximately 50,000 sites of genetic variations to about 1,000 SNP positions with unique mutation profiles.

The filtered and cleaned data were grouped into datasets (Table 1) based on laboratory findings on drug-resistance status of the corresponding organisms (for example, within dataset 2 we examined genetic markers of drug resistance to ofloxacin assuming that all strains in the current dataset are sensitive to the second-line injectable drugs).

Table 1. Datasets formed from the original samples based on resistance status

Dataset	Target drug	Size	Conditions of strain selection
1	All 1st-line drugs and ofloxacin	132	All available
2	Ofloxacin	48	Susceptible to aminoglycosides
3	Aminoglycosides	54	Susceptible to ofloxacin
4	Ofloxacin	23	Susceptible to aminoglycosides, but resistant to rifampicin
5	Capreomycin	63	Susceptible to amikacin

(Continued)

Table 1. (*Continued*)

Dataset	Target drug	Size	Conditions of strain selection
6	Aminoglycosides	122	All available
7	Aminoglycosides	29	Susceptible to ofloxacin, but resistant to rifampicin
8	Non-aminoglycoside 2nd-line drugs	48	Susceptible to aminoglycosides
9	Non-aminoglycoside 2nd-line drugs	122	All available

3.2   Population Structure

We examined the population structure of the sequenced organisms to identify associations with well-known taxonomic groups. Methods of phylogenetic analysis, digital spoligotyping and principal component analysis were used to determine relatedness among strains.

Our genome-wide phylogenetic tree was built from all positions which contained at least one passing SNP call. Using RAxML package [10] we constructed maximum likelihood phylogenetic tree using the GTR model with *M. canettii* as an outgroup.

For digital spoligotyping reads were mapped against 43 spacer sequences and frequency was tallied. Background null model for the expected coverage was made from total sequencing data using an exponential distribution under the assumption that more than 90 % of reads align. The Benjamini-Hochberg correction was applied to p-values calculated for frequency and when significant ($p < 0.01$), the marker was reported as present.

Principal components analysis (PCA) was used to visualize the data and segregate strains for subsequent revision of statistical test outcomes. We used phylogeny and spoligotyping results to plot affiliations of *M. tuberculosis* samples with most frequent genotype lineages and principal genetic groups. We applied method implemented in EIGENSTRAT software [11] for running PCA and visualizing population stratification along the principal components (Fig. 1). The Kaiser criteria showed 28 principal components as significant for inferring axes of genetic variation from a set of 7,000 randomly selected SNPs for 132 tested organisms.

3.3   Single-Marker Drug-Association Analysis

Most single-marker methods rely on contingency table analysis comparing allele frequencies in sets of resistant and susceptible samples. Tables are constructed independently for any combination of tested drug and mutation. These allows using sufficient number of observations to calculate test statistics and estimate parameters. However, classical implementations of single-marker methods do not consider pairwise and higher-order interactions between genetic variants. We applied Cochran-Mantel-Haenszel (CMH) test implemented in PLINK package [12] and Pearson chi-squared statistics with EIGENSTRAT correction calculated using the method of principal

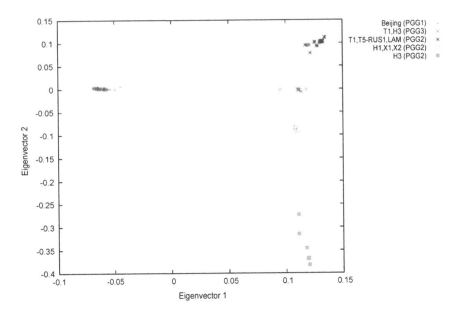

Fig. 1. Top two axes of variation of *M.tuberculosis* samples represented in Belarus

components [11]. This makes it possible to neutralize spurious correlations caused by strains segregation based on membership in genetically more homogeneous groups, which is known as founder effect. Calculated p-values were adjusted for multiple hypotheses testing using Benjamin-Hochberg correction.

3.4 Multi-marker Drug-Association Analysis

More sophisticated multi-marker methods can realistically model the multiplicity of genotypic factors, while bringing a number of other challenges associated with analyses of high-dimensional data in GWAS experiments: number of SNPs (parameters) is significantly greater than the number of sequences (observations). To overcome these difficulties we experimented with regularized logistic regression, linear mixed model (LMM) [13] and mode-oriented stochastic search (MOSS) [14].

According to the classical logistic regression model a posterior probability of drug-resistant phenotype is given by the formula $P\{y_i = 1|x_i\} = 1/(1 + \exp(-\beta^T x_i))$, where $y_i \in \{0, 1\}$ indicates the phenotype (susceptible/resistant) of the i th organism, x_i - genotype vector and β - vector of parameters which can be interpreted as significance of mutations for the development of drug resistance. Elastic net showed most relevant results as a regularization method encouraging a grouping effect, when strongly correlated predictors tend to occur together in a produced sparse model. The elastic net solves an optimization problem $\sum_{i=1}^{n} l(x_i, y_i, \beta) + \lambda_1 ||\beta||_{l_1} + \lambda_2 ||\beta||_{l_2} - > \min_{\beta}$, where $l(x_i, y_i, \beta)$ refers to the misclassification loss function for i th sample. A stage-wise LARS-EN algorithm was used to solve the elastic net problem [15].

Unlike logistic regression, linear mixed model allows explicit correction for population structure due to the random effect of the model that is calculated based on kinship/relatedness matrix. In general, linear mixed model can be represented in a matrix form as $y = X\beta + u + \varepsilon$, where y is a vector of phenotypes, X - matrix of genotypes, β - vector of parameters, $\varepsilon \sim N_n(0, \sigma_\varepsilon^2 I)$ - noise and $u \sim N_n(0, \lambda \sigma_\varepsilon^2 K)$ - vector of random effects. Parameter $\lambda = \sigma_u^2 / \sigma_\varepsilon^2$ represents the ratio of variances explained by internal (genetic) and environmental factors respectively. Kinship matrix K can be evaluated as $K = XX^T / m$ for m organisms. Model parameters are estimated by the maximum likelihood or restricted maximum likelihood methods.

The Wald test was used to investigate statistical significance of the individual genetic markers, where null hypothesis $H_0 : \beta_j = 0$ was tested against the alternative $H_1 : \beta_j \neq 0$ for all $j = \overline{1, m}$.

The MOSS algorithm proved to be an interesting model search technique, which investigates models allowing to identify combinations of the best predictive SNPs associated with the response. Let M be a subset of searched hierarchical log-linear models with $k \in [2, 5]$ variables. For each element $\mu \in M$ define a priori probability and a neighboring function that returns its environment (a subset of the previous and subsequent models). The algorithm automatically removes models with low posterior probability from consideration giving priority to the most promising candidates. The procedure ends with a set of the most suitable models $M(c) = \left\{ \mu \in M : P\langle \mu \mid X, y \rangle \geq c \cdot \text{``} \max_{\mu' \in M} P(\mu | X, y) \text{''} \right\}$, where $P(\mu | X, y)$- a posteriori probability of model $\mu \in M$ and $c \in (0, 1)$ is a hyperparameter that influences the amount of enumerated candidate models and the resulting solution set.

3.5   Correction for Consistency of the Resulting SNP Sets

There are situations when drug-association tests include in the resulting set pairs of correlated mutations with high and low association scores within the same pair. Relevance feature network (RFN) allows refining the resulting sets of significant genetic markers to smooth the scores. This is based on building a graph of a special structure to take into account correlations between the mutated sites so that any highly correlated SNPs should be either significant or non-significant. Pearson coefficients of correlation were used to estimate linkages between loci. Algorithm searches for the minimum cut in the graph that, eventually, splits SNPs into subsets of significant and non-significant [16]. We used F-measure to check the accuracy of classification after the FRN correction.

4   Results

Principal component analysis (PCA) has shown that all strains from Belarusian patients can be segregated into five groups. Phylogeny and spoligotyping established the most prevalent sublineages: Beijing (63.6 %), T1 (18.9 %), H3 (5.6 %) and T5 (2.8 %).

Within the descriptive analysis, we discovered pairwise correlations of drug susceptibility testing results to investigate cross-resistance between drugs (Table 2). As

Table 2. Correlations between the results of TB drug susceptibility testing

	EMB	INH	RIF	PZA	STM	CYCL	ETH	PARA	AMIK	CAPR	OFLO
EMB	1.00	0.90	0.90	1.00	0.80	0.36	0.34	0.25	0.53	0.59	0.55
INH	0.90	1.00	1.00	0.91	0.89	0.36	0.31	0.23	0.48	0.52	0.53
RIF	0.90	1.00	1.00	0.91	0.89	0.37	0.31	0.22	0.48	0.53	0.53
PZA	1.00	0.91	0.91	1.00	0.72	0.38	0.38	0.18	0.38	0.46	0.49
STM	0.80	0.89	0.89	0.72	1.00	0.33	0.27	0.20	0.42	0.45	0.47
CYCL	0.36	0.36	0.37	0.38	0.33	1.00	0.33	0.32	0.27	0.35	0.23
ETH	0.34	0.31	0.31	0.38	0.27	0.33	1.00	0.06	0.33	0.46	0.14
PARA	0.25	0.23	0.22	0.18	0.20	0.32	0.06	1.00	0.07	0.13	0.23
AMIK	0.53	0.48	0.48	0.38	0.42	0.27	0.33	0.07	1.00	0.90	0.57
CAPR	0.59	0.52	0.53	0.46	0.45	0.35	0.46	0.13	0.90	1.00	0.61
OFLO	0.55	0.53	0.53	0.49	0.47	0.23	0.14	0.23	0.57	0.61	1.00

anticipated, the highest levels of correlations were detected between amikacin, capreomycin and inside groups of first-line drugs.

Interesting results showed analysis of the proportion of phenotypic variance explained (PVE) by SNP genotypes (Table 3), which can be summarized as $k^2_{SNP} = \sigma^2_G/(\sigma^2_G + \sigma^2_E)$, where σ^2_G is variance due to genotypic markers and σ^2_E is influence of the environmental factors.

Table 3. PVE values for anti-TB drugs analyzed

Drug	PVE %	Standard error %	Drug	PVE %	Standard error %
1st-line drugs			*2nd-line drugs*		
INH (Izoniazid)	99.997	0.021	CYCL (Cycloserine)	75.716	12.386
RIF (Rifampicin)	99.997	0.021	CAPR (Capreomycin)	73.903	11.048
PZA (Pyrazinamide)	99.997	0.049	AMIK (Amikacin)	69.831	11.925
STM (Treptomycin)	99.997	0.036	OFLO (Ofloxacin)	58.682	12.922
EMB (Ethambutol)	97.119	1.695	ETH (Ethionamide)	45.680	24.906
			PARA (Para-aminosalicyclic acid)	29.998	17.010

According to these results, SNPs do not fully explain resistance to most of second-line TB drugs in our datasets, possibly, due to issues of DST protocols [17] or other factors, which should be considered when interpreting GWAS results.

Genotype/phenotype association tests resulted in high-confidence mutation lists that were ordered according to mutation significance values for each drug and annotated using NCBI databases. We provide an overview of the predictions quality obtained by some association analysis methods (Table 4) applied to our datasets. We used cross-validation (CV) for parameters tuning and calculated generally accepted metrics to evaluate the quality of predictions: precision, recall, F-measure (the weighted harmonic mean of precision and recall), accuracy.

Table 4. Comparative analysis of prediction quality scores produced by multi-marker drug association tests using cross-validation for parameters tuning

Drug	Resistant samples	Susceptible samples	Method	Precision	Recall	F1	Accuracy
OFLO	69	63	MOSS	0.929	0.752	0.831	0.840
			LMM	0.557	1	0.715	0.583
			Logistic regression	0.971	0.986	0.978	0.977
EMB	102	30	MOSS	0.962	0.981	0.972	0.955
			LMM	1	0.108	0.195	0.311
			Logistic regression	0.990	1	0.995	0.992
INH	106	26	MOSS	1	0.981	0.990	0.985
			LMM	1	1	1	1
			Logistic regression	1	1	1	1
PZA	28	6	MOSS	1	1	1	1
			LMM	0.966	1	0.983	0.971
			Logistic regression	1	1	1	1
RIF	106	26	MOSS	1	0.869	0.930	0.895
			LMM	1	1	1	1
			Logistic regression	1	1	1	1
STM	110	22	MOSS	1	0.954	0.977	0.962
			LMM	1	0.991	0.995	0.992
			Logistic regression	1	1	1	1
AMIK	59	63	MOSS	1	0.847	0.917	0.926
			LMM	0.484	1	0.652	0.484
			Logistic regression	1	0.932	0.965	0.967
CAPR	66	51	MOSS	1	0.731	0.845	0.848
			LMM	0.564	1	0.721	0.564
			Logistic regression	1	1	1	1
CYCL	46	70	MOSS	0.502	0.453	0.477	0.604
			LMM	0.397	1	0.568	0.397
			Logistic regression	1	0.978	0.989	0.991

Lists of significant mutations differ in the amount of selected features depending on the method. MOSS and LMM algorithms provided the smallest number of significant variations with sufficiently good classification quality. Regularized logistic regression showed the best results but produced much larger lists of significant SNPs, which may indicate an overfitting and inclusion of noise characteristics in the resulting sets. We performed a second run of logistic regression using genetic markers selected at the first run but this did not reduce the output list of SNPs significantly. Single-marker methods also provided sufficiently large number of mutations associated with drug-resistance. However, all methods agreed in assigning the highest scores to the genetic markers used in GenoType MTBDRplus/MTBDRsl assays.

After RFN-based selection procedure, long mutation lists could be at least halved for most genotype/phenotype association tests. Figure 2 shows the number of significant SNPs and F-measure vary depending on the RFN parameter while searching for ofloxacin resistance markers. In most cases, a serious reduction of the resulting mutation set slightly degraded classification quality.

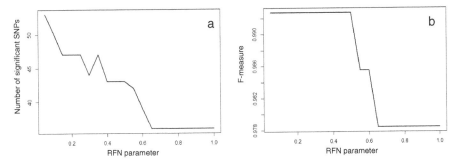

Fig. 2. Plot that illustrates (**a**) changes in the number of significant genomic markers and (**b**) alteration of F-measure depending on RFN parameter variation while adjusting logistic regression results for ofloxacin

The association tests were run for a second time for each drug where known high-confidence resistance mutations were excluded from the analysis. In such a way we intended to determine more realistic scores for mutations which appeared not very significant but correlated to the most significant SNPs in some models. Results showed that the exclusion of a few significant SNPs did not change the other scores dramatically. For example, the MOSS algorithm populated resulting sets of most promising log-linear models with a bulk of middle-quality models and lower SNP-inclusion probabilities. Correction of the significance scores using relevance feature network after the second run demonstrated a notable loss in predictive power of the remained SNPs in comparison to the first run.

5 Conclusions

Analysis of international databases showed that *M. tuberculosis* study is one of the fastest growing areas. Despite high publication activity in mutation analysis of *M. tuberculosis*, there are few genome-wide projects directed at comprehensive discovery of MDR and XDR tuberculosis.

Here we proposed a methodology of genome-wide association study of pathogenic microorganisms, which allows evaluating the contribution of genetic variations in drug-resistance and adjusting results in case of contradictory. We applied this approach to the analysis of 132 *M. tuberculosis* isolated from patients in Belarus with various forms of pulmonary tuberculosis. We implemented this methodology in laboratory software tools using third-party libraries and original scripts in R language.

6 Availability

Elements of this approach are used in current to establish the Belarus tuberculosis portal (http://tuberculosis.by) and conduct comprehensive study of obtained MDR and XDR TB strains. All sequences have been exposed publically available through GeneBank bioproject under accession PRJNA200335.

Acknowledgements. The authors are grateful to the Republican Scientific and Practical Center of Pulmonology and Tuberculosis of Ministry of Health of Belarus for cooperation and assistance in providing data. We express our thanks to the colleagues in Broad Institute of MIT and Harvard for collaboration in genome sequencing.

References

1. World Health Organization: Global Tuberculosis Report 2015. World Health Organization Press, Geneva (2015)
2. Skrahina, A., et al.: Multidrug-resistant tuberculosis in Belarus: the size of the problem and associated risk factors. Bull. World Health Organ. **91**(1), 36–45 (2013)
3. Farhat, M.R., et al.: Genomic analysis identifies targets of convergent positive selection in drug-resistant Mycobacterium tuberculosis. Nat. Genet. **45**(10), 1183–1189 (2013)
4. Zhang, H., et al.: Genome sequencing of 161 Mycobacterium tuberculosis isolates from China identifies genes and intergenic regions associated with drug resistance. Nat. Genet. **45**(10), 1255–1260 (2013)
5. Walker, T.M., et al.: Whole-genome sequencing for prediction of Mycobacterium tuberculosis drug susceptibility and resistance: a retrospective cohort study. Lancet Infect. Dis. **15**(10), 1193–1202 (2015)
6. Sandgren, A., et al.: Tuberculosis drug resistance mutation database. PLoS Med. **6**(2), 132–136 (2009)
7. GenoType MTBDRplus - your test system for a fast and reliable way to detect MDR-TB. Hain Lifesciences [Electronic resource]. http://www.hain-lifescience.de/en/products/microbiology/mycobacteria/genotype-mtbdrplus.html

8. GenoType MTBDRsl - your important assistance for detection of XDR-TB. Hain Lifesciences [Electronic resource]. http://www.hain-lifescience.de/en/products/microbiology/mycobacteria/genotype-mtbdrsl.html

9. Walker, B.J., et al.: Pilon: an integrated tool for comprehensive microbial variant detection and genome assembly improvement. PLoS ONE (2014). doi:10.1371/journal.pone.0112963

10. Stamatakis, A.: RAxML version 8: a tool for phylogenetic analysis and post-analysis of large phylogenies. Bioinformatics (2014). doi:10.1093/bioinformatics/btu033

11. Price, A.L., et al.: Principal components analysis corrects for stratification in genome-wide association studies. Nat. Genet. **38**(8), 904–909 (2006)

12. Purcell, S., et al.: PLINK: a tool set for whole-genome association and population-based linkage analyses. Am. J. Hum. Genet. **81**(3), 559–575 (2007)

13. Zhou, X., Stephens, M.: Genome-wide efficient mixed-model analysis for association studies. Nat. Genet. **44**, 821–824 (2012)

14. Dobra, A., et al.: The mode oriented stochastic search (MOSS) for log-linear models with conjugate priors. Stat. Methodol. **7**, 240–253 (2010)

15. Zou, H., Hastie, T.: Regularization and variable selection via the elastic net. J. Roy. Stat. Soc. Series B **67**, 301–320 (2010)

16. Kolmogorov, V.: What energy functions can be minimized via graph cuts? IEEE Trans. Pattern Anal. Mach. Intell. **26**(2), 147–159 (2004)

17. Horne, D.J., et al.: Diagnostic accuracy and reproducibility of WHO-endorsed phenotypic drug susceptibility testing methods for first-line and second-line antituberculosis drugs. J. Clin. Microbiol. **51**(2), 393–401 (2013)

Haplotype Inference for Pedigrees
with Few Recombinations

B. Kirkpatrick$^{(\boxtimes)}$

Intrepid Net Computing, Dillon, MT, USA
bbkirk@intrepidnetcomputing.com

Abstract. Pedigrees, or family trees, are graphs of family relationships that are used to study inheritance. A fundamental problem in computational biology is to find, for a pedigree with n individuals genotyped at every site, a set of Mendelian-consistent haplotypes that have the minimum number of recombinations. This is an NP-hard problem and some pedigrees can have thousands of individuals and hundreds of thousands of sites.

This paper formulates this problem as a optimization on a graph and introduces a tailored algorithm with a running time of $O(n^{(k+2)}m^{6k})$ for n individuals, m sites, and k recombinations. Since there are generally only 1-2 recombinations per chromosome in each meiosis, k is small enough to make this algorithm practically relevant.

Keywords: Pedigrees · Haplotype inference · Minimum recombination haplotype configuration (MRHC)

Full Manuscript. Pre-print publication of the full manuscript is available at arXiv [10].

1 Introduction

The study of pedigrees is of fundamental interest to several fields: to computer science due the combinatorics of inheritance [8,17], to epidemiology due to the pedigree's utility in disease-gene finding [15,18] and recombination rate inference [3], and to statistics due to the connections between pedigrees and graphical models in machine learning [11]. The central calculation on pedigrees is to compute the likelihood, or probability with which the observed data observed are inherited in the given genealogy. This likelihood serves as a key ingredient for computing recombination rates, inferring haplotypes, and hypothesis testing of disease-loci positions. State-of-the-art methods for computing the likelihood, or sampling from it, have exponential running times [1,2,6,7,16].

The likelihood computation with uniform founder allele frequencies can be reduced to the combinatorial MINIMUM RECOMBINATION HAPLOTYPE CONFIGURATION (MRHC) first introduced by Li and Jiang [12]. A solution to MRHC is a set of haplotypes that appear with maximum probability. The MRHC problem is NP-hard, and as such is unlikely to be solvable in polynomial time.

© Springer International Publishing Switzerland 2016
A. Bourgeois et al. (Eds.): ISBRA 2016, LNBI 9683, pp. 269–283, 2016.
DOI: 10.1007/978-3-319-38782-6_23

The MRHC problem differs from more general haplotype phasing approaches [13] that attempt to phase unrelated or partially related individuals. The MRHC problem applies specifically to individuals in a family with known relationships, and this problem has a variation with mutations [14,19]. Xiao, et al. considered a bounded number of recombinations in a probabilistic phasing model [20].

This paper gives an exponential algorithm for the MRHC problem with running time tailored to the required recombinations $O(n^{(k+2)}m^{6k})$ having exponents that only depend on the minimum number of recombinations k which should be relatively small (i.e. one or two recombinations per chromosome per individual per generation). This is an improvement on previous formulation that rely on integer programming solvers rather than giving an algorithm which is specific to MRHC [12]. We also define the minimum-recombination (MR) graph, connect the MR graph to the inheritance path notation and discuss its properties.

The remainder of this paper is organized as follows. Section 2 introduces the combinatorial model for the pedigree analysis. Section 3 provides a construction of the MR graph. Finally, Sect. 4 gives a solution to the MRHC problem based on a coloring of the minimum recombination graph. Due to space constraints, several algorithms and proofs have been deferred to the extended version of the paper.

2   Pedigree Analysis

This section gives the background for inferring haplotype configurations from genotype data of a pedigree. We use the Iverson bracket notation, so that $[E]$ equals 1 if the logical expression E is true and 0 otherwise [9].

A *pedigree* is a directed acyclic graph P whose vertex set $I(P)$ is a set of *individuals*, and whose directed arcs indicate genetic inheritance from parent to child. A pedigree is *diploid* if each of its individuals has either no or two incoming arcs; for example, human, cow, and dog pedigrees are diploid. For a diploid pedigree P, every individual without incoming arcs is a *founder* of P, and every other individual i is a *non-founder* for which the vertices adjacent to its two incoming arcs are its *parents* $p_1(i), p_2(i)$, mother and father, respectively. Let $F(P)$ denote the set of founders of P.

In this paper, every individual has genetic data of importance to the haplotype inference problem. We abstract this data as follows. A *site* is an element of an ordered set $\{1, \ldots, m\}$. For two sites s, t in the interval $[1, m]$, their *distance* is $\mathsf{dist}(s, t) = |s - t|$. For a pedigree P, let $n = |I(P)|$ be the number of its individuals. A *haplotype* h is a string of length m over $\{0, 1\}$ whose elements represent binary *alleles* that appear together on the same chromosome. We use p_1 and p_2 to indicate maternal and paternal chromosomes, respectively, and let $h^{p_1}(i), h^{p_2}(i)$ be binary strings that denote the maternal and paternal haplotypes of individual i. For a site s, the maternal (resp. paternal) haplotype of individual i at site s is the allele $h^{p_1}(i, s)$ (resp. $h^{p_2}(i, s)$) of the string $h^{p_1}(i)$ (resp. $h^{p_2}(i)$) at position s. A *haplotype configuration* is a matrix H with m columns and n rows, whose entry H_{rc} at row r and column c is the vector $\binom{h^{p_1}(r,c)}{h^{p_2}(r,c)}$.

Haplotype data is expensive to collect; thus, we observe genotype data and recover the haplotypes by inferring the parental and grand-parental origin of each allele. The genotype of each individual i at each site s is the conflation $g(i, s)$ of the alleles on the two chromosomes: formally,

$$g(i, s) = \begin{cases} h^{p_1}(i, s), & \text{if } h^{p_1}(i, s) = h^{p_2}(i, s), \\ 2, & \text{otherwise.} \end{cases} \tag{1}$$

Genotype $g(i, s)$ is *homozygous* if $g(i, s) \in \{0, 1\}$ and *heterozygous* otherwise. Let G be the matrix of genotypes with entry $g(i, s)$ at row i and column s. We have defined the genotypes in the generative direction from the haplotypes. We are interested in the inverse problem of recovering the haplotypes given the genotypes. For a matrix G having η heterozygous sites across all individuals, there are $2^{\eta-1}$ possible configurations satisfying *genotype consistency* given by (1).

Throughout, we assume that Mendelian inheritance at each site in the pedigree proceeds with recombination and without mutation. This assumption imposes Mendelian consistency rules on the haplotypes (and genotypes) of the parents and children. For $\ell \in \{1, 2\}$, a haplotype $h^{p_\ell}(i)$ is *Mendelian consistent* if, for every site s, the allele $h^{p_\ell}(i, s)$ appears in $p_\ell(i)$'s genome as either the grand-maternal allele $h^{p_1}(p_\ell(i), s)$ or grand-paternal allele $h^{p_2}(p_\ell(i), s)$. Mendelian consistency is a constraint imposed on our haplotype configuration that is in addition to genotype consistency in (1). From now on, we will define a haplotype configuration as *consistent* if it is both genotype and Mendelian consistent.

For each non-founder $i \in I(P) \setminus F(P)$ and $\ell \in \{1, 2\}$, we indicate the *origin* of each allele of $p_\ell(i)$ by the binary variable $\sigma^{p_\ell}(i, s)$ defined by

$$\sigma^{p_\ell}(i, s) = \begin{cases} p_1, & \text{if } h^{p_\ell}(i, s) = h^{p_1}(p_\ell(i), s), \\ p_2, & \text{if } h^{p_\ell}(i, s) = h^{p_2}(p_\ell(i), s). \end{cases} \tag{2}$$

In words, $\sigma^{p_\ell}(i, s)$ equals p_1 if $h^{p_1}(i, s)$ has grand-maternal origin and equals p_2 otherwise. The set $\sigma(s) = \{(\sigma^{p_1}(i, s), \sigma^{p_2}(i, s)) \mid i \in I(P)\}$ is the *inheritance path for site s*. A *recombination* is a change of allele between consecutive sites, that is, if $\sigma^{p_\ell}(i, s) \neq \sigma^{p_\ell}(i, s+1)$ for some $\ell \in \{1, 2\}$ and $s \in \{1, \ldots, m-1\}$. For a haplotype configuration H, 2^ζ inheritance paths satisfy (2), where ζ is the number of homozygous sites among all parent individuals of the pedigree. This means that for a genotype matrix G, we have at most $O(2^{\eta-1}2^\zeta)$ possible tuples (H, σ), and this defines the search space for the MRHC problem where the goal is to choose a tuple (H, σ) with a minimum number of recombinations represented in σ.

For a pedigree P and observed genotype data G, the formal problem is:

MINIMUM RECOMBINATION HAPLOTYPES (MRHC)
Input: A pedigree P with genotype matrix G
  Task: Find $h^{p_\ell}(i, s)$ for $i \in I(P), s \in \{1, \ldots, m\}, \ell \in \{1, 2\}$ minimizing
     the number of required recombinations, i.e., compute
     $\text{argmin}_{(H,\sigma)} \sum_{i \in I(P) \setminus F(P)} \sum_{s \geq 1}^{m-1} \sum_{\ell=1}^{2} [\sigma^{p_\ell}(i, s) \neq \sigma^{p_\ell}(i, s+1)]$

---

3   Minimum Recombination Graph

We now fix a pedigree P and describe a vertex-colored graph $R(P)$, the minimum recombination graph (MR graph) of P, which allows us to reduce the MRHC problem on P to a coloring problem on $R(P)$. The concept of the MR graph was introduced by Doan and Evans [4] to model the phasing of genotype data in P. However, our graph definition differs from theirs, because, as we will argue later, their definition does not model all recombinations of all haplotypes consistent with the genotype data.

3.1   Definition of the Minimum Recombination Graph

Intuitively, the minimum recombination graph represents the Mendelian consistent haplotypes and the resulting minimum recombinations that are required for inheriting those haplotypes in the pedigree: vertices represent genome intervals, vertex colors represent haplotypes on those intervals, and edges represent the potential for inheritance with recombination.

Formally, the *minimum recombination graph* of P is a tuple $(R(P), \phi, \mathcal{S})$, where R is an undirected multigraph, ϕ is a coloring function on the vertices of $R(P)$, and $\mathcal{S}$ is a collection of "parity constraint sets". The vertex set $V(R(P))$ of $R(P)$ consists of one vertex i_{st} for each individual $i \in I(P)$ and each genomic interval $1 \le s < t \le m$, plus one *special* vertex b. A vertex i_{st} is *regular* if sites s and t are contiguous heterozygous sites in individual i, and *supplementary* otherwise. A vertex i_{st} is *heterozygous* (*homozygous*) if i has heterozygous (homozygous) genotypes at both s, t.

Vertex-Coloring. The coloring function ϕ assigns to each regular or supplementary vertex i_{st} a color $\phi(i_{st}) \in \{\mathsf{gray}, \mathsf{blue}, \mathsf{red}, \mathsf{white}\}$. The color of vertex i_{st} indicates the different "haplotype fragments" that are Mendelian consistent at sites s and t in the genome of individual i. A *haplotype fragment* $f(i_{st})$ of a vertex i_{st} at sites s and t is an (unordered) set of two haplotypes which we will write horizontally with sites s and t side-by-side and the two haplotypes stacked on top of each other. Let $\Phi(i_{st})$ be the set of haplotype fragments generated by the color assignment of vertex i_{st}. The colors are defined in Table 1. The *haplotype pair of individual i at sites s and t* is the $\{0,1\}$-valued 2×2-matrix $H(i,s,t) = \begin{pmatrix} h^{p_1}(i,s) & h^{p_1}(i,t) \\ h^{p_2}(i,s) & h^{p_2}(i,t) \end{pmatrix}$. We denote unordered (set) comparison of the haplotype fragments and haplotype pairs by $H(i,s,t) \doteq f(i_{st})$. Similarly, for set comparison of sets, we write $\{H(i,s,t) | \ \forall H\} \doteq \Phi(i_{st})$ where the first set considers all consistent haplotype configurations H. Then the color and genotype of i_{st} precisely represent its haplotype fragments, as defined in Table 1. Figure 1 gives an example of the genotypes, haplotypes, and vertex colorings.

For a heterozygous vertex i_{st}, its color $\phi(i_{st})$ indicates the relative paternal origin of the heterozygous alleles at sites s and t and corresponds to a haplotype configuration (red and blue have a one-to-one correspondence with the two possible haplotypes for the sites of i_{st}). But these haplotypes are fragmented, and,

Table 1. Rules for coloring vertex i_{st} of the minimum recombination graph. The $\doteq$ symbol denotes a set comparison operation (i.e., an unordered comparison of elements).

$g(i,s)$	$g(i,t)$	$\{H(i,s,t)\vert\forall H\} \doteq \Phi(i_{st})$	$\phi(i_{st})$
2	2	$\{\begin{pmatrix}0\,1\\1\,0\end{pmatrix},\begin{pmatrix}0\,0\\1\,1\end{pmatrix}\}$	gray
2	2	$\begin{pmatrix}0\,1\\1\,0\end{pmatrix}$	red
2	2	$\begin{pmatrix}0\,0\\1\,1\end{pmatrix}$	blue
0	0	$\begin{pmatrix}0\,0\\0\,0\end{pmatrix}$	blue
1	1	$\begin{pmatrix}1\,1\\1\,1\end{pmatrix}$	blue
0	1	$\begin{pmatrix}0\,1\\0\,1\end{pmatrix}$	red
1	0	$\begin{pmatrix}1\,0\\1\,0\end{pmatrix}$	red
otherwise		$\{\begin{pmatrix}0\,0\\0\,1\end{pmatrix},\begin{pmatrix}0\,0\\1\,0\end{pmatrix},\begin{pmatrix}1\,0\\1\,1\end{pmatrix},\begin{pmatrix}0\,1\\1\,1\end{pmatrix}\}$	white

hence, may or may not be consistent with a single haplotype configuration. Note that colors may or may not indicate Mendelian consistent haplotype fragments.

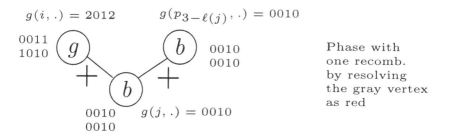

Fig. 1. The genotypes and the haplotypes are given for three individuals and four sites. Here $(s,t) = (1,4)$ since those are the heterozygous sites in the left parent. From Table 1, we see that the left parent is gray, line 1, that the right parent is blue, line 4, and that the child is blue, line 4. This figure is an instance of Table 2, case 1 (in the full manuscript's Appendix [10], it is the first case). There are no parity constraints in this example. A better phasing with zero recombinations would be to resolve the gray parent as blue.

Parity Constraint Sets. We now describe the collection $\mathcal{S}$ of parity constraint sets. The collection $\mathcal{S}$ contains one set S for each gray heterozygous supplementary vertex. A *parity constraint set* is a tuple (S, ρ_S) with parity color $\rho_S \in \{\text{red, blue}\}$ for even parity, and a set S consisting of a heterozygous supplementary vertex i_{st} and all regular heterozygous vertices i_{pq} such that $s \leq p < q \leq t$. Here, $\rho_S = \text{red}$ (resp. $\rho_S = \text{blue}$) indicates even parity of the red (resp. blue) vertices. As sites s, t are heterozygous and i_{st} is supplementary, the set S contains at least two regular vertices i_{sp} and either i_{pt} or i_{qt}.

Given $\mathcal{S}$, every Mendelian consistent haplotype configuration *induces* a vertex coloring $\phi_{\mathcal{S}}$ of $R(P)$, defined by

$$\phi_{\mathcal{S}}(i_{st}) = \begin{cases} \phi(i_{st}), & \text{if } \phi(i_{st}) \neq \text{gray}, \\ \text{red}, & \text{if } \phi(i_{st}) = \text{gray} \wedge \exists \, H(i,s,t) \doteq \left(\begin{smallmatrix} 0 & 1 \\ 1 & 0 \end{smallmatrix}\right), \\ \text{blue}, & \text{otherwise}. \end{cases}$$

However, we need further constraints to guarantee that the coloring $\phi_{\mathcal{S}}$ has a corresponding Mendelian consistent haplotype configuration. Intuitively, these constraints ensure that the collection of overlapping haplotype fragments selected by coloring the gray vertices are consistent with two longer haplotypes. Examples of red parity constraint sets are given in Fig. 2.

For coloring $\phi_{\mathcal{S}}$, the number of ρ_S-colored vertices in each parity constraint set $(S, \rho_S) \in \mathcal{S}$ must be even. When $\rho_S = $ red it properly models that the gray vertices i_{pq}, $s < p < q < t$ with $\phi(i_{pq}) = $ gray and $\phi_{\mathcal{S}}(i_{st}) = $ red indicate alternating alleles 0-1 along the chromosome. For now, we focus on the case where $\rho_S = $ red which is the default color for ρ_S. Informally, we want the red-colored gray vertices in the parity constraint set to indicate alternating 0-1 pattern along the haplotype. Therefore, the color of the unique supplementary vertex in each set S must agree with the pattern indicated by the regular vertices in S. Later we will see that $\rho_S = $ blue only for particular cases where the blue vertices are adjacent to red vertices on edges without recombination, meaning that these red vertices indicate alternative allele 0-1 along the chromosome.

We call a parity constraint set S *satisfied* by $\phi_{\mathcal{S}}$ if S contains an even number of vertices i_{pq}, $s < p < q < t$ with $\phi(i_{pq}) = $ gray and color $\phi_{\mathcal{S}}(i_{st}) = \rho_S$; and we call $\mathcal{S}$ *satisfiable* if there exists a coloring $\phi_{\mathcal{S}}$ induced by $\mathcal{S}, \phi, H$ such that each set $S \in \mathcal{S}$ is satisfied. By definition, a coloring $\phi_{\mathcal{S}}$ induced by a Mendelian consistent haplotype configuration satisfies all sets $(S, \phi_{\mathcal{S}}) \in \mathcal{S}$. The converse is also true:

Observation 1. *Any assignment $\phi_{\mathcal{S}}$ of colors* red *and* blue *to vertices i_{pq}, $s < p < q < t$ with $\phi(i_{pq}) = $* gray *that satisfies all sets of the form $(S, \phi_{\mathcal{S}}) \in \mathcal{S}$ represents a Mendelian consistent haplotype configuration H.*

In other words, there is a bijection between haplotype configurations and colorings that satisfy the parity constraint sets. For $\phi_{\mathcal{S}} = $ red, the justification follows from the 0-1 alternating alleles of gray vertices in any genotype consistent haplotype. We will see later that in the instances where we have $\rho_S = $ blue, the bijection will also hold.

Edge Creation. It remains to describe the edge set $E(R(P))$ of $R(P)$, which requires some preparation. Consider a haplotype configuration H and a minimum recombination inheritance path for those haplotypes. Let r be a recombination that occurs during the inheritance from an individual i to its child j between contiguous sites q and $q + 1$. Let $\ell \in \{1, 2\}$ indicate whether $i = p_\ell(j)$ is the maternal or paternal parent of j. Then the recombination r of i's haplotypes is indicated in the inheritance path by $\sigma^{p_\ell}(j, q) \neq \sigma^{p_\ell}(j, q + 1)$. Fixing all recombinations $r' \neq r$ in the inheritance path, r can be shifted to the right or to the left in j's inheritance path to produce a new inheritance path which is also consistent with the haplotype configuration H. The *maximal genomic interval* of r is the unique maximal set $[s,t] = \{s, s+1, \ldots, t-1, t\}$ of sites such that r can placed between any contiguous sites $q, q+1$ in the interval with the resulting inheritance path being consistent with H. Since all genotype data is observed, the maximal genomic interval $[s,t]$ of r always means that both s, t are heterozygous sites in the parent i, and therefore $[s,t]$ is determined only by the recombination position q and the pair $\{s,t\}$, independent of H. This interval $[s,t]$ is pertinent to which haplotype fragments are represented in $R(P)$, and it is elucidated by the "min-recomb property" defined below.

The set $E(R(P))$ will be the disjoint union of the set E^+ of *positive* edges and the set E^- of *negative* edges. An edge $\{u,v\} \in E(R(P))$ will be called *disagreeing* if either $\{u,v\} \in E^+$ and vertices u, v are colored differently, or if $\{u,v\} \in E^-$ and vertices u, v have the same color. Our goal is to create edges such that $R(P)$ satisfies the "min-recomb property".

Definition 1. *Let P be a pedigree with $I(P)$ its set of individuals. A graph with vertex set $I(P)$ has the* min-recomb property *if for every individual $j \in I(P)$ with parents $p_1(j), p_2(j)$, and every haplotype configuration H for the genotype data, for $\ell \in \{1,2\}$, a recombination between $i = p_\ell(j)$ and j in the maximal genomic interval $[s,t]$ is in some minimum recombination inheritance path for H if and only if the recombination is represented in the graph by a disagreeing edge incident to vertex $i_{st} = p_\ell(i)_{st}$.*

Let i_{st} be a regular vertex of $R(P)$ with $g(i,s) = g(i,t) = 2$ and let $j \in I(P) \setminus F(P)$ be such that $i = p_\ell(j)$. Then $\phi(i_{st}) \in \{\text{gray, blue, red}\}$, and we create edges incident to i_{st} and j depending on their colors and genotypes, according to Table 2. Figure 1 gives an example of the first case in this table. Note that $R(P)$ is a multigraph, but there is at most one negative edge $\{i_{st}, p_{3-\ell}(j)\}$ for any tuple $(j, i_{st} = p_\ell(j), p_{3-\ell}(j))$.

Table 2. Rules for creating edges of the minimum recombination graph.

Case	$\phi(p_{3-\ell}(j))$	$\phi(j)$	Edges to create
1	$\{\text{gray, blue, red}\}$	$\{\text{gray, blue, red}\}$	$\{i_{st}, j_{st}\}, \{p_{3-\ell}(j)_{st}, j_{st}\} \in E^+$
2	white	$\{\text{gray, blue, red}\}$	$\{i_{st}, j_{st}\} \in E^+$
3	$\{\text{gray, blue, red}\}$	white	$\{i_{st}, p_{3-\ell}(j)_{st}\} \in E^-$
4	white	white	(see text)

It remains to describe the edges to create in Case 4, when $\phi(p_{3-\ell}(j)) = \phi(j) = \text{white}$. This will be done according to the following subcases:

4(a) If $p_{3-\ell}(j)$ and j have a common heterozygous site, that is, if $g(p_{3-\ell}(j), s) = g(j,s) = 2$ or $g(p_{3-\ell}(j), t) = g(j,t) = 2$, then there is a unique site $z \in \{s,t\}$

that is heterozygous in both individuals j and $p_{3-\ell}(j)$. Let $q(j) \in \{s, s + 1, ..., t-1, t\} \setminus \{z\}$ be the heterozygous site in j that is closest to z, or $q(j) = +\infty$ if no such site exists. Similarly, let $q(p_{3-\ell}(j)) \in \{s, s+1 \ldots, t\} \setminus \{z\}$ be the heterozygous site in $p_{3-\ell}(j)$ that is closest to z, or $q(p_{3-\ell}(j)) = +\infty$ if no such site exists. If $\min\{q(j), q(p_{3-\ell}(j))\} = +\infty$ then vertex i_{st} remains isolated; otherwise, let $z_{\min} = \min\{z, q\}, z_{\max} = \max\{z, q\}, \bar{z} = \{s, t\} \setminus \{z\}$, and create edges incident to i_{st} according to Table 3.

4(b) If j and $p_{3-\ell}(j)$ do not have a heterozygous site at the same position, then either $g(p_{3-\ell}(j), s) = g(j, t) = 2$ or $g(j, s) = g(p_{3-\ell}(j), t) = 2$. Let $z \in \{s, t\}$ be such that $g(p_{3-\ell}(j), z) \neq 2$ and let $\bar{z} \in \{s, t\}$ be such that $g(j, \bar{z}) \neq 2$. If $g(p_{3-\ell}(j), z) = g(j, \bar{z})$, create the edge $\{i_{st}, b\} \in E^-$, else create the edge $\{i_{st}, b\} \in E^+$.

Table 3. Case 4(a): rules for creating edges incident to a vertex i_{st} with $\min\{q(j), q(p_{3-\ell}(j))\} < +\infty$.

$\phi(j_{z_{\min} z_{\max}})$	$g(i, \min\{q(j), q(p_{3-\ell}(j))\})$	edge to create
$\{\text{blue}, \text{red}, \text{gray}\}$	$= g(p_{3-\ell}(j), \bar{z})$	$\{i_{st}, j_{z_{\min} z_{\max}}\} \in E^+$
$\{\text{blue}, \text{red}, \text{gray}\}$	$\neq g(p_{3-\ell}(j), \bar{z})$	$\{i_{st}, j_{z_{\min} z_{\max}}\} \in E^-$
white	$= g(p_{3-\ell}(j), \bar{z})$	$\{i_{st}, p_{3-\ell}(j)_{z_{\min} z_{\max}}\} \in E^-$
white	$\neq g(p_{3-\ell}(j), \bar{z})$	$\{i_{st}, p_{3-\ell}(j)_{z_{\min} z_{\max}}\} \in E^+$

Graph Cleanup. To complete the construction of $R(P)$, we pass through its list of supplementary vertices to remove some of their edges: this is necessary as some edges adjacent to a supplementary vertex might over-count the number of recombinations; see the example in Fig. 2.

Let $\{i_{st}, j_{st}\}$ be an edge adjacent to a supplementary gray vertex i_{st} where i is the parent of j. Let $(S(i_{st}), \rho_{S(i_{st})}) \in \mathcal{S}$ be the set containing i_{st}. If all regular vertices i_{pq} in $S(i_{st})$, for $s \leq p < q \leq t$, are incident to an edge $\{i_{pq}, j_{pq}\}$ then the supplementary edge $\{i_{st}, j_{st}\}$ over-counts. We remove $\{i_{st}, j_{st}\}$ and replace the set $S(i_{st})$ by a set $S(j_{st})$, which has vertices with the same indices as those in $S(i_{st})$ and where the parity constraint is to have an even number of $\bar{\rho}_{S(i_{st})}$ vertices where $\bar{\rho}_{S(i_{st})} = \text{blue}$ if $\rho_{S(i_{st})} = \text{red}$ and $\bar{\rho}_{S(i_{st})} = \text{red}$ if $\rho_{S(i_{st})} = \text{blue}$. Notice that j_{st} must also be a supplementary vertex, for the condition to be satisfied.

Note that this edge-removal rule does not apply to edges in Case 4, and does not apply to negative edges, as a negative edge $\{i_{st}, j_{st}\}$ adjacent to a supplementary vertex i_{st} has at least one regular vertex i_{pq}, $s \leq p < q \leq t$ in the parity constraint set $S(i_{st})$ for which there is no edge $\{i_{pq}, j_{pq}\}$.

Observation 2. *Any assignment ϕ_S of colors* red *and* blue *to vertices i_{st} with $\phi(i_{st}) = $ gray *that satisfies all parity constraint sets $(S, \rho_S) \in \mathcal{S}$ represents a Mendelian consistent haplotype configuration H.*

Comparing the MR graph $R(P)$ as defined in this section, with the graph $D(P)$ defined by Doan and Evans [4], we find that $D(P)$ fails to properly model the phasing of genotype data; see Sect. 3.4 for details.

3.2 Algorithms

Our motivation for introducing the ϕ-colored MR graph and parity constraint sets $\mathcal{S}$ is to model the existence of Mendelian consistent haplotypes for the genotypes in P; we formalize this in Lemma 1. Complete algorithms will be given in the extended version of this paper.

Lemma 1. Given $(R(P), \phi, \mathcal{S})$, there exists a Mendelian consistent haplotype configuration H for the genotypes if and only if there exists a coloring $\phi_{\mathcal{S}}$ that satisfies all parity constraint sets in $\mathcal{S}$.

Proof. Given a haplotype configuration H, let $\phi_{\mathcal{S}}$ be a coloring of regular and supplementary vertices in $I(P)$ defined as follows. For any vertex $i_{st} \in I(P)$ with $\phi_{\mathcal{S}}(i_{st}) \neq$ gray, set $\phi_{\mathcal{S}}(i_{st}) = \phi(i_{st})$. For any vertex $i_{st} \in I(P)$ with $\phi_{\mathcal{S}}(i_{st}) =$ gray and $H(i, s, t) \doteq \left(\begin{smallmatrix} 0 & 1 \\ 1 & 0 \end{smallmatrix} \right)$, set $\phi_{\mathcal{S}}(i_{st}) =$ red. For any vertex $i_{st} \in I(P)$ with $\phi_{\mathcal{S}}(i_{st}) =$ gray and $H(i, s, t) \doteq \left(\begin{smallmatrix} 0 & 0 \\ 1 & 1 \end{smallmatrix} \right)$, set $\phi_{\mathcal{S}}(i_{st}) =$ blue. Then $\phi_{\mathcal{S}}$ satisfies the parity constraint sets in $\mathcal{S}$, since each haplotype in H is a contiguous sequence of alleles.

Conversely, let $\phi_{\mathcal{S}}$ satisfy the parity constraint sets in $\mathcal{S}$. We generate the haplotype sequences for all individuals by the MR Haplotype algorithm, which results in the haplotypes from the colored minimum recombination graph. For individual i and site s, given its genotype $g(i, s)$ the algorithm arbitrarily selects an $\ell \in \{1, 2\}$ and obtain haplotype $h^{p_\ell}(i)$ from the graph. Recall that the haplotype fragments are unordered, so the symmetry between the first haplotype fragments is broken by arbitrarily selecting the zero allele of the first locus. Since the haplotype fragments of all following vertices overlap with the fragments of the previous vertex, all other symmetries are broken by the original choice. Then the algorithm sets $h^{p_{3-\ell}}(i) = g(i, s) - h^{p_\ell}(i)$. Let h_{is} be the haplotype allele for i at site s. For the smallest heterozygous site s_0 of i, setting $h(i, t) = 0$ allows to arbitrarily select one of the haplotypes of i. To obtain the rest of the haplotype alleles, the loop iterates along the genome setting the alleles as indicated by the colors. All gray vertices are used, and since the parity constraints are satisfied by the supplementary vertices, the alleles set by the regular gray vertices and the supplementary gray vertices are identical.

We defined the minimum recombination graph $(R(P), \phi, \mathcal{S})$ in terms of the minimum recombination property, proved that such a graph exists and satisfies the coloring property.

In the rest of this section we discuss how to construct a minimum recombination graph in polynomial time from the genotype data for all individuals in the pedigree P. We make three claims: (1) that the white vertices are irrelevant, (2) that the algorithms we give construct the minimum recombination graph of P, and (3) that the algorithms run in polynomial time.

First, consider the white vertices of $(R(P), \phi, \mathcal{S})$. These are not connected to any other vertex of $R(P)$ and are therefore not involved in any recombinations. They never change their color and are therefore not involved in specifying the haplotype configuration. Thus, removing the white vertices from $R(P)$ yields a graph that still satisfies the minimum recombination property and the coloring property. Our algorithms therefore do not create any white vertices.

Second, we claim that the MR Graph algorithm constructs the minimum recombination graph from the given genotype data for all individuals in the pedigree P. Considering the color $\phi(i)$ of any heterozygous vertex created. If Mendelian consistency requires vertex i to have a particular color $c \in \{\text{red}, \text{blue}\}$, then $\phi(i)$ is set to c. By definition of $(R(P), \phi, \mathcal{S})$, any heterozygous vertex is colored a particular color if every Mendelian consistent haplotype configuration has the appropriate corresponding haplotypes. The analysis of all genotype and haplotype possibilities in the proof of Lemma 2 shows that Mendelian consistency criterion is necessary and sufficient to obtain these colors. The cases show that when considering this vertex as the parent, there are haplotype configurations for both colors of the vertex, regardless of the genotypes of the children. However, when this vertex is the child, there are instances where the vertex has a determined color. These cases in the tables are marked with bold; the disallowed genotype combinations are indicated with MI and by a slash through the offending color with the only feasible color in bold. Since the table shows all Mendelian consistent genotype possibilities, it follows that any vertex constrained to be a particular color must be constrained by one of the Mendelian compatibility instances in the table. Therefore these Mendelian consistency cases are necessary and sufficient for initially coloring the heterozygous vertices.

Note that the parity constraint sets add no further coloring constraints to the heterozygous vertices beyond those given by the Mendelian consistency constraints. To see this, suppose, for the sake of contradiction, that there is a parity constraint set $S \in \mathcal{S}$ with exactly one vertex i_{st} of color $\phi(i_{st}) = \text{gray}$. Then in every haplotype configuration H, the color $\phi_{\mathcal{S}}$ is uniquely determined. Therefore, of all possible haplotype cases in the proof of Lemma 2, since the only ones having a determined color for a heterozygous vertex are Mendelian consistency cases, then this single gray vertex color must be determined by Mendelian consistency.

It remains to verify that the edges of $R(P)$ are created according to the rules given above. It is possible to write an MR Trio algorithm that satisfies this, this algorithm is given in the extended version of this paper.

Third, we claim that the MR Graph algorithm runs in time polynomial in $|P|$. Its running times is determined by the number of vertices that are processed. Let $n = |I(P)|$ be the number of individuals in P, let m be the number of sites, and c be the maximum number of individuals j for any i with $p_\ell(j) = i$. Then the MR Graph algorithm runs in time $O(cnm)$, since for each individual $i \in I(P)$ there are at most m vertices for contiguous heterozygous sites. For each of those vertices, MR Trio algorithm is called at most c times, and performs a constant-time edge-creation operation. All these algorithms are given in the extended version of this paper.

3.3   Properties of the Minimum Recombination Graph

We prove basic properties of the minimum recombination graph $(R(P), \phi, \mathcal{S})$.

First, there can be multiple colorings of gray vertices by red or blue that satisfy those parity constraints corresponding to a particular choice of haplotypes for all individuals in P; this is formalized in Lemma 2.

Lemma 2. *Given* $(R(P), \phi, \mathcal{S})$, *a coloring* ϕ' *of regular and supplementary vertices of* $R(P)$ *satisfies all parity constraint set in* $\mathcal{S}$ *if*

$$\phi'(i_{st}) \in \begin{cases} \{\phi(i_{st})\}, & \text{if } \phi(i_{st}) \neq \text{gray, and regular} \\ \{\text{red, blue}\}, & \text{if } \phi(i_{st}) = \text{gray, and regular} \\ parity(\rho_s) & \text{if supplementary} \end{cases} \qquad (3)$$

Proof. By definition of ϕ, for any regular vertex i_{st} with $\phi(i_{st}) = \text{gray}$ there exist two haplotype configurations, one in which i_{st} has the red haplotype fragments, $\left(\begin{smallmatrix} 0 & 1 \\ 1 & 0 \end{smallmatrix}\right)$, and one in which i_{st} has the blue haplotype fragments, $\left(\begin{smallmatrix} 0 & 1 \\ 1 & 0 \end{smallmatrix}\right)$. In both cases, there exists a haplotype configuration, one represented by blue and the other by red. After coloring all the regular vertices, we can select the color of the supplementary vertices to satisfy parity. Thus, any coloring ϕ' obtained from the haplotype fragments that appear in the haplotype configuration and subject to (3) satisfies the parity constraint sets.

Second, we show that each edge in the graph is necessary, in that there exists a haplotype configuration with the indicated recombination.

Theorem 3. *For any edge* $e = \{i_{st}, j_{pq}\} \in E(R(P))$ *there exists a haplotype configuration* H *having a minimum recombination inheritance path with the recombination indicated by* e. *(Proof in the extended version of the paper.)*

Third, we prove that $(R, \phi, \mathcal{S})$ satisfies the min-recomb property.

Theorem 4. *Let* H *be a Mendelian consistent haplotype configuration, let* $i, j \in I(P)$ *be such that* $i = p_\ell(j)$, *and let* s, t *be sites such that* $s < t$. *Then a recombination between* i *and* j *in the maximal genomic interval* $[s, t]$ *is in some minimum recombination inheritance path of* H *if and only if it is represented in* $R(P)$ *by a disagreeing edge incident to* i_{st}.

Theorem 4 proves that the edge construction cases result in an MR graph, since those particular edges satisfy the min-recomb property.

Corollary 1. *For a Mendelian consistent haplotype configuration* H, *let* ϕ' *be the coloring induced on* $R(P)$ *by* H, *and let* $E' = \{\{i_{st}, j_{pq}\} \in E^- \mid \phi'(i_{st}) = \phi'(j_{pq})\} \cup \{\{i_{st}, j_{pq}\} \in E^+ \mid \phi'(i_{st}) \neq \phi'(j_{pq})\}$. *Then the minimum number of recombinations required for any inheritance of those haplotypes equals* $|E'|$. *(Proof in the extended version of the paper.)*

Note that similar to the proof of Theorem 4, from $R(P)$ and ϕ, we can exploit the edge cases for the disagreeing edges to obtain a minimum recombination inheritance path from $R(P)$ in time $O(|E(R(P))|)$ time. The running time is due to a constant number of cases being considered for each disagreeing edge. From each of the cases, a feasible inheritance path is an immediate consequence.

Corollary 2. *A solution to the* MRHC *problem corresponds to a coloring* ϕ_S *that satisfies* S *and has a minimum number of disagreeing edges.*

3.4   Comparison of the MR Graph with the Doan-Evans Graph

We now compare the MR graph $R(P)$, as defined in Sect. 3, with the graph $D(P)$ defined by Doan and Evans [4]. We claim that the graph $D(P)$ fails to properly model the phasing of genotype data.

First, in $D(P)$ any vertex that represents two heterozygous sites is colored gray. However, as some of the gray vertices are constrained by Mendelian consistency to be either red or blue, D represents Mendelian inconsistent haplotype configurations. For example, in some instances where both parents are white, i.e. $\left(\begin{smallmatrix} 0 & 0 \\ 0 & 1 \end{smallmatrix}\right)$ and $\left(\begin{smallmatrix} 0 & 0 \\ 1 & 0 \end{smallmatrix}\right)$, the heterozygous child must be colored red.

Second, $D(P)$ violates the minimum recombination property: in Fig. 1(c) of their paper [4], there exists haplotypes for the two parents and child such that H indicates a different number of recombinations than required by the haplotypes. Specifically, let the left parent have haplotypes 0101 and 1110, the right parent have haplotypes 0010 and 1111, and the child have haplotypes 0111 and 1111. Then $D(P)$ indicates one recombination, whereas the minimum number of recombinations required by the haplotypes is two.

Third, the parity constraint sets defined by Doan and Evans [4] can overcount the number of recombinations. For example, consider the pedigree P with $n = 5$ individuals consisting of an individual i, its parents, and its paternal grand-parents, see Fig. 2.

$$(S_1, \rho_1) = (\{\phi(i_{1,3}) = red, \phi(i_{1,2}), \phi(i_{2,3})\}, red)$$
$$(S_2, \rho_2) = (\{\phi(j_{1,3}), \phi(j_{1,2}) = blue, \phi(j_{2,3})\}, red)$$

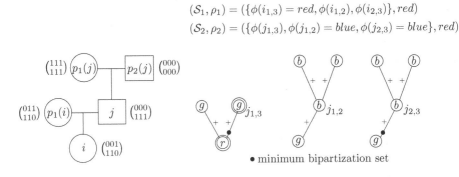

• minimum bipartization set

Fig. 2. The specified haplotypes induce two disagreeing edges in $D(P)$, but only one recombination is required to inherit the haplotypes. The supplementary gray vertices are indicated with double circles. Their parity constraint sets are given at the top of the figure.

4   Coloring the MR Graph by Edge Bipartization

In this section, we solve a variant of an edge bipartization problem on a perturbation of the minimum recombination graph. The solution to this problem is in one-to-one correspondence with a Mendelian consistent haplotype configuration for the genotype data, because of Observation 2.

First, we perturb the graph $(R(P), \phi, \mathcal{S})$ by substituting each of the positive edges in $R(P)$ by two negative edges. That is, bisect every positive edge $\{i_{st}, j_{st}\} \in E^+$ with a new gray vertex x and add the resulting two edges $\{i_{st}, x\}, \{x, j_{st}\}$. Once this step has been completed for all positive edges of $R(P)$, call the resulting graph $R(P)^-$. Observe that $R(P)^-$ is not a minimum recombination graph, since the new gray vertices do not represent a maximal genomic interval. Further, colorings of $R(P)$ and $R(P)^-$ are in one-to-one correspondence, as the color of i_{st} in $R(P)$ equals the color of i_{st} in $R(P)^-$. Similarly, $R(P)^-$ has the same number of disagreeing edges of a given coloring of $R(P)$, and thus preserves the number of recombinations of any coloring. Thus, by Observation 2, $R(P)$ has a bipartization set of size k if and only if $R(P)^-$ has.

Second, we perturb the graph $(R(P)^-, \phi, \mathcal{S})$ by turning $R(P)^-$ into an uncolored graph $\overline{R(P)}$. The graph $\overline{R(P)}$ has the same vertex set as $R(P)^-$ (with colors on the vertices removed), plus two additional vertices v_r and v_b. The graph contains all edges of $R(P)^-$, plus a *parity edge* for every vertex colored red connecting it to v_b and a parity edge for every vertex colored blue connecting it to v_r. This way, color constraints are preserved. For a graph, a subset B of its edges is called a *bipartization set* if removing the edges in B from the graph yields a bipartite graph.

A bipartization set is *minimal* if it does not include a bipartization set as proper subset. A bipartization set is *respectful* if it also satisfies the parity constraint sets. We claim that respectful bipartization sets of $R(P)^-$ are respectful bipartization set of $\overline{R(P)}$. Those bipartization sets of $\overline{R(P)}$ that are not bipartization sets of $R(P)^-$ contain at least one parity edge. Here we need to compute a bipartization set B (with size at most k) of non-parity edges such that the graph $R(P) - B$ satisfies all parity constraint sets in $\mathcal{S}$; we call such a set B *respectful (with respect to $\mathcal{S}$)*.

4.1   The Exponential Algorithm

A MRHC problem instance has parameters n for the number of individuals, m for the number of sites, and k for the number of recombinations.

The algorithm considers in brute-force fashion the number of recombinations $\{0, 1, 2, ..., k\}$ and stops on the first k such that there exists some set S of k edges whose removal from the graph produces (1) a bipartite graph and (2) satisfies the parity constraints. For each selection of k edges, the two checks require (1) traversing the graph in a depth-first search in time $O(n^2 m^4)$ and (2) computing the parity of all the parity constraint sets in time $O(nm^3)$.

The number of sets S with k recombination edges is $|E|^k$ where $E = E(\overline{R(P)})$ is the edge set of $\overline{R(P)}$ and where $|E| = O(nm^2)$. So, the running time of the whole algorithm is $O(n^{(k+2)}m^{6k})$.

5   Discussion

This paper gives an exponential to compute minimum recombination haplotype configurations for pedigrees with all genotyped individuals, with only polynomial dependence on the number m of sites (which can be very large in practice) and small exponential dependence on the minimum number of recombinations k. This algorithm significantly improves, and corrects, earlier results by Doan and Evans [4,5]. An open question is how this algorithm performs when implemented and applied to data. Another open question is how to handle missing alleles in the data.

Acknowledgments. BK thanks M. Mnich at the Cluster of Excellence, Saarland University, Saarbrücken, Germany for critical reading of the manuscript. BK thanks arXiv for pre-print publication of the full manuscript [10].

References

1. Abecasis, G., Cherny, S., Cookson, W., Cardon, L.: Merlin-rapid analysis of dense genetic maps using sparse gene flow trees. Nat. Genet. **30**, 97–101 (2002)
2. Browning, S., Browning, B.: On reducing the statespace of hidden Markov models for the identity by descent process. Theoret. Popul. Biol. **62**(1), 1–8 (2002)
3. Coop, G., Wen, X., Ober, C., Pritchard, J.K., Przeworski, M.: High-resolution mapping of crossovers reveals extensive variation in fine-scale recombination patterns among humans. Science **319**(5868), 1395–1398 (2008)
4. Doan, D.D., Evans, P.A.: Fixed-parameter algorithm for haplotype inferences on general pedigrees with small number of sites. In: Moulton, V., Singh, M. (eds.) WABI 2010. LNCS, vol. 6293, pp. 124–135. Springer, Heidelberg (2010)
5. Doan, D., Evans, P.: An FPT haplotyping algorithm on pedigrees with a small number of sites. Algorithms Mol. Biol. **6**, 1–8 (2011)
6. Fishelson, M., Dovgolevsky, N., Geiger, D.: Maximum likelihood haplotyping for general pedigrees. Hum. Hered. **59**, 41–60 (2005)
7. Geiger, D., Meek, C., Wexler, Y.: Speeding up HMM algorithms for genetic linkage analysis via chain reductions of the state space. Bioinformatics **25**(12), i196 (2009)
8. Geiger, D., Meek, C., Wexler, Y.: Speeding up HMM algorithms for genetic linkage analysis via chain reductions of the state space. Bioinformatics **25**(12), i196–i203 (2009)
9. Iverson, K.E.: A Programming Language. Wiley, New York (1962)
10. Kirkpatrick, B.: Haplotype inference for pedigrees with few recombinations. arXiv 1602.04270 (2016). http://arxiv.org/abs/1602.04270
11. Lauritzen, S.L., Sheehan, N.A.: Graphical models for genetic analysis. Stat. Sci. **18**(4), 489–514 (2003)

12. Li, J., Jiang, T.: Computing the minimum recombinant haplotype configuration from incomplete genotype data on a pedigree by integer linear programming. J. Comput. Biol. **12**(6), 719–739 (2005)

13. O'Connell, J., Gurdasani, D., et al.: A general approach for haplotype phasing across the full spectrum of relatedness. PLoS Genet **10**(4), e1004234 (2014)

14. Pirola, Y., Bonizzoni, P., Jiang, T.: An efficient algorithm for haplotype inference on pedigrees with recombinations and mutations. IEEE/ACM Trans. Comput. Biol. Bioinform. **9**(1), 12–25 (2012)

15. Risch, N., Merikangas, K.: The future of genetic studies of complex human diseases. Science **273**(5281), 1516–1517 (1996)

16. Sobel, E., Lange, K.: Descent graphs in pedigree analysis: applications to haplotyping, location scores, and marker-sharing statistics. Am. J. Hum. Genet. **58**(6), 1323–1337 (1996)

17. Steel, M., Hein, J.: Reconstructing pedigrees: a combinatorial perspective. J. Theoret. Biol. **240**(3), 360–367 (2006)

18. Thornton, T., McPeek, M.: Case-control association testing with related individuals: a more powerful quasi-likelihood score test. Am. J. Hum. Genet. **81**, 321–337 (2007)

19. Wang, W.B., Jiang, T.: Inferring haplotypes from genotypes on a pedigree with mutations, genotyping errors and missing alleles. J. Bioinform. Comput. Biol. **9**, 339–365 (2011)

20. Xiao, J., Lou, T., Jiang, T.: An efficient algorithm for haplotype inference on pedigrees with a small number of recombinants. Algorithmica **62**(3), 951–981 (2012)

Improved Detection of 2D Gel Electrophoresis Spots by Using Gaussian Mixture Model

Michal Marczyk[(✉)]

Data Mining Group, Institute of Automatic Control, Silesian University
of Technology, Akademicka 16, 44-100 Gliwice, Poland
Michal.Marczyk@polsl.pl

Abstract. 2D gel electrophoresis is the most commonly used method in bio-medicine to separate even thousands of proteins in a complex sample. Although the technique is quite known, there is still a need to find an efficient and automatic method for detection of protein spots on gel images. In this paper a mixture of 2D normal distribution functions is introduced to improve the efficiency of spot detection using the existing software. A comparison of methods is based on simulated datasets with known true positions of spots. Fitting a mixture of components to the gel image allows for achieving higher sensitivity in detecting spots, better overall performance of the spot detection and more accurate esti-mates of spot centers. Efficient implementation of the algorithm enables parallel computing capabilities that significantly decrease the computational time.

Keywords: 2D gel electrophoresis · Spot detection · Gaussian mixture model

1 Introduction

2D gel electrophoresis (2DGE) is a powerful tool for separation and fractionation of complex protein mixtures from tissues, cells or other biological samples. It allows for the separation of thousands of proteins in a single gel. Modeling and image analysis are crucial in extracting the biological information from a 2D gel electrophoresis experi-ment [1]. Goals of such an analysis are the rapid identification of proteins located on a single gel and/or differentially expressed proteins between samples run on a set of 2D gels. Applications of 2DGE include whole proteome analysis, detection of biomarkers and disease markers, drug discovery, cancer research, purity checks, microscale protein purification and product characterization [2, 3]. The lack of efficient, effective and reliable methods for 2D gel analysis has been a major factor limiting the contribution of 2DGE to the biomedical research on a wide scale.

A biological experiment consists of two separation steps: first dimension and second dimension. In the first dimension, protein molecules are resolved depending on their isoelectric point. The separation of proteins under a pH gradient allows intense band recovering using various tactics. In the second dimension, protein separation is performed based on the molecular weight using sodium dodecyl sulfate (SDS) buffers. Due to the fact that it is improbable that different protein molecules may have the same physico-chemical properties, proteins are efficiently separated by 2DGE. A result of applying 2DGE on a protein sample is a 2D grayscale image with dark spots corresponding to

© Springer International Publishing Switzerland 2016
A. Bourgeois et al. (Eds.): ISBRA 2016, LNBI 9683, pp. 284–294, 2016.
DOI: 10.1007/978-3-319-38782-6_24

particular proteins. There are also additional artifacts visible, like the random noise and the background signal. The background signal in 2DGE images is not uniform, but consists of local regions of elevated pixel-intensities. The local background intensity is often correlated with the local density of protein spots. The noise in 2D gel images is usually seen as a common white noise, high-intensity spikes with sharp boundaries covering a few pixels and larger artifacts resulting from the dust and other pollutions acquired during the experimental setup.

The detection and quantification of data features, such as spots in two-dimensional images, is a particularly important component of low-level analysis, because it works to reduce the data size and to gather only true information about the analyzed sample. The goal of this step is to find spot positions with the surrounding boundary and determine their quantities. The simplest approach is to find local extrema in the image and then apply some filtering criteria. A commonly used method called Pinnacle was introduced by Morris et al. [4]. Pinnacle's spot detection is performed on a denoised average image of a properly aligned 2D gel image set and is achieved by detecting local minima and combining them within a defined proximity. Another approach is to model a spot's intensity as a Gaussian normal distribution and derive spots quantity and boundaries from the model. The use of a Gaussian is motivated by the 3D shape of spots and by general considerations on diffusion processes in the gel [5]. Spot models can be used in the spot quantification, with overlapping spots being represented as the sum of multiple single-spot models.

In this paper improvements in the detection of protein spots estimated by existing software, by use of the mixture of 2D functions, are presented. The choice of the detection algorithm is based on the comparative study described in [6]. Pinnacle is a quick and automatic non-commercial method that has been shown to yield better spot detection than commonly used solutions. It is simple to implement and has intuitive tunable parameters. A natural choice for the component function is a 2D Gaussian distribution with circular shape. Comparison of the created algorithm to Pinnacle software is based on a large number of artificially generated datasets, demonstrating improvements in the performance of the spot detection and the accuracy of spot location estimation achieved by the use of the Gaussian mixture modeling.

2 Materials and Methods

2.1 Synthetic Data Simulation

The main goal of the spot detection algorithm is to determine if a particular detected feature is a protein spot or just a consequence of noise. Spot detection methods may be graded by creating synthetic images with a known location and quantification of protein spots. In the literature it is stated that to create a proper 2DGE signal we must assume that the observed image is an effect of accumulating background, random noises and real information about protein abundance [7, 8]. In order to imitate the real 2DGE image more accurately and retain its characteristics, the information about distributions of all image elements is taken from the publicly available dataset with annotated spots called GelA [9]. The noise signal from GelA image is extracted using the Median Modified Wiener Filter [10], that preserves the morphology of close-set spots, and avoids spot and spike fuzzyfication. The Gaussian distribution was fitted to estimate parameters of the noise distribution. The background signal was found using a rolling ball algorithm and a log-normal distribution was fitted to estimate parameters of its distribution. The location of spots in the image was modeled separately in both axes, using a beta distribution. The spots intensity was modeled by a log-normal distribution of pixel intensities. Since a spread of the protein spot is strictly connected with its intensity (higher intensity spots are wider on the gel image) it was found that a second order polynomial may describe such a connection.

By using the algorithm for creating synthetic images, five simulation scenarios with a different number of true proteins, varying from 1000 to 2000, were generated. Each dataset contains 10 images defined over the same equally spaced grid of 1200 points in both directions. For each image spots are generated using a spot model based on diffusion principles that occur in 2DGE experiment [11], given as

$$f(x,y) = \frac{C_0}{2}\left[\operatorname{erf}\left(\frac{a'+r'}{2}\right) + \operatorname{erf}\left(\frac{a'-r'}{2}\right)\right] + \left[\exp\left(-\left(\frac{a'+r'}{2}\right)^2\right) + \exp\left(-\left(\frac{a'-r'}{2}\right)^2\right)\right]$$

(1)

where $r' = \sqrt{\frac{(x-x_0)^2}{D_x} + \frac{(y-y_0)^2}{D_y}}$, C_0 is a height-defining parameter, D_X and D_y are the diffusion width parameters, a' is the area of the disc from which the diffusion process starts and x_0, y_0 are the spot coordinates. All parameters of the spot model (besides area of the disc from which the diffusion process starts, that was set to 3 as in [7]) are drawn from theoretical distributions developed on a real GelA image. 0.5 % of the highest intensity data was cropped to introduce the effect of the spot saturation. At the last step the Gaussian distribution noise and the background signal, defined by smooth spatial stochastic process, were introduced with parameters estimated from the real image. Finally, 50 synthetic images with a total number of 75,000 spots were created (Fig. 1).

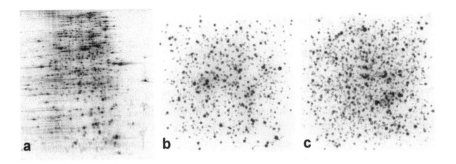

Fig. 1. Example of 2DGE images: (a) GelA image, (b) synthetic image with 1000 spots, (c) synthetic image with 2000 spots

2.2   2D Mixture Modeling

To detect spots in a 2DGE image a similar idea that was presented in Polanski et al. [12] for analysis of 1D spectra is applied. The partition of the gel image into fragments is augmented by the use of marker-controlled watershed segmentation to separate touching objects in the image. First, morphological techniques called "opening-by-reconstruction" and "closing-by-reconstruction" are used to clean up the image and flat maxima inside each object that can be separated. Next, thresholding by use of the Otsu method is performed to establish background regions. Computing the skeleton by influence zones of the foreground can thin the background. It is done by computing the watershed transform of the distance transform of background signal and looking for the watershed ridge lines.

The mixture of 2D Gaussian components can be defined as:

$$f(x, y) = \sum_{k=1}^{K} \alpha_k f_k(x, y, \mu_k, \Sigma_k) \tag{2}$$

where K stands for the number of Gaussian components, α_k, $k = 1,2,...K$ are component weights, which sum up to 1, f_k is the probability density function of the k-th Gaussian component, μ_k is the location of the component and Σ_k is the component covariance. Mixture models of 2DGE images are fitted locally to fragments of the image by use of a modified version of the expectation-maximization (EM) algorithm, which stabilizes successive estimates of parameters of the mixture distribution. If a standard EM algorithm for 2D data modeling is used we must create a $2 \times nI$ matrix of input values, where nI is a number of all pixels n multiplied by total intensity values in modeled image I. Particularly, for each pixel the same m vectors of size 2×1 with its coordinates are created, where m is a signal intensity in that pixel. For images with more than 8 bit color depth it will create huge input matrices, what reduces efficiency of the algorithm drastically. In the proposed modification two variables are created: $2 \times n$ input matrix of pixel coordinates and $1 \times n$ vector of pixel intensities. Additionally, pixels with zero intensity values are removed to reduce the data size and increase the calculation speed. The vector of spot intensities is used as a weight for pixel coordinates in a maximization step. Such modification gives a better fit of the model when intensity values are not integers, and exactly the same solution as the standard EM in other cases. In all scenarios, memory and computational time are drastically reduced using the modified EM. The introduced method of the spot detection is called 2DGMM in further text.

An important element of the EM algorithm is the choice of initial conditions. Theoretical positions of spots may be found by applying the method based on local maxima. In this paper Pinnacle software is used to find spots, which may serve as initial conditions for parameters of Gaussian components (mean and covariance) and number of model components K. To provide the regularization of the GMM model a Bayesian information criterion is used. The final number of model components K is estimated in a backward elimination scheme. Mixture models created locally on obtained image fragments are aggregated into the mixture model of the whole image.

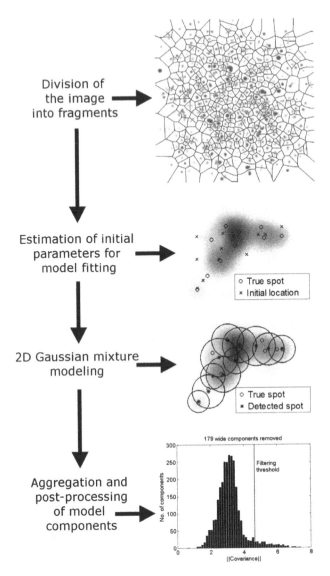

Fig. 2. Flowchart of the algorithm with exemplary images.

The mixture model is well fitted to 2DGE image and it may still represent a small amount of the background signal and noise. The post-processing of model components is needed to remove signal artifacts that failed to have been corrected. In the proposed approach distributions of some parameters of model components are analyzed. Too wide components are filtered out by setting a threshold value on the component covariance. By analyzing the distribution of height of the components, too small components are detected and removed. To find the threshold for removing components an outlier detection method, that was created to deal with skewed distributions [13], is used. The modeling method with enabled post-processing of model components is called 2DGMM-proc in further text. Flowchart of the algorithm is presented in Fig. 2.

3 Results and Discussion

The comparison of methods of the spot detection is based on simulated datasets, where true positions of spots are known. Structure of the simulated data is changed by assuming a different number of true spots in a gel. In this paper the detection power of different algorithms is analyzed. Spots detected by Pinnacle and GMM-based methods are treated as a true positive finding if the distance to the true location is lower than three pixels. Several performance indexes are computed to compare results obtained by different algorithms. False discovery rate (FDR) is the number of spots among those detected by the procedure which do not correspond to true spots, divided by the number of all spots detected by the procedure. Sensitivity index (Sens) is the number of true spots detected by the procedure divided by the number of all true spots in the sample. FDR and Sens performance measures are aggregated into the one index F1, which is defined as the harmonic mean of 1-FDR and Sens. Higher values of F1 score imply better performance and lower values - poorer performance of the evaluated method. Also, the number of spots detected by a spot detection algorithm is reported.

Pinnacle algorithm has few tunable parameters, which should be chosen prior to their application. Two main parameters are tuned: minimum spot intensity (q_a) and minimum distance to another spot (δ). In this analysis the q_a parameter is changed in a range from 0.6 to 0.98 and δ parameter in a range from 2 to 20. F1 index is used as a base for optimizing above mentioned parameters and for each of the three algorithms examined, the best parameters are found. For all scenarios the proposed range of parameters searching provides finding a global maximum of F1 score. All parameters for methods of image fragmentation and model building were selected automatically or were set as a constant in all simulation scenarios.

For each scenario with a given number of spots and for each spot detection method optimal values of two parameters are found by averaging F1 score calculated for 10 images within each scenario. Results of the spot detection after applying all algorithms with their optimal parameters are shown in Fig. 3. Sensitivity of GMM-based algorithms is higher than the one obtained after using Pinnacle. All methods show the same patterns of change; sensitivity of detecting true spots decreases with increase of the number of true spots. In the plots of FDR index it can be seen that Pinnacle method gives a lower number of false positives. It may be caused by the fact that in 2DGMM one additional large spot, which corresponds to background level, is introduced for each gel fragment. Also, it is desirable in the model fitting procedure to find more spots for each fragment of a gel to represent full image pattern, including remaining artifacts. Filtering of components gives indispensable decrease in FDR, which however may still be improved. For all methods a slight increase in FDR can be noticed when gel complexity grows. F1 score, which is treated as an index of overall detection performance, shows that using 2DGMM-proc method gives the best results for the whole range of numbers of true spots. When the number of spots is greater than 1500

2DGMM method gives better results than Pinnacle. It may be caused by an increase in the number of overlapping spots, which are better represented using the mixture model. Setting optimal parameters for Pinnacle method gives a lower number of detected spots than the number of true spots in all scenarios. For GMM-based methods these numbers are comparable.

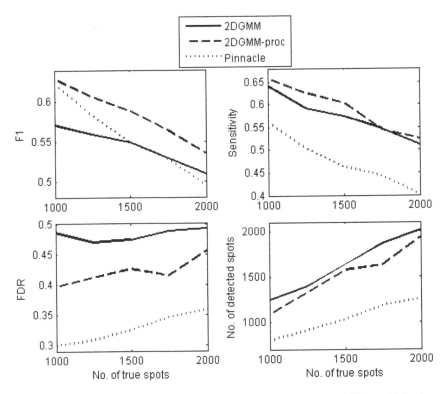

Fig. 3. Performance of the spot detection measured by F1 score (upper left), sensitivity (upper right), false discovery rate (lower left) and the number of detected spots (lower right).

In previous calculations use of optimal parameters for all methods is assumed. To check if applying GMM improves the efficiency of detecting spots, a percentage difference in F1 score, sensitivity, FDR and the number of detected spots between Pinnacle and 2DGMM-proc is calculated for all parameter settings. Additionally, it is checked how the post-processing of model components influences the performance of the spot detection by comparing differences between results of applying 2DGMM and 2DGMM-proc. In Table 1 median values of detection quality scores are presented. In both comparisons using 2DGMM-proc leads to the increase of F1 score (5.95 % to 2DGMM and 17.69 % to Pinnacle, in average). Filtering about 13 % of 2DGMM model components gives a small decrease in spot detection sensitivity (–0.3 % in average) and fine decrease in finding false positive spots (–14.06 % in average). About

17.69 increase of sensitivity is observed, when 2DGMM_proc method is used on the initial conditions found by Pinnacle. Due to the factors described in previous paragraph about 5.41 % increase in FDR is observed. To obtain a full potential of applying mixture models to 2DGE images analysis, a more sophisticated method for discrimination between those components, which are likely to correspond to protein spots and those, which were rather generated by noise in the signal and/or by residuals of baseline, is needed. For example, distributions of different model components parameters, like component weight or coefficient of variation, may be analyzed or the algorithm described in [14] may be used to find filtering thresholds.

Table 1. A median percentage difference in the spot detection performance indices between different methods.

No. of true spots	2DGMM_proc vs 2DGMM				2DGMM_proc vs Pinnacle			
	F1	Sensitivity	FDR	No. of spots	F1	Sensitivity	FDR	No. of spots
1000	6.35	−0.16	−12.29	−14.02	3.85	14.29	7.65	32.19
1250	6.99	−0.35	−16.18	−14.64	8.93	17.61	4.58	30.51
1500	6.36	−0.48	−16.16	−13.94	12.54	19.31	1.20	29.61
1750	5.59	−0.33	−12.94	−11.59	7.15	16.64	9.55	24.52
2000	4.50	−0.45	−12.75	−11.07	13.76	20.62	4.11	28.84

The accuracy of estimating spot locations is found by calculating the Euclidean distance between the center of a detected spot and the true location. In Fig. 4 improvement in the accuracy of estimation of spot positions, achieved by application of GMM-based methods, is highlighted. In Pinnacle spot center is localized by indicating the particular pixel with locally highest intensity. In that case the average distance to true location is about square root of 2, which corresponds to the distance of 1 pixel in both directions. When the mixture model is introduced spot center locations may be found among image pixels, as a result of the shape of the whole spot. Such ability leads to reducing the accuracy of the spot location comparing to Pinnacle software. When the number of spots in an image grows, finding of a proper spot center gets harder. In such case the accuracy is slightly decreased.

Computations for all datasets were performed by use of the computational server with two six core Intel Xeon X5680 processors (3.4 GHz in normal work, 3.6 GHz in turbo mode) and 32 GB DDR3 1333 MHz RAM memory. Since GMM modeling is performed for each image fragment separately and there is no communication between analyzed fragments, it is possible to introduce a perfectly parallel computation. The average processing time of 2DGMM-proc method with parallel and single core computations is presented in Table 2. By average, a 5.43× decrease in the computational time was observed for 1200 × 1200 pixels images with the number of spots varying from 1000 to 2000. When the number of spots grows parallel implementation gives more benefits. Analyzing a single image in parallel mode takes about 18 s.

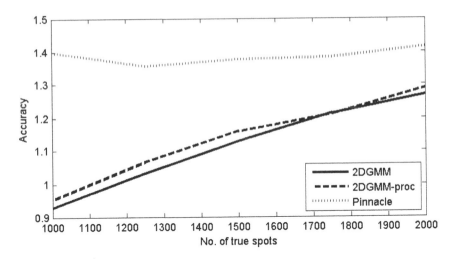

Fig. 4. A mean distance between the center of detected spots and true locations.

Table 2. A mean computational time of 2DGMM-proc in seconds

No. of true spots	Parallel	Single core	Boost
1000	14.33	69.92	4.88×
1250	16.73	87.46	5.23×
1500	17.80	103.51	5.81×
1750	18.23	105.03	5.76×
2000	23.23	127.14	5.47×

3.1　Comparison to Compound Fitting Algorithm

Fitting two dimensional Gaussian function curves for the extraction of data from 2DGE images was recently studied by Brauner et al. [7]. They propose a Compound Fitting (CF) algorithm based on simultaneous fitting of neighboring groups of spots detected by any spot detection method (built-in algorithm is also available). The main difference between CF and 2DGMM-proc algorithm is that they use nonlinear least-square error type of fit and assume some bounds for the model parameters. By default, the initial peak position can be refined by the fit only by 2 pixels. In that case, fitting a Gaussian curve is used rather as a spot quantification method than the spot detection method. In our algorithm there are no constraints on estimating peak positions, so the performance of spot detection and quantification can be improved simultaneously.

CF algorithm is semi-automatic, thus it requires from the user to mark the center of two clearly separable spots with the smallest distance between them. Based on this information the input parameter w is calculated. After the spot detection step the user must set the parameter t, that is the number of standard deviations a pixel has to be higher than the local background to be accepted as a spot, by visual inspection of the gel image with highlighted spots. The influence of both parameters on the simulated data was analyzed. Setting $w = 10$ and $t = 2$ gave the best visual results. Other parameters, like the compound area edge length and the maximum number of pixel for which a peak position can be refined by the fit, was set as default.

Comparing the overall performance of the spot detection (Table 3) CF algorithm gives higher values of F1 score. Sensitivity is higher when 2DGMM_proc is used. As previously, both methods show the same patterns of change. Good performance of CF algorithm is provided by a low value of FDR. The accuracy of estimating the location of a spot is comparable between methods. These findings show that 2DGMM_proc method finds more true spots than CF, but still there are too many false positives. Post-processing of model components must be improved. It will be also interesting too check the peak detection method used in CF as an initial condition step in 2DGMM_proc application.

Table 3. A comparison of the spot detection performance and accuracy between two methods.

No. of true spots	2DGMM_proc				Compound fitting			
	F1	Sensitivity	FDR	Accuracy	F1	Sensitivity	FDR	Accuracy
1000	0.63	**0.65**	0.39	**0.71**	**0.74**	0.64	**0.12**	0.81
1250	0.61	**0.62**	0.41	**0.87**	**0.69**	0.58	**0.12**	0.88
1500	0.59	**0.60**	0.42	1.02	**0.65**	0.52	**0.13**	0.97
1750	0.56	**0.54**	0.41	1.08	**0.61**	0.47	**0.13**	1.04
2000	0.53	**0.52**	0.45	1.19	**0.55**	0.41	**0.15**	1.10

4   Conclusions

The idea of proposed algorithm for the detection of protein spots in 2DGE images is simple. A full image is cut into fragments and an attempt is made to improve the already discovered pattern of spots by application of a mixture model. Aggregation of obtained results leads to a precise mixture model of gel image. Some mixture components obtained in the iterative EM algorithm do not correspond to protein spots. They are too wide or too low, so they are filtered out from the final model.

In the cases where there are clusters of overlapping spots, like in scenarios with the highest number of true spots, the mixture model enables detecting components hidden behind others. Model components are well characterized by the accurate spot position and shape, while in some spot detection methods the information on shape is missing. Fitting the mixture model to synthetic 2DGE image allows for achieving higher sensitivity in detecting spots and better overall performance of the spot detection than applying the Pinnacle algorithm. Parallel implementation of the algorithm gives significant decrease in the computational time.

Acknowledgments. This work was financially supported by the internal grant for young researchers from Silesian University of Technology number BKM/514/RAu1/2015/34. All calculations were carried out using GeCONiI infrastructure funded by the project number POIG.02.03.01-24-099/13.

References

1. Magdeldin, S., Enany, S., Yoshida, Y., Xu, B., Zhang, Y., Zureena, Z., Lokamani, I., Yaoita, E., Yamamoto, T.: Basics and recent advances of two dimensional-polyacrylamide gel electrophoresis. Clin. Proteomics **11**, 16 (2014)
2. Magdeldin, S., Li, H., Yoshida, Y., Enany, S., Zhang, Y., Xu, B., Fujinaka, H., Yaoita, E., Yamamoto, T.: Comparison of two dimensional electrophoresis mouse colon proteomes before and after knocking out Aquaporin 8. J. Proteomics **73**, 2031–2040 (2010)
3. Wu, W., Tang, X., Hu, W., Lotan, R., Hong, W.K., Mao, L.: Identification and validation of metastasis-associated proteins in head and neck cancer cell lines by two-dimensional electrophoresis and mass spectrometry. Clin. Exp. Metastasis **19**, 319–326 (2002)
4. Morris, J.S., Clark, B.N., Gutstein, H.B.: Pinnacle: a fast, automatic and accurate method for detecting and quantifying protein spots in 2-dimensional gel electrophoresis data. Bioinformatics **24**, 529–536 (2008)
5. Berth, M., Moser, F.M., Kolbe, M., Bernhardt, J.: The state of the art in the analysis of two-dimensional gel electrophoresis images. Appl. Microbiol. Biotechnol. **76**, 1223–1243 (2007)
6. Morris, J.S., Clark, B.N., Wei, W., Gutstein, H.B.: Evaluating the performance of new approaches to spot quantification and differential expression in 2-dimensional gel electrophoresis studies. J. Proteome Res. **9**, 595–604 (2010)
7. Brauner, J.M., Groemer, T.W., Stroebel, A., Grosse-Holz, S., Oberstein, T., Wiltfang, J., Kornhuber, J., Maler, J.M.: Spot quantification in two dimensional gel electrophoresis image analysis: comparison of different approaches and presentation of a novel compound fitting algorithm. BMC Bioinform. **15**, 181 (2014)
8. Shamekhi, S., Miran Baygi, M.H., Azarian, B., Gooya, A.: A novel multi-scale Hessian based spot enhancement filter for two dimensional gel electrophoresis images. Comput. Biol. Med. **66**, 154–169 (2015)
9. Raman, B., Cheung, A., Marten, M.R.: Quantitative comparison and evaluation of two commercially available, two-dimensional electrophoresis image analysis software packages, Z3 and Melanie. Electrophoresis **23**, 2194–2202 (2002)
10. Cannistraci, C.V., Montevecchi, F.M., Alessio, M.: Median-modified Wiener filter provides efficient denoising, preserving spot edge and morphology in 2-DE image processing. Proteomics **9**, 4908–4919 (2009)
11. Bettens, E., Scheunders, P., Van Dyck, D., Moens, L., Van Osta, P.: Computer analysis of two-dimensional electrophoresis gels: a new segmentation and modeling algorithm. Electrophoresis **18**, 792–798 (1997)
12. Polanski, A., Marczyk, M., Pietrowska, M., Widlak, P., Polanska, J.: Signal partitioning algorithm for highly efficient Gaussian mixture modeling in mass spectrometry. PLoS ONE **10**, e0134256 (2015)
13. Hubert, M., Van der Veeken, S.: Outlier detection for skewed data. J. Chemometr. **22**, 235–246 (2008)
14. Marczyk, M., Jaksik, R., Polanski, A., Polanska, J.: Adaptive filtering of microarray gene expression data based on Gaussian mixture decomposition. BMC Bioinform. **14**, 101 (2013)

Abridged Track 2 Abstracts

Predicting Combinative Drug Pairs via Integrating Heterogeneous Features for Both Known and New Drugs

Jia-Xin Li[1], Jian-Yu Shi[1], Ke Gao[2], Peng Lei[3], and Siu-Ming Yiu[4]

[1] School of Life Sciences, Northwestern Polytechnical University, Xi'an, China
lijiaxin0932@mail.nwpu.edu.cn, jianyushi@nwpu.edu.cn
[2] School of Computer Science, Northwestern Polytechnical University,
Xi'an, China
vipgaoke@163.com
[3] Department of Chinese Medicine, Shaanxi Provincial People's Hospital,
Xi'an, China
leipengml@163.com
[4] Department of Computer Science, The University of Hong Kong,
Pok Fu Lam, Hong Kong
smyiu@cs.hku.hk

An ordinary disease caused by the anomaly of expression level of an individual gene, can be treated by a specific drug, which regulates the gene's expression. However, this single-drug treatment has very low effectiveness on complex diseases [1], which usually involve multiple genes and pathways of the metabolic network. Drug combination, as one of multiple-target treatments, has demonstrated its effectiveness in treating complex diseases, such as HIV/AIDS [2] and colorectal cancer [3]. However, it is still costly and time-consuming in clinical trials to find an effective combination of individual drugs. Fortunately, both the number of approved drug combinations [4] and the amount of available heterogeneous information about drugs are increasing. It became feasible to develop computational approaches to predict potential candidates of drug pairs for the treatments of complex diseases [5, 6].

Current computational approaches can be roughly categorized into two groups, disease-driven and drug-driven. The former, developed for a specific disease, relies heavily on disease-associated genes and targets in pathways [5]. Disease-driven approaches are able to predict multiple drug combinations for a specific disease. However, current approaches use only genotype information, but have not yet integrated other information, such as pharmacology or clinic phenotype.

In contrast, drug-driven approaches focus on drugs, not diseases, and can be applied to all drugs in a large scale manner. They characterize each drug as a feature vector capturing various attributes of the drug [6], and directly apply supervised learning to predict unknown drug pairs, based on the assumption that combinative drug pairs (positives) are similar to each other and different from conflicting or ineffective drug pairs (negatives). In these approaches, heterogeneous features are usually

This work was supported by the Fundamental Research Funds for the Central Universities of China (Grant No. 3102015ZY081).

A. Bourgeois et al. (Eds.): ISBRA 2016, LNBI 9683, pp. 297–298, 2016.
DOI: 10.1007/978-3-319-38782-6

combined into a high-dimensional vector in a straight-forward manner, which however leads to a time-consuming training as well as over-parameterized or over-fitted classifier model. In addition, existing drug-driven approaches mainly are only applicable to "known" drugs which were already used in some approved combinative drug pairs, but not to new drugs.

In summary, there are two issues that have been not addressed by former approaches. (i) Existing approaches do not have an effective method to integrate or make full use of heterogeneous features, in particular, none of the approaches have used information from pharmaceutical drug-drug interaction (DDI). (ii) They are not applicable or effective for "new" drugs that were not approved to be combined with other drugs. Predicting potential drug combination among "new" drugs remains difficulty.

In this paper, we focus on developing a novel drug-driven approach that tackles the above two issues. (1) Instead of combining heterogeneous features simply as a high-dimensional vector, we define features based on heterogeneous data, including pharmaceutical DDI, ATC codes, drug-target interactions (DTI) and side effects (SE), where we, in particular, make use of pharmaceutical DDI information not used by existing approaches. In addition, we design an efficient fusion scheme to integrate four heterogeneous features with the advantage of avoiding high-dimensional feature vector, which usually causes time-consuming training as well as over-parameterized or over-fitted classifier model. (2) More importantly, we propose the appropriate schemes of cross validation for three different predicting scenarios, including predicting potential combinative pairs (S1) between known drugs, (S2) between new drugs and known drugs, and (S3) between any two new drugs. Experiments on real data show that our proposed method is effective, not only for predicting unknown pairs among known drugs (AUC = 0.954, AUPR = 0.821 in S1), but also for providing the first tool to predict potential pairs between known drugs and new drugs (AUC = 0.909, AUPR = 0.635 in S2) and potential pairs among new drugs (AUC = 0.809, AUPR = 0.592 in S3). In addition, we provide a detailed analysis on how each kind of heterogeneous data is related to the formation of combinative drug pairs that motivates the design of our approach.

References

1. Jia, J., Zhu, F., Ma, X., Cao, Z., Li, Y., Chen, Y.Z.: Mechanisms of drug combinations: interaction and network perspectives. Nat. Rev. Drug Discov. **8**, 111–128 (2009)
2. Henkel, J.: Attacking AIDS with a 'cocktail' therapy? FDA Consum **33**, 12–17 (1999)
3. Feliu, J., Sereno, M., De Castro, J., Belda, C., Casado, E., Gonzalez-Baron, M.: Chemotherapy for colorectal cancer in the elderly: who to treat and what to use. Cancer Treat. Rev. **35**, 246–254 (2009)
4. Liu, Y., Hu, B., Fu, C., Chen, X.: DCDB: drug combination database. Bioinformatics **26**, 587–588 (2010)
5. Pang, K., Wan, Y.W., Choi, W.T., Donehower, L.A., Sun, J., Pant, D., Liu, Z.: Combinatorial therapy discovery using mixed integer linear programming. Bioinformatics **30**, 1456–1463 (2014)
6. Huang, H., Zhang, P., Qu, X.A., Sanseau, P., Yang, L.: Systematic prediction of drug combinations based on clinical side-effects. Sci. Rep. **4**, 7160 (2014)

SkipCPP-Pred: Promising Prediction Method for Cell-Penetrating Peptides Using Adaptive k-Skip-n-Gram Features on a High-Quality Dataset

Wei Leyi and Zou Quan[✉]

School of Computer Science and Technology, Tianjin University, Tianjin,
People's Republic of China
zouquan@nclab.net

Cell-penetrating peptides (CPPs) are short peptides usually comprising 5–30 amino acid residues. Also known as protein transduction domains (PTDs), membrane translocating sequences (MTSs), and Trojan peptides, CPPs can directly enter cells without significantly damaging the cell membrane [1, 2]. This unique ability of CPPs could be exploited to improve the cellular uptake of various bioactive molecules, which is inherently poor because bioactive cargoes tend to become trapped in the endosomes. When transported by CPPs, these cargoes are immediately freed in the cytosol to reach their intracellular targets (immediate bioavailability). CPPs are considered as very promising tools for non-invasive cellular import of cargoes, and have been successfully applied in in vitro and in vivo delivery of therapeutic molecules (e.g., small chemical molecules, nucleic acids, proteins, peptides, liposomes and particles). They also offer great potential as future therapeutics [2, 3] such as gene therapy and cancer treatments. The medical applicability of CPPs would be further enhanced by correct classification of peptides into CPPs or non-CPPs.

Predicting CPPs by traditional experimental methods is time-consuming and expensive. Thus, there is an urgent demand for fast prediction by computational methods. Most of the recent computational methods are based on machine-learning algorithms, which can automatically predict the cell-penetrating capability of a peptide. Although machine-learning-based methods have intrinsic advantages (time- and cost-saving) over experimental methods, they are less reliable than experimental methods. Therefore, they can play only a complementary role to experimental methods. Consequently, improving the predictive ability of computational predictors has been the major concern in this field.

In recent years, the predictive performance of computational methods has improved. However, such improvements seem doubtful because the datasets used for model training are non-representative. Most of the benchmark datasets used in the literature are too small to yield statistical results. For example, each of the four datasets constructed by Sanders et al. [4] contains fewer than 111 true CPPs. Besides being statistically insufficient, existing benchmark datasets are highly redundant, which biases the prediction results. For instance, the sequences in the current largest dataset proposed by Gautam et al. [5] share high sequence similarity. However, the high

© Springer International Publishing Switzerland 2016
A. Bourgeois et al. (Eds.): ISBRA 2016, LNBI 9683, pp. 299–300, 2016.
DOI: 10.1007/978-3-319-38782-6

performance of their method on their proposed dataset (> 90 % accuracy) is probably not generalizable to other datasets. Therefore, a representative benchmark dataset is essential for robust CPP prediction by computational methods.

Motivated by the aforementioned limitations of existing benchmark datasets, we propose a high-quality updated benchmark dataset of CPPs with no more than 80 % similarity. Moreover, the new CPP dataset is sufficiently large to build prediction models. Considering the importance of negative samples on predictive performance [6], the collected negative samples (non-CPPs) are strictly based on the distribution of true CPPs in the dataset. To our knowledge, the proposed dataset is the most stringent benchmark dataset in the literature. Using this dataset, we then train a novel CPP prediction method called SkipCPP-Pred, which integrates the features of the proposed adaptive k-skip-2-gram and the RF classifier. As demonstrated by jackknife results on the proposed new dataset, the accuracy (ACC) of SkipCPP-Pred is 3.6 % higher than that of state-of-the-art methods. The proposed SkipCPP-Pred is freely available from an online server (http://server.malab.cn/SkipCPP-Pred/Index.html), and is anticipated to become an efficient tool for researchers working with CPPs.

Acknowledgments. The work was supported by the National Natural Science Foundation of China (No. 61370010).

References

1. Kilk, K.: Cell-penetrating peptides and bioactive cargoes. Strategies and mechanisms (2004)
2. Milletti, F.: Cell-penetrating peptides: classes, origin, and current landscape. Drug Discov. Today **17**(15), 850–860 (2012)
3. Heitz, F., Morris, M.C., Divita, G.: Twenty years of cell-penetrating peptides: from molecular mechanisms to therapeutics. Br. J. Pharmacol. **157**(2), 195–206 (2009)
4. Sanders, W.S., Johnston, C.I., Bridges, S.M., Burgess, S.C., Willeford, K.O.: Prediction of cell penetrating peptides by support vector machines. PLoS Comput. Biol. **7**(7), e1002101 (2011)
5. Gautam, A., Chaudhary, K., Kumar, R., Sharma, A., Kapoor, P., Tyagi, A., Raghava, G.P.: In silico approaches for designing highly effective cell penetrating peptides. J. Transl. Med. **11**(1), 74 (2013)
6. Wei, L., Liao, M., Gao, Y., Ji, R., He, Z., Zou, Q.: Improved and promising identification of human microRNAs by incorporating a high-quality negative set. IEEE/ACM Trans. Comput. Biol. Bioinf. **11**(1), 192–201 (2014)

CPredictor2.0: Effectively Detecting Both Small and Large Complexes from Protein-Protein Interaction Networks

Bin Xu[1(✉)], Jihong Guan[1], Yang Wang[3], and Shuigeng Zhou[2]

[1] Department of Computer Science and Technology, Tongji University,
Shanghai 201804, China
{1110423,jhguan}@tongji.edu.cn

[2] Shanghai Key Lab of Intelligent Information Processing, School of Computer
Science, Fudan University, Shanghai 200433, China
sgzhou@fudan.edu.cn

[3] School of Software, Jiangxi Normal University, Nanchang 330022, China
yang1995t@163.com

1 Introduction

In organisms, most biological functions are not performed by a single protein, but by complexes that consist of multiple proteins [1, 2]. Protein complexes are formed by proteins with physical interactions. With the advance of biological experimental technologies and systems, huge amounts of high-throughput *protein-protein interaction* (PPI) data are available. PPI data are usually represented as a *protein-protein interaction network* (PIN) where nodes are proteins and edges indicate interactions between the corresponding proteins (nodes). Existing methods have demonstrated their abilities to predict protein complexes from PINs. However, up to now there is no approach that is able to provide a unified framework to detect protein complexes of various sizes from PINs with an acceptable performance. In the complex dataset of MIPS [3], there are 61 size-two complexes, 42 size-three complexes and 170 larger complexes; while in CYC2008 [4], there are 156, 66 and 127 size-two complexes, size-three complexes and larger complexes, respectively. Small complexes and large complexes both account for a large proportion of the total complexes. A size-two complex is actually a single edge in the PIN. A size-three complex consists of three proteins with two or three protein interactions. Traditional graph clustering method is not effective in detecting such small-size complexes. Thus, it is important and challenging to detect protein complexes of all sizes.

In this study, a new method named CPredictor2.0, which considers both protein functions and protein interactions, is proposed to detect both small and large complexes from a given PIN.

This work was partially supported by the Program of Shanghai Subject Chief Scientist under grant No. 15XD1503600, and NSFC under grant No. 61272380.

A. Bourgeois et al. (Eds.): ISBRA 2016, LNBI 9683, pp. 301–303, 2016.
DOI: 10.1007/978-3-319-38782-6

2   Methods

Our approach mainly consists of three steps:

Step 1: According to the function scheme of MIPS, protein functions are described by terms of various levels in a hierarchy. In order to compare functions among proteins, we first take the functional annotations specified by the terms of the first N levels in the hierarchy, where N is a user-specified parameter. Then, we assign proteins into the same group if they possess similar annotations.

Step 2: A network is built upon each protein group. Nodes are proteins in the group; Two proteins are linked if they interact with each other according to PPI data. Then, in each network, we employ the Markov Clustering Algorithm (MCL) [5] to detect preliminary clusters.

Step 3: Highly overlapping clusters will be merged in case of redundancy. The derived clusters are regarded as predicted protein complexes.

3   Results

We use three PPI datasets of Saccharomyces cerevisiae, including Gavin *et al*, Krogan *et al.* and Collins *et al.* Protein complex datasets MIPS and CYC2008 are used as benchmark datasets. We consider complexes with two or three proteins as small complexes, and those with at least four proteins as large complexes. We compare our method with several existing methods. Protein complexes are detected from the three PPI datasets, and compare with MIPS and CYC2008 as benchmark datasets respectively. The *F-measure* values are shown in Table 1. Results of small complexes and large complexes, indicated by S and L, are presented separately. It is obvious that the proposed method CPredictor2.0 significantly outperforms the other methods in most cases.

Table 1. *F-measure* comparison between CPredictor2.0 and six existing methods.

(a) MIPS as benchmark set

Method	Gavin *et al.*		Krogan *et al.*		Collins *et al.*	
	S	L	S	L	S	L
MCODE	0.08	0.39	0.06	0.37	0.08	0.50
RNSC	0.10	0.45	0.11	0.39	0.21	0.50
DPClus	0.11	0.34	0.10	0.27	0.20	0.45
CORE	0.16	0.21	0.15	0.08	0.21	0.35
ClusterONE	0.14	0.44	0.10	0.23	0.23	0.50
CPredictor	0.16	0.46	0.13	0.46	0.25	0.52
CPredictor2.0	**0.21**	**0.59**	**0.23**	**0.54**	**0.27**	**0.62**

(b) CYC2008 as benchmark set

Method	Gavin *et al.*		Krogan *et al.*		Collins *et al.*	
	S	L	S	L	S	L
MCODE	0.04	0.50	0.04	0.41	0.05	0.60
RNSC	0.22	0.50	0.24	0.50	0.35	**0.63**
DPClus	0.08	0.44	0.12	0.33	0.32	0.54
CORE	0.21	0.21	0.24	0.13	0.35	0.45
ClusterONE	0.17	0.53	0.16	0.28	0.35	0.56
CPredictor	0.18	0.55	0.20	**0.53**	0.32	**0.63**
CPredictor2.0	**0.30**	**0.59**	**0.34**	**0.53**	**0.39**	0.62

References

1. Gavin, A.C., Bösche, M., Krause, R., Grandi, P., Marzioch, M., Bauer, A., Schultz, J., Rick, J.M., Michon, A.M., Cruciat, C.M., et al.: Functional organization of the yeast proteome by systematic analysis of protein complexes. Nature **415**(6868), 141–147 (2002)

2. Gavin, A.C., Aloy, P., Grandi, P., Krause, R., Boesche, M., Marzioch, M., Rau, C., Jensen, L.J., Bastuck, S., Dümpelfeld, B., et al.: Proteome survey reveals modularity of the yeast cell machinery. Nature **440**(7084), 631–636 (2006)
3. Mewes, H.W., Frishman, D., Güldener, U., Mannhaupt, G., Mayer, K., Mokrejs, M., Morgenstern, B., Münsterkötter, M., Rudd, S., Weil, B.: Mips: a database for genomes and protein sequences. Nucleic Acids Res. **30**(1), 31–34 (2002)
4. Pu, S., Wong, J., Turner, B., Cho, E., Wodak, S.J.: Up-to-date catalogues of yeast protein complexes. Nucleic Acids Res. **37**(3), 825–831 (2009)
5. Enright, A.J., Van Dongen, S., Ouzounis, C.A.: An efficient algorithm for large-scale detection of protein families. Nucleic Acid Res. **30**(7), 1575–1584 (2002)

Structural Insights into Antiapoptotic Activation of Bcl-2 and Bcl-xL Mediated by FKBP38 and tBid

Valery Veresov and Alexander Davidovskii

Department of Cell Biophysics, Institute of Biophysics and Cell Engineering,
NAS of Belarus, Ackademicheskaya Str 27, 220072 Minsk, Belarus
{veresov,davidovskii}@ibp.org.by

The recruitment of antiapoptotic proteins Bcl-2 and Bcl-xL to the mitochondrial outer membrane and their activation by BH3-only protein tBid are essential for their anti-apoptotic function. In a number of neuronal tissues the mitochondrial targeting of Bcl-2 and Bcl-xL is executed by FKBP38, an integral mitochondrial outer-membrane protein [1]. In the absence of detailed structural information, the molecular mechanisms of the underlying interactions remain elusive and FKBP38 activity in apoptosis regulation is contradictory. Here, computational structural biology tools were applied to gain structural insights into mechanisms of interactions between FKBP38, antiapoptotic proteins Bcl-2 and Bcl-xL, and BH3-only protein tBid.

Because the atomic-level structure of FKBP38 is absent, we determined the 3D-structure of this membrane protein by computational modeling. To do this, the FKBP38 protein was divided, in accordance with experimental data, into the membrane (residues 390-412) and cytosol (residues 1-389) parts and the 3D-structures of two overlapping segments, those of residues 1-390 and 360-412, were modeled using two different protocols. In the former case, homology modeling program PHYRE [2] and the iterative threading assembly refinement (I-TASSER) protocol [3] were applied. In the latter case the MEMOIR system of membrane proteins homology modeling was used [4]. Thereafter two structures obtained by these simulations were superimposed within the segment of the 360-390 residues and were next integrated into one structure using the Rosetta Loop Closure protocol [5]. The OPM database ("Orientations of proteins in membranes database") [6] was used to predict the orientations of FKBP38 in the membrane.

The prediction of the 3D-structures of the FKBP38/Bcl-2, FKBP38/Bcl-xL, FKBP38/Bcl-xL/tBid and FKBP38/tBid protein-protein complexes was performed in a stepwise fashion with an initial rigid-body global search and subsequent steps to refine initial predictions. To do this, a four - stage molecular docking protocol PIPER [7] – ROSETTADOCK$_1$ [8] – HADDOCK [9] – ROSETTADOCK$_2$ (abbreviated by PRHR) was used. Clustering of structures and evaluation of energy funnels were used to improve the ability of finding the correct structures of the complexes. In the present work, the ranking by binding affinities among different complexes was based on the ROSETTADOCK$_2$ interface energy score (I_sc). This ranking was favored over that using the ROSETTADOCK binding score (RDBS) [10], as the one which correlates more universally with experimental binding affinity data and is free from the problems

© Springer International Publishing Switzerland 2016
A. Bourgeois et al. (Eds.): ISBRA 2016, LNBI 9683, pp. 304–306, 2016.
DOI: 10.1007/978-3-319-38782-6

associated with global minimization of pulled apart individual components; these problems are especially grave when membrane proteins or protein complexes are docked.

The PRHR docking strategy has resulted in the structure of the FKBP38/Bcl-2 complex possessing good shape complementarity ($\Delta VDWS_{RD} = -148.9$, BSA = 2079 Å^2) where $\Delta VDWS_{RD}$ is the ROSETTADOCK weighted interface Lennard-Jones score, which is calculated by subtracting the corresponding score of complexes from those of the individual chains, and BSA is 'Buried Surface Area'. Four salt bridges and one hydrogen bond took place between FKBP38 and Bcl-2. Taken together, this resulted in a low ROSETTADOCK interface energy score (I_sc) of -8.7 for the highest-ranked ROSETTADOCK$_2$ structure of the complex, suggesting a high binding affinity of FKBP38 towards Bcl-2. In the case of the FKBP38/Bcl-xL complex, a fine shape complementarity ($\Delta VDWS_{RD} = -162.5$; BSA = 3390.4 Å^2) took place. Besides, one salt bridge and four hydrogen bonds were formed. Together, this resulted in a low ROSETTADOCK interface energy score value of -8.8, suggesting high binding affinity between Bcl-xL and FKBP38. With binding of tBid to FKBP38, the PRHR approach has resulted in the structure of the FKBP38/tBid complex with fine shape complementarity ($\Delta VDWS_{RD} = -247.6$; BSA = 3579.0 Å^2). Besides, three salt bridges and one hydrogen bond took place between protein components. Taken together, this resulted in a very low ROSETTADOCK interface binding score (Isc) of -8.9 for the highest-ranked ROSETTADOCK structure of the complex. Collectively, these results suggest that tBid has higher binding affinity towards FKBP38 as compared with that of Bcl-2 or Bcl-xL, thus being capable of displacing these antiapoptotic proteins from sequestration by FKBP38 and making them free for further antiapoptotic association with proapoptotic proteins.

References

1. Edlich, F., Lücke, C.: From cell death to viral replication: the diverse functions of the membrane-associated FKBP38. Curr. Opin. Pharmacol. **11**, 348–353 (2011)
2. Bennett-Lovsey, R.M., Herbert, A.D., Sternberg, M.J., Kelley, L.A.: Exploring the extremes of sequence/structure space with ensemble fold recognition in the program Phyre. Proteins. **70**, 611–625 (2008)
3. Roy, A., Kucukural, A., Zhang, Y.: I-TASSER: a unified platform for automated protein structure and function prediction. Nat. Protoc. **5**, 725–738 (2010)
4. Ebejer, J.P., Hill, J.R., Kelm, S., Shi, J., Deane, C.M.: Memoir: template-based structure prediction for membrane proteins. Nucleic Acids Res. **41**(Web Server issue), W379–383 (2013)
5. Mandell, D.J., Coutsias, E.A., Kortemme, T.: Sub-angstrom accuracy in protein loop reconstruction by robotics-inspired conformational sampling. Nat. Methods **6**, 551–552 (2009)
6. Lomize, M.A., Lomize, A.L., Pogozheva, I.D., Mosberg, H.I.: OPM: orientations of proteins in membranes database. Bioinformatics **22**, 623–625 (2006)
7. Kozakov, D., Brenke, R., Comeau, S.R., Vajda, S.: PIPER: an FFT-based protein docking program with pairwise potentials. Proteins **65**, 392–406 (2006)

8. Gray, J.J., Moughon, S., Wang, C., Schueler-Furman, O., Kuhlman, B., Rohl, C.A., Baker, D.: Protein-protein docking with simultaneous optimization of rigid-body displacement and side-chain conformations. J. Mol. Biol. **331**, 281–299 (2003)
9. de Vries, S.J., van Dijk, M., Bonvin, A.M.J.J.: The HADDOCK web server for data-driven biomolecular docking. Nat. Protoc. **5**, 883–897 (2010)
10. Veresov, V.G., Davidovskii, A.I.: Structural insights into proapoptotic signaling mediated by MTCH2, VDAC2, TOM40 and TOM22. Cell. Signal. **26**, 370–382 (2014)

VAliBS: A Visual Aligner
for Bisulfite Sequences

Min Li[1], Xiaodong Yan[1], Lingchen Zhao[1], Jianxin Wang[1],
Fang-Xiang Wu[2], and Yi Pan[1,3]

[1] School of Information Science and Engineering, Central South University,
Changsha 410083, China
{limin,yanxiaodong,jxwang}@csu.edu.cn
[2] Division of Biomedical Engineering and Department of Mechanical
Engineering, University of Saskatchewan, Saskatoon, SK S7N 5A9, Canada
faw341@mail.usask.ca
[3] Department of Computer Science, Georgia State University,
Atlanta, GA 30302-4110, USA
pan@cs.gsu.edu

Methylation is a common modification of DNA. It has been a very important and hot topic to study the correlation between methylation and diseases in medical science. Because of special process with bisulfite treatment, traditional mapping tools do not work well with such methylation experimental reads. Traditional aligners are not designed for mapping bisulfite-treated reads, where the unmethylated 'C's are converted to 'T's. In this paper, we develop a reliable and visual tool, named VAliBS, for mapping bisulfate sequences to a genome reference. VAliBS works well even on large scale data or high noise data. By comparing with other state of the art tools (BisMark [1], BSMAP [2], BS-Seeker2 [3]), VAliBS can improve the accuracy of bisulfite mapping. Moreover, VAliBS is a visual tool which makes its operations more easily and the alignment results are shown with colored marks which makes it easier to be read. VAliBS provides fast and accurate mapping of bisulfite-converted reads, and provides a friendly window system to visualize the detail of mapping of each read.

VAliBS uses three stages to map bisulfite reads: pre-processing, mapping, and post-processing. The schematic diagrams of VAliBS is shown in Fig. 1.

For masking the base difference from methylation, we use the wild useful three-letter strategy. It means, masking the difference from C and T artificially (in the other strand is G and A). Concretely, for every reference, we do two copies for it, one converting all C to T, the other one converting all G to A; for every read, we do the same process. In the mapping stage, we employee BWA [21] and Bowtie2 [20] to map converted reads and references. In the post-processing, we have implemented a filter procedure for wiping off most of mapping mistakes from base conversion. We also consider the mismatches with SNP tolerant by inputting SNP files to avoid filtering correct results.

This work was supported in part by the National Natural Science Foundation of China under Grants (No.61232001, No. 61379108, and No.61428209) and the Program for New Century Excellent Talents in University (NCET-12-0547).

A. Bourgeois et al. (Eds.): ISBRA 2016, LNBI 9683, pp. 307–308, 2016.
DOI: 10.1007/978-3-319-38782-6

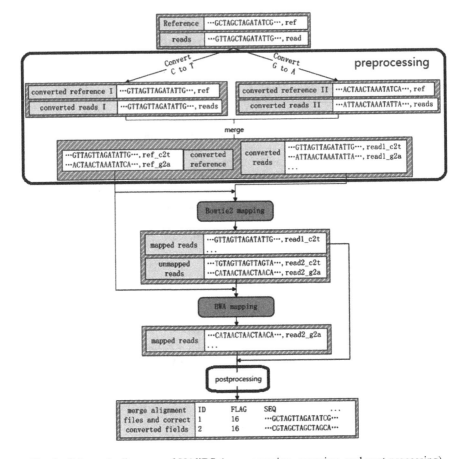

Fig. 1. Schematic diagrams of VAliBS (pre-processing, mapping, and post-processing).

References

1. Krueger, F., Andrews, S.R.: Bismark: a flexible aligner and methylation caller for Bisulfite-Seq applications. Bioinformatics **27**(11), 1571–1572 (2011)
2. Xi, Y., Li, W.: BSMAP: whole genome bisulfite sequence MAPping program. BMC Bioinform. **10**(1), 1 (2009)
3. Guo, W., Fiziev, P., Yan, W., et al.: BS-Seeker2: a versatile aligning pipeline for bisulfite sequencing data. BMC Genomics **14**, 774 (2013)
4. Langmead, B., Trapnell, C., Pop, M., Salzberg, S.L.: Ultrafast and memory-efficient alignment of short DNA sequences to the human genome. Genome Biol. **10**(3), R25 (2009)
5. Li, H., Durbin, R.: Fast and accurate long-read alignment with Burrows-Wheeler transform. Bioinformatics **26**(5), 589–595 (2010)

MegaGTA: A Sensitive and Accurate Metagenomic Gene-Targeted Assembler Using Iterative de Bruijn Graphs

Dinghua Li[1], Yukun Huang[1], Henry Chi-Ming Leung[1],
Ruibang Luo[1,2,3], Hing-Fung Ting[1], and Tak-Wah Lam[1,2]

[1] Department of Computer Science, The University of Hong Kong,
Hong Kong, China
{dhli,ykhuang,cmleung2,rbluo,hfting,twlam}@cs.hku.hk
[2] L3 Bioinformatics Limited, Hong Kong, China
[3] United Electronics Co., Ltd., Beijing, China

Metagenomic gene-targeted assemblers generate better gene contigs of a specific gene family both in quantity and quality than *de novo* assemblers. The recently released gene-targeted metagenomic assembler, named Xander [1], has demonstrated that profile Hidden Markov Model (HMM [2]) guided traversal of *de Bruijn* graph is remarkably better than other gene-targeted or *de novo* assembly methods, albeit, great improvements could be made in both quality and computation. Xander is restricted to use only one k-mer size for graph construction, which results in compromised sensitivity or accuracy. The use of Bloom-filter representation of *de Bruijn* graph in order to maintain a small memory footprint, also brings in false positives, the side-effect of which remains unclear. The multiplicity of k-mers, which is a proven effective measure to differentiate erroneous k-mers and correct k-mers, was not recorded also because of the use of Bloom-filter.

This paper presents a new gene-target assembler named MegaGTA, which succeeded in improving Xander in both quality and computation (non-gene-targeted assemblers not included as their performances on this dataset are not comparable with both MegaGTA and Xander [1]). Quality-wise, it utilizes iterative *de Bruijn* graphs [3] to take advantage of multiple k-mer sizes to make the best of both sensitivity and accuracy. It introduces a penalty score for low coverage error-prone k-mers to enhance the HMM model. Computation-wise, it adopts succinct *de Bruijn* graphs (SdBG [4]) to achieve a small memory footprint. Unlike Bloom filter, SdBG is an exact representation and it enables MegaGTA to be false positive k-mers free. The highly efficient parallel algorithm for constructing SdBG [5] results in an order of magnitude higher assembly speed.

We compared MegaGTA and Xander on an HMP-defined mock metagenomic dataset, and found MegaGTA excelled in both sensitivity and accuracy (Table 1). In terms of sensitivity, MegaGTA achieved same or higher gene fractions for every benchmarked microorganism. Notably, MegaGTA was able to recover two genes almost to its full-length in a single run, while Xander gave fragmented results even by merging the contigs generated with multiple k-mers sizes in multiple runs. In terms of accuracy, MegaGTA produced less misassemblies, mismatches, and had a lower duplication ratio.

© Springer International Publishing Switzerland 2016
A. Bourgeois et al. (Eds.): ISBRA 2016, LNBI 9683, pp. 309–311, 2016.
DOI: 10.1007/978-3-319-38782-6

Table 1. Assembly results of MegaGTA (using iterative *de Bruijn* graph) and Xander (merging contigs of three *k*-mer sizes) on rplB genes of the HMP mock dataset. *The rplB gene of *Streptococcus mutans* was in fact covered by two fragmented contigs of Xander

	MegaGTA (iterates on $k = 30$, 36, 45)	Xander (Union of $k = 30$, 36, 45)
# contigs	10	19
# genes recovered	10	10
duplication ratio	1	1.79
# misassembled contigs	0	1
# partially unaligned contigs	1	2
# mismatches per 100kbp	13.52	453.05
Time (second)	277	2927
The gene fraction of each recovered rplB genes (%)		
Acinetobacter_baumannii	98.77	84.77
Bacteroides_vulgatus	82.48	82.48
Deinococcus_radiodurans	99.64	99.64
Escherichia_coli	81.39	81.39
Propionibacterium_acnes	78.14	78.14
Rhodobacter_sphaeroides	98.21	98.21
Staphylococcus_aureus	99.64	99.64
Staphylococcus_epidermidis	99.64	99.64
Streptococcus_mutans	99.29	99.29*
Streptococcus_pneumoniae	63.31	62.23

The performance of MegaGTA on real data was evaluated with a large rhizosphere soil metagenomic sample [1] (327 Gbps). MegaGTA produced 9.7–19.3 % more contigs than Xander, which were assigned to 10–25 % more gene references. On this dataset, 0.02 %, 0.39 % and 10.52 % of contigs generated by Xander contain false positive *k*-mers, when using a Bloom filter size of 256 GB, 128 GB and 64 GB, respectively. In comparison, MegaGTA required 242 GB memory without any false *k*-mers. We found that most of the false *k*-mers located amid a contig, which may lead to misassemblies. Therefore, one should be careful with the size of the Bloom filter in Xander.

MegaGTA advances in both assembly quality and efficiency. Depends on the number of *k*-mers used, it is about two to ten times faster than Xander. The source code of MegaGTA is available at https://github.com/HKU-BAL/megagta.

References

1. Wang, Q., et al.: Xander: employing a novel method for efficient gene-targeted metagenomic assembly. Microbiome **3**(1), 1–13 (2015)
2. Eddy, S.R.: What is a hidden Markov model? Nat. Biotech. **22**(10), 1315–1316 (2004)

3. Peng, Y., Leung, H.C.M., Yiu, S.M., Chin, F.Y.L.: IDBA – a practical iterative de Bruijn graph de novo assembler. In: Berger, B. (ed.) RECOMB 2010. LNCS, vol. 6044, pp. 426–440. Springer, Heidelberg (2010)
4. Bowe, A., Onodera, T., Sadakane, K., Shibuya, T.: Succinct de Bruijn Graphs. In: Raphael, B., Tang, J. (eds.) WABI 2012. LNCS, vol. 7534, pp. 225–235. Springer, Heidelberg (2012)
5. Li, D., et al.: MEGAHIT: an ultra-fast single-node solution for large and complex metagenomics assembly via succinct de Bruijn graph. Bioinformatics (2015)

EnhancerDBN: An Enhancer Prediction Method Based on Deep Belief Network

Hongda Bu[1(✉)], Yanglan Gan[2], Yang Wang[4], Jihong Guan[1], and Shuigeng Zhou[3]

[1] Department of Computer Science and Technology,
Tongji University, Shanghai 201804, China
jhguan@tongji.edu.cn
[2] School of Computer, Donghua University, Shanghai, China
ylgan@dhu.edu.cn
[3] Shanghai Key Lab of Intelligent Information Processing and
School of Computer Science, Fudan University,
Shanghai 200433, China
sgzhou@fudan.edu.cn
[4] School of Software, Jiangxi Normal University, Nanchang 330022, China
yang1995t@163.com

1 Introduction

Transcriptional enhancers are important regulatory elements that play critical roles in regulating gene expression. As enhancers are independent of the distance and orientation to their target genes, predicting distal enhancers is still a challenging task for bioinformatics researchers. Recently, with the development of high-throughout ChiP-seq technology, some computational methods, including RFECS [1] and EnhancerFinder [2] were developed to predict enhancers using genomic or epigenetic features. However, the unsatisfactory prediction performance and the inconsistency of these computational models across different cell-lines call for further exploration in this area.

2 Methods

We proposed EnhancerDBN, a deep belief network (DBN) based method for predicting enhancers. The method was trained on the VISTA Enhancer Browser data set that contains biologically validated enhancers, with three types of features, including DNA sequence compositional features, histone modifications and DNA methylation. Our method mainly consists of the following three steps:

(1) Feature calculation. Three types of features were used to represent enhancers, including 169 DNA sequence compositional features, 106 histone modification features and DNA methylation feature.

This work was partially supported by the Program of Shanghai Subject Cheif Scientist under grant No. 15XD1503600, and NSFC under grant No. 61272380.

H. Bu — The authors wish it to be known that, in their opinion, the first two authors should be regarded as joint first authors.

A. Bourgeois et al. (Eds.): ISBRA 2016, LNBI 9683, pp. 312–313, 2016.
DOI: 10.1007/978-3-319-38782-6

(2) Training the EnhanerDBN classifier for enhancer prediction. EnhancerDBN uses a two-step scheme to turn the prediction problem into a binary classification task that decides whether any DNA region is an enhancer candidate or not. The first step is to construct the DBN by training a series of Restricted Boltzmann Machines (RBMs); the second step is to train and optimize the EnhancerDBN classifier by using the trained DBN and an additional output layer with the backpropagation (BP) algorithm [3].

(3) Enhancer prediction and performance evaluation. 10-fold validation was conducted to evaluate the proposed method. We first evaluated the predictive power of different types of features in terms of prediction error rate, then compared our method with existing methods in terms of AUC value or accuracy.

3   Results

Figure 1 shows the ROCs of our method and five existing methods, from which we can see that our method clearly outperforms the other five methods. In summary, our experimental results show that (1) EnhancerDBN achieves higher prediction performance than 13 existing methods, and (2) DNA methylation and GC content can serve as highly relevant features for enhancer prediction.

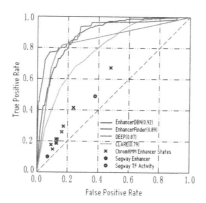

Fig. 1. Performance comparison with five existing methods in ROC space.

References

1. Rajagopal, N., Xie, W., Li, Y., et al.: RFECS: A random-forest based algorithm for enhancer identification from chromatin state. PLoS Comput Biol 9(3)(2013) e1002968
2. Erwin, G.D., Oksenberg, N., Truty, R.M., et al.: Integrating diverse datasets improves developmental enhancer prediction. PLoS Comput Biol 10(6)(2014) e1003677
3. Hinton, G.E., Osindero, S., Teh, Y.W.: A fast learning algorithm for deep belief nets. Neural Comput. **18**(7), 1527–1554 (2006)

An Improved Burden-Test Pipeline for Cancer Sequencing Data

Yu Geng[1,6], Zhongmeng Zhao[1,4]($\boxtimes$), Xuanping Zhang[1,4], Wenke Wang[1],
Xiao Xiao[5], and Jiayin Wang[2,3,4]($\boxtimes$)

[1] Department of Computer Science and Technology, Xi'an Jiaotong University,
Xi'an 710049, Shaanxi, China
zmzhao@mail.xjtu.edu.cn
[2] School of Management, Xi'an Jiaotong University, Xi'an 710049, Shaanxi, China
wangjiayin@mail.xjtu.edu.cn
[3] Institute of Data Science and Information Quality, Xi'an Jiaotong University,
Xi'an 710049, Shaanxi, China
[4] Shaanxi Engineering Research Center of Medical and Health Big Data,
Xi'an Jiaotong University, Xi'an 710049, Shaanxi, China
[5] State Key Laboratory of Cancer Biology, Xijing Hospital of Digestive Diseases,
Xi'an 710032, Shaanxi, China
[6] Jinzhou Medical University, Jinzhou 121001, Liaoning, China

Along with the rapid development of sequencing technology, several cancer-genome projects, such as TCGA and ICGC, have been achieved. In such projects, at least two separate samples are sequenced for each patient, one from tumor tissue and the other from adjacent normal tissue or blood (not in the case of leukemia) as a control. The alignments and variation callings of the two sets of sequencing data are compared to identify plausible differences, many of which may represent genuine germline variants and somatic mutational events.

Identifying novel deleterious germline variation and somatic mutational events is one of the essential topics in cancer genomics. As rare germline variants are often uniquely inherited in a population, while highly recurrent somatic mutations only make up a small proportion of total rare somatic events, burden-test, also known as collapsing-based strategy, becomes a major technique for association analysis, which is considered to relieve the underpowered issues caused by low minor allelic frequencies (MAFs). These approaches collapse given variants to one or multiple virtual loci, whose MAF(s) is/are obviously increased, and the statistical tests are then applied to these virtual loci. Because of this approach, many rare variants have been reported as having hereditary effects not only on cancers but also on a number of complex diseases and traits.

However, existing approaches are challenged by multiple issues and are still expected to improve for cancer sequencing data. In traditional rare variant association studies, interactions among rare variants are suggested to be either non-existent or too weak to be observed significantly. In contrast, several models of interacting germline and somatic hotspots have been proposed and supported in cancer research; e.g., the two-hit hypothesis serves as a classic genetic model for DNA repair and tumor suppressor genes, some germline variants are reported to significantly increase the mutation rates of somatic events in local genomic

A. Bourgeois et al. (Eds.): ISBRA 2016, LNBI 9683, pp. 314–315, 2016.
DOI: 10.1007/978-3-319-38782-6

regions, and so on. Moreover, several computational approaches, such as the significant mutated gene test, clonal expansion analysis and allelic imbalance test, are developed to observe such interactions. Second, due to technical limitations, sample contamination (purity), sequencing bias and branch effect are ineluctable and may reveal differences among individuals. To obtain better results, an association pipeline for cancer sequencing data should carefully consider these issues; otherwise, the burden-test results may be weakened by decreasing the statistical powers and introducing false positives.

We propose a novel pipeline, *RareProb-C* for association analysis on cancer sequencing data. Our approach extends and improves the existing framework of *RareProb*, while the old model becomes part of one out of four components. *RareProb-C* has three new components, which are detecting interacting variants, estimating the significant mutated regions with higher probability of harboring deleterious variants, and inferring and removing singular variants/cases. During implementation, all these components are executed simultaneously within a hidden Markov random field model.

We compare *RareProb-C* to several widely used approaches, including *RWAS*, *LRT* and *RareCover*. We first apply *RareProb-C* to a real cancer sequencing dataset. This dataset consists of 429 TCGA serous ovarian cancer (OV) cases. Each case has one tumor sample with whole exome sequencing data and one normal sample with whole exome sequencing data. The control cohort is from the NHLBI Women's Health Initiative (WHI), which consists of 557 samples with whole exome sequencing data. All of the data are aligned to human reference build37 using BWA, and variants are identified using VarScan, GATK, and Pindel, with stringent downstream filtering to standardize specificity. Variant annotation is based on Ensembl release 70_37_v5. The variant list for association analysis contains 3050 germline truncation variants and 4724 somatic truncation mutations. *RareProb-C* successfully identifies the known highlighted variants, which are reported to be associated with enriched disease susceptibilities in literatures. The comparison approaches, on the other hand, identify part of those genes/variants. *RareProb-C* also achieves higher statistical power than the existing approaches on multiple artificial datasets with different configurations.

Acknowledgments. This work is supported by the National Science Foundation of China (Grant No: 61100239, 81400632), Shaanxi Science Plan Project (Grant No: 2014JM8350) and the Ph.D. Program Foundation by Ministry of Education of China (Grant No: 20100201110063).

Modeling and Simulation of Specific Production of Trans10, cis12-Conjugated Linoleic Acid in the Biosynthetic Pathway

Aiju Hou[1], Xiangmiao Zeng[1], Zhipeng Cai[2], and Dechang Xu[1(✉)]

[1] School of Food Science and Engineering, Harbin Institute of Technology,
Harbin 150090, People's Republic of China
dcxu@hit.edu.cn
[2] Department of Computer Science, Georgia State University, Atlanta, GA, USA

Abstract. A mathematical model of CLA metabolism was established to study the regulation mechanism in CLA metabolic pathway. The simulation results indicate that if there was only t10c12-CLA pathway left in CLA metabolic pathway, the LA could completely be converted into t10c12-CLA within less than 6 h. In addition, the HY emzyme activity has a significant effect on the forming rate of t10c12-CLA.

Keywords: Conjugated linoleic acid · Biosynthesis · Mathematical modeling

1 Introduction

CLAs (Conjugated linoleic acids) have isomer-specific, health-promoting properties. However, the content of CLA in natural foods is too small to exert its physiological activity, which drives scientists to focus on the development of efficient methods for CLA production. Although some species of bacteria were found having CLA-producing capacity, their specific and efficient production was not established, especially for the production of t10c12-CLA, which was the only one of the isomers with fat reducing activity. The regulation of metabolic pathway of LA (Linoleic acid) in bacteria might be a promising strategy to produce specific isomers of CLA. However, the complexity of metabolism of LA makes it difficult and laborious to regulate each parameter involved in the process through wet experiments. Therefore, a model based on the biochemical reactions involved in metabolic pathway of LA was built up to find out the key parameters in the pathway through. This model could be useful for directional regulating the production of specific CLA isomers.

2 Methods

A relatively complete metabolic pathway of CLA was constructed based on previous studies [1, 2]. 14 reactions were involved totally (Fig. 1). According to the Michaelis-Menten equation, the enzymatic reaction kinetics model and the differential equation model of CLA metabolism were established and simulated.

© Springer International Publishing Switzerland 2016
A. Bourgeois et al. (Eds.): ISBRA 2016, LNBI 9683, pp. 316–318, 2016.
DOI: 10.1007/978-3-319-38782-6

Fig. 1. The metabolic pathway of conjugated linoleic acid

3 Results and Conclusions

3.1 The Preliminary Simulation of CLA Metabolic Model

The data obtained from the preliminary simulation shows that the conversion rate of t10c12-CLA is only ~1 % and that of total CLA is ~14 %. There was an accumulation of oleic acid (OA) during the process and most of the fatty acids converted to saturated fatty acid (SA) in the end. The simulation results are consistent with the previous studies.

3.2 The Influence of Bypasses on the Yield of t10c12-CLA

10-OH, cis12-C18:1 has three metabolic directions: R_2, R_3 and R_6 (Fig. 1). If the reaction rate of R_3 and R_6 is adjusted to zero, higher yield of t10c12-CLA should be expected. However, the yield of t10c12-CLA didn't show a significant increase when R_6 and R_3 was blocked sequentially. Originally, c9c12-LA also participated in the $\Delta12$ hydroxylation (R_4), which was converted into SA. So the R_4 bypass was also blocked. Surprisingly, the yield of t10c12-CLA was still at a very low level while the accumulation of OA reached a very high level. The accumulated OA finally converted into t10c12-CLA after the reaction time was prolonged 20 times compared to the previous simulations. From this result, we see that if there was only t10c12-CLA pathway left, the LA could totally be converted into t10c12-CLA, but the output time to reach the steady state was 20 times long than that without any operation.

3.3 The Influence of Parameters on the Yield of t10c12-CLA

The limiting step related parameter $k_{cat-HY}^{OH10-cis12}$, $K_{m-HY}^{NAD^+}$, $K_{m-HY}^{OA_{OH10-cis12}}$, $K_{s-HY}^{NAD^+}$ was studied respectively to observe their influences on the production of t10c12-CLA through amplifying or reducing 10 times of themselves. Among these parameters, $k_{cat-HY}^{OH10-cis12}$ has great effect on the yield of t10c12-CLA. When $k_{cat-HY}^{OH10-cis12}$ was expanded to 10 times, the accumulated peak of 10-OH, cis12-C18:1 was reduced by ~30 % and the generating rate of t10c12-CLA was increased 10 times roughly.

Altogether, among the factors that affect the production of t10c12-CLA, the catalytic constant of related enzyme affects not only the final yield, but also decides the generating rate of t10c12-CLA. Blocking some bypasses also contributes to the yield.

References

1. Shigenobu, K., Michiki, T., Si-Bum, P., Akiko, H., Nahoko, K., Jun, K., Hiroshi, K., Ryo, I., Yosuke, I., Makoto, A., Hiroyuki, A., Kazumitsu, U., Jun, S., Satomi, T., Kenzo, Y., Sakayu, S., Jun, O.: Polyunsaturated fatty acid saturation by gut lactic acid bacteria affecting host lipid composition. PNAS, 1–6 (2013)
2. Michiki, T., Shigenobu, K., Kaori, T., Akiko, H., Si-Bum, P., Sakayu, S., Jun, O.: Hydroxy fatty acid production by Pediococcus sp. Eur. J. Lipid Sci. Technol. **115**, 386–393 (2013)

Dynamic Protein Complex Identification in Uncertain Protein-Protein Interaction Networks

Yijia Zhang, Hongfei Lin, Zhihao Yang, and Jian Wang

The College of Computer Science and Technology,
Dalian University of Technology, Dalian 116023, China
{zhyj,hflin,yangzh,wangjian}@dlut.edu.cn

Protein complexes play critical roles in many biological processes and most proteins are only functional after assembly into protein complexes [1]. One of the most important challenges in the post-genomic era is to computationally analyze the large PPI networks and accurately identify protein complexes. Most studies on protein complexes identification have been focused on the static PPI networks without accounting for the dynamic properties of biological systems. However, cellular systems are highly dynamic and responsive to cues from the environment [2]. In reality, the PPI network in a cell is not static but dynamic, which is changing over time, environments and different stages of cell cycles.

Some studies have projected the additional information such as gene expression data onto PPI networks to reveal the dynamic behavior of PPIs. *de* Lichtenberg *et al.* [3] integrate data on protein interactions and gene expression to analyze the dynamics of protein complexes during the yeast cell cycle, and reveal most complexes consist of both periodically and constitutively expressed subunits. Srihari *et al.* [4] exploit gene expression data to incorporate 'time' in the form of cell-cycle phases into the prediction of complexes, and study the temporal phenomena of complex assembly and disassembly across cell-cycle phases. On the other hand, both static PPI networks and dynamic static PPI networks are mainly based on the high-throughput PPI data. Some studies have indicated that the high-throughput PPI data contains high false positive and false negative rates, due to the limitations of the associated experimental techniques and the dynamic nature of PPIs [5]. This makes the accuracy of the PPI networks based on such high-throughput experimental data is far from satisfactory. So far, it is still impossible to construct an absolutely reliable PPI networks. In fact, each protein and PPI is associated with an uncertainty value in the PPI networks. It is more reasonable to use uncertain graph model to deal with PPI networks.

In this paper, we use the uncertain model to identify protein complexes in the dynamic uncertain PPI networks (DUPN). Firstly, we use the three-sigma method to calculate the active time point and the existence probability of each protein based on the gene expression data. DUPN are constructed to integrate the dynamic information of gene expression and the topology information of high-throughput PPI data. Due to the noises and incompletion of the high-throughput data and gene expression, each node and edge is associated with an existence probability on a DUPN. Then, we propose CDUN algorithm to identify protein complexes on DUPN based on the

© Springer International Publishing Switzerland 2016
A. Bourgeois et al. (Eds.): ISBRA 2016, LNBI 9683, pp. 319–320, 2016.
DOI: 10.1007/978-3-319-38782-6

uncertain model. We test our method on the different PPI datasets including Gavin [6], DIP [7], and STRING [8] datasets, respectively. We download the gene expression data GSE3431 [9] from Gene Expression Omnibus. The gold standard data are CYC2008 [10], which consist of 408 protein complexes. Precision, recall and F-score are used to evaluate the performance of our method. It is encouraging to see that our approach is competitive or superior to the current protein complexes identification methods in performance on the different yeast PPI datasets, and achieves the state-of-the-art performance.

Acknowledgments. This work is supported by grant from the Natural Science Foundation of China (No. 61300088, 61572098 and 61572102), the Fundamental Research Funds for the Central Universities (No. DUT14QY44).

References

1. Butland, G., Peregrín-Alvarez, J.M., Li, J., Yang, W., Yang, X., Canadien, V., Starostine, A., Richards, D., Beattie, B., Krogan, N.: Interaction network containing conserved and essential protein complexes in Escherichia coli. Nature **433**, 531–537 (2005)
2. Przytycka, T.M., Singh, M., Slonim, D.K.: Toward the dynamic interactome: it's about time. Brief. Bioinform. bbp057 (2010)
3. de Lichtenberg, U., Jensen, L.J., Brunak, S., Bork, P.: Dynamic complex formation during the yeast cell cycle. Science **307**, 724–727 (2005)
4. Srihari, S., Leong, H.W.: Temporal dynamics of protein complexes in PPI networks: a case study using yeast cell cycle dynamics. BMC Bioinform. **13**, S16 (2012)
5. Li, X., Wu, M., Kwoh, C.-K., Ng, S.-K.: Computational approaches for detecting protein complexes from protein interaction networks: a survey. BMC Genom. **11**, S3 (2010)
6. Gavin, A.-C., Aloy, P., Grandi, P., Krause, R., Boesche, M., Marzioch, M., Rau, C., Jensen, L.J., Bastuck, S., Dümpelfeld, B.: Proteome survey reveals modularity of the yeast cell machinery. Nature **440**, 631–636 (2006)
7. Xenarios, I., Salwinski, L., Duan, X.J., Higney, P., Kim, S.-M., Eisenberg, D.: DIP, the database of interacting proteins: a research tool for studying cellular networks of protein interactions. Nucleic Acids Res. **30**, 303–305 (2002)
8. Franceschini, A., Szklarczyk, D., Frankild, S., Kuhn, M., Simonovic, M., Roth, A., Lin, J., Minguez, P., Bork, P., von Mering, C.: STRING v9. 1: protein-protein interaction networks, with increased coverage and integration. Nucleic Acids Res. **41**, D808–D815 (2013)
9. Tu, B.P., Kudlicki, A., Rowicka, M., McKnight, S.L.: Logic of the yeast metabolic cycle: temporal compartmentalization of cellular processes. Science **310**, 1152–1158 (2005)
10. Pu, S., Wong, J., Turner, B., Cho, E., Wodak, S.J.: Up-to-date catalogues of yeast protein complexes. Nucleic Acids Res. **37**, 825–831 (2009)

Predicting lncRNA-Protein Interactions Based on Protein-Protein Similarity Network Fusion (Extended Abstract)

Xiaoxiang Zheng[1], Kai Tian[1], Yang Wang[2], Jihong Guan[3], and Shuigeng Zhou[1]

[1] Shanghai Key Lab of Intelligent Information Processing, and School of Computer Science, Fudan University, Shanghai 200433, China
{xxzheng,ktian14,sgzhou}@fudan.edu.cn
[2] School of Software, Jiangxi Normal University, Nanchang 330022, China
yang1995t@163.com
[3] Department of Computer Science and Technology, Tongji University, Shanghai 201804, China
jhguan@tongji.edu.cn

Introduction. Long non-coding RNAs (lncRNAs in short), one type of non-protein coding transcripts longer than 200 nucleotides, play important roles in complex biological processes, ranging from transcriptional regulation, epigenetic gene regulation to disease identification. Researches have shown that most lncRNAs exert their functions by interfacing with multiple corresponding RNA binding proteins. Therefore, predicting *lncRNA-protein interactions* (lncRPIs) is an important way to study the functions of lncRNAs.

In this paper, to boost the performance of lncRPI prediction we propose to fuse multiple protein-protein similarity networks (PPSNs), and integrate the fused PPSN with known lncRPIs to construct a more informative heterogeneous network, on which new lncRPIs are inferred.

Method. Figure 1 illustrates the pipeline of our method. In the left side of the figure, the rectangles represent lncRNAs and the circles represent proteins. In the right side of the figure, four small networks represent different protein-protein similarity networks (PPSNs), which are first constructed with protein sequences, protein domains, protein functional annotations from GO and the STRING database, respectively. These four PPSNs are then fused by using the Similarity Network Fusion (SNF) algorithm [2] to get a more informative PPSN. The resulting PPSN is combined with the lncRNA-protein bipartite constructed by known lncRPIs to obtain a heterogeneous lncRNA-protein network. On this constructed heterogeneous network, the HeteSim algorithm [1] is finally applied to predicting new lncRPIs.

Results. The AUC values of the baseline [3] and our method with different settings of fusing two, three and four PPSNs are illustrated in Fig. 2. We can see that under all settings, our method outperforms the baseline; and as more PPSNs are fused, our method performs better and better.

This work was partially supported by the Program of Shanghai Subject Chief Scientist under grant No. 15XD1503600, and NSFC under grant No. 61272380.

A. Bourgeois et al. (Eds.): ISBRA 2016, LNBI 9683, pp. 321–322, 2016.
DOI: 10.1007/978-3-319-38782-6

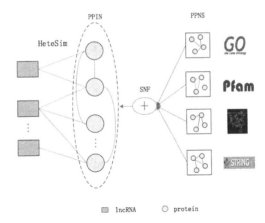

Fig. 1. Illustration of the pipeline of our method.

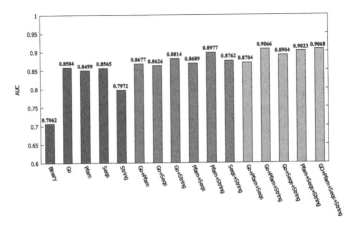

Fig. 2. The AUC values of the baseline and our method under different settings.

References

1. Shi, C., Kong, X., Huang, Y., Yu, P., Wu, B.: HeteSim: A general framework for relevance measure in heterogeneous networks. IEEE Trans. Knowl. Data Eng. **26**(10), 2479–2492 (2014)
2. Wang, B., Mezlini, A.M., Demir, F., Fiume, M., Tu, Z., Brudno, M., Haibe-Kains, B., Goldenberg, A.: Similarity network fusion for aggregating data types on a genomic scale. Nature Methods **11**(3), 333–337 (2014)
3. Yang, J., Li, A., Ge, M., Wang, M.: Prediction of interactions between lncRNA and protein by using relevance search in a heterogeneous lncRNA-protein network. In: Proceedings of 34th Chinese Control Conference (CCC 2015), pp. 8540–8544. IEEE (2015)

DCJ-RNA: Double Cut and Join for RNA Secondary Structures Using a Component-Based Representation

Ghada Badr[1,2] and Haifa Al-Aqel[3]

[1] King Saud University College of Computer and Information Sciences,
Riyadh, Kingdom of Saudi Arabia
ghbadr@ksu.edu.sa
[2] IRI - The City of Scientific Research and Technological Applications,
University and Research District, P.O. 21934
New Borg Alarab, Alexandria, Egypt
[3] Imam Mohammad ibn Saud Islamic University,
College of Computer and Information Sciences,
Riyadh, Kingdom of Saudi Arabia
h.alaqel@ccis.imamu.edu.sa

Abstract. Genome rearrangements are essential processes for evolution and are responsible for existing varieties of genome architectures. RNA is responsible for transferring the genetic code from the nucleus to the ribosome to build proteins. In contrast to representing the genome as a sequence, representing it as a secondary structure provides more insight into the genome's function. This paper proposes a double cut and join(DCJ-RNA) algorithm. The main aim is to suggest an efficient algorithm that can help researchers compare two RNA secondary structures based on rearrangement operations. DCJ-RNA calculates the distance between structures and reports a scenario based on the minimum rearrangement operation. The results, which are based on real datasets, show that DCJ-RNA is able to clarify the rearrangement operation, as well as a scenario that can increase the similarity between the structures. Preliminary work has been presented as a poster in [1].

1 DCJ-RNA Algorithm

The RNA component-based rearrangement algorithm uses a component-based representation [2] that allows for the unique description of any RNA pattern and shows the main features of the pattern efficiently. The proposed algorithm also uses the DCJ algorithm to describe rearrangement operations. The DCJ-RNA algorithm is described as follows:

Input: The input consists of two RNA secondary structures given in a component-based representation. A: (a1, a. . . an) and B: (b1, b2. . . bm)

Output: This consists of the number of rearrangement (DCJ) operations and one evolutionary scenario.

The DCJ-RNA algorithm completes three main steps.

© Springer International Publishing Switzerland 2016
A. Bourgeois et al. (Eds.): ISBRA 2016, LNBI 9683, pp. 323–325, 2016.
DOI: 10.1007/978-3-319-38782-6

Step 1: Alignment of similar components based on their component lengths and stem lengths.

In this step, calculate the similarity between components in terms of their component lengths and stem lengths [3]. Similar components are assigned together, beginning with those with the greatest similarity by using the similarity measure:

d1(fai,fbj) = ComponentLength(fai,fbi).StemLength(fai,fbi)

Step 2: Permutation generation

In this step, a corresponding permutation is generated for each of the two structures. This is completed by determining the components to be inserted or deleted, as well as the order of the similar components using the alignment that is generated from step 1. A two-dimensional array of 3 X in size is constructed and identified as SortArray. The first row contains the desired structure, the second row contains the deleted components, and the third row contains the inserted components. An index value of zero for the rows is reserved for the number of components in the same row.

Step 3: Applying the DCJ algorithm.

The component numbers are used to determine the permutations in the DCJ algorithm [4]. Each permutation has two chromosomes:

For the first permutation: The first chromosome is the actual structure of the components, and the second chromosome is the inserted components.

For the second permutation: The first chromosome is the desired structure, and the second chromosome consists of the deleted components.

Each permutation is represented by its adjacencies and telomeres. Finally, the DCJ algorithm is applied to the first and second permutations as input.

2   Results and Conclusion

To test and validate the DCJ-RNA algorithm, three different experiments are applied to three different datasets collected using real data from the NCBI GenBank [6] and Rfam Database [7][1]. The worst time for the algorithm is O(N log N). The total space requirement is $O(N^2)$ (N is the number of components). The DCJ-RNA algorithm is optimal because the DCJ algorithm on which it is based is also optimal.

References

1. Badr, G., Alaqel, H.: Genome rearrangement for RNA secondary structure using a component-based representation: an initial framework. Poster presentation at RECOMB-CG, New York, United States (2014)
2. Badr, G., Turcotte, M.: Component-based matching for multiple interacting RNA sequences. In: Chen, J., Wang, J., Zelikovsky, A. (eds.) ISBRA 2011. LNCS, vol. 6674, pp. 73–86. Springer, Heidelberg (2011)

---

[1] The experiments are obtained using real and simulated data in [5].

3. Alturki, A., Badr, G., Benhidour, H.: Component-based pair-wise RNA secondary structure alignment algorithm. Master Project, King Saud University, Riyadh (2013)
4. Zhang, K., Shasha, D.: Simple fast algorithms for the editing distance between trees and related problems. SIAM J. Comput. **18**, 1245–1262 (1989)
4. Al-aqel, H., Badr, G.: Genome rearrangement for RNA secondary structure using a component-based representation. Master Project, King Saud University, Riyadh (2015)
5. Benson, D.A., Clark, K., Karsch-Mizrachi, I., Lipman, D.J., Ostell, J., et al.: GenBank. Nucleic Acids Res. **41**, D36–D42 (2012)
6. Burge, S.W., Daub, J., Eberhardt, R., Tate, J., Barquist, L., Nawrocki, E.P., et al.: Rfam 11.0: 10 years of RNA families. Nucleic Acids Res. 1–7 (2012)

Improve Short Read Homology Search Using Paired-End Read Information

Prapaporn Techa-Angkoon, Yanni Sun, and Jikai Lei

Department of Computer Science and Engineering, Michigan State University,
East Lansing, MI 48824, USA
yannisun@msu.edu

1 Introduction

Homology search is still an important step in sequence-based functional analysis for genomic data. By comparing query sequences with characterized protein or domain families, functions and structures can be inferred. Profile Homology search has been extensively applied in genome-scale protein domain annotation in various species. However, when applying the state-of-the-art tool HMMER [1] for domain analysis in NGS data lacking reference genomes, such as RNA-Seq data of non-model species and metagenomic data, its sensitivity on short reads deteriorates quickly.

In order to quantify the performance of HMMER on short reads, we applied HMMER to 9,559,784 paired-end Illumina reads of 76 bp sequenced from a normalized cDNA library of *Arabidopsis Thaliana* [3]. The read mapping and domain annotations are used to quantify the performance. The results of this experiment were shown in Table 1.

Table 1. Performance of HMMER and our tool on classifying reads in a RNA-Seq data. 2,972,809 read pairs can be uniquely mapped to 3,577 annotated domain families in the reference genome. However, HMMER missed at least half of the reads when aligning these reads to the domains. There are three cases of alignments. Case 1: only one end can be aligned to a domain. Case 2: both ends can be aligned to a domain. Case 3: No end can be aligned to any domain in the *Arabidopsis Thaliana* RNA-Seq dataset. Each number in the table represents the percentage of a case. "HMMER w/o filtration": turning off all filtration steps and run full Forward/Backward. "HMMER GA cutoff": running HMMER with gathering thresholds.

Case	HMMER, E-value 10	HMMER, w/o filtration, E-value 10	HMMER, GA	Ours
Case 1	34.51%	32.83%	22.51%	0.42%
Case 2	28.42%	31.58%	8.84%	62.51%
Case 3	37.07%	35.59%	68.65%	37.07%

© Springer International Publishing Switzerland 2016
A. Bourgeois et al. (Eds.): ISBRA 2016, LNBI 9683, pp. 326–329, 2016.
DOI: 10.1007/978-3-319-38782-6

2   Method

We introduce the first profile-based homology search method designed for short reads by taking advantage of the paired-end sequencing properties. By applying an approximate Bayesian approach incorporating alignment scores and fragment length distribution of paired-end reads, our method can rescue missing end using the most sensitive mode of HMMER to align the other end to the identified protein families. Then, we compute the posterior probability of the alignments between a read pair and a domain family. For each aligned read pair, we estimate the "alignment quality" using an approximate Bayesian approach [2, 5]. The alignment quality is the probability that a read pair is aligned accurately to its native domain family. As a read pair could be aligned to multiple protein families and some of them are not in ground truth, we use the calculated probability to rank all alignments and discard ones with small probability.

3   Experimental Results

Our homology search method is designed for NGS data that lacks reference genomes. We applied our tool to align short reads in a RNA-Seq dataset and a metagenomic dataset.

3.1   Profile-Based Homology Search in RNA-Seq Dataset of *Arabidopsis Thaliana*

In this experiment, we used the RNA-Seq dataset as described in Sect. 1. As shown in Sect. 1, HMMER can miss one end or both ends of at least half of the read pairs in this experiment. By using our method, the percentage of paired-end reads with both ends being aligned increases from 28.42 % to 62.51 %. Detailed comparison is presented in Table 1.

Moreover, we quantified the performance of homology search for each read by comparing its true domain family membership and predicted membership. We calculated the sensitivity and FP rate for each read pair and then report the average of all read pairs using ROC curves in Figs. 1 and 2. Our method can be used to remove falsely aligned families and thus achieved better tradeoff between sensitivity and FP rate.

3.2   Protein Domain Analysis in a Metagenomic Dataset of Synthetic Communities

The second dataset is a metagenomic data set is sequenced from synthetic communities of *Archaea* and *Bacteria* [4]. There were 52,486,341 paired-end reads of 101 bp. In this experiment, we align all these reads to a catalog of single copy genes, including nearly all ribosomal proteins and tRNA synthases found in nearly all free-living bacteria. The 111 domains were downloaded from TIGR-FAMs and Pfam database.

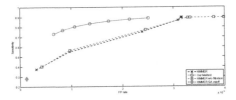

Fig. 1. ROC curves of short read homology search for RNA-Seq data of *Arabidopsis*. Our tool and HMMER are compared on **case 1**, where only one end is aligned by HMMER (default E-value). Note that there is one data point for HMMER with GA cutoff.

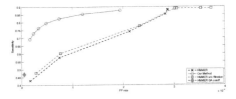

Fig. 2. ROC curves of short read homology search for RNA-Seq data of *Arabidopsis*. Our tool and HMMER are compared on **case 2**, where both ends can be aligned by HMMER (default E-value). Note that there is one data point for HMMER with GA cutoff.

As the reads are longer than the first experiment, HMMER is able to align more reads to correct domain families. Even so, there are still one third of read pairs with at most one end being aligned to the underlying domains. By using our method, the percentage of case 2 (both ends aligned) is improved from 65.82 % to 88.71 %. Table 2 presents the percentage of three cases by HMMER and our tool.

Table 2. The percentage of three cases of pair-end read alignments by our tool and HMMER for the metagenomic data. Case 1: only one end. Case 2: both ends. Case 3: none of end.

Case	HMMER, E-value 10	HMMER, w/o filtration, E-value 10	HMMER, GA	Ours
Case 1	23.15 %	21.63 %	3.76 %	0.26 %
Case 2	65.82 %	68.46 %	2.46 %	88.71 %
Case 3	11.03 %	9.91 %	93.77 %	11.03 %

Furthermore, we computed the sensitivity and FP rate for each read pair and then present the average of all read pairs using ROC curves. Our method showed better overall performance compared to the state-of-the-art methodology of homology search.

References

1. Eddy, S.R.: Accelerated profile HMM searches. PLoS Comput. Biol. **7**(10), e1002195 (2011)
2. Lunter, G., Goodson, M.: Stampy: a statistical algorithm for sensitive and fast mapping of illumina sequence reads. Genome Res. **21**(6), 936–939 (2011)
3. Marquez, Y., Brown, J.W., Simpson, C., Barta, A., Kalyna, M.: Transcriptome survey reveals increased complexity of the alternative splicing landscape in Arabidopsis. Genome Res. **22**(6), 1184–1195 (2012)
4. Shakya, M., Quince, C., Campbell, J.H., Yang, Z.K., Schadt, C.W., Podar, M.: Comparative metagenomic and rRNA microbial diversity characterization using archaeal and bacterial synthetic communities. Environ. Microbiol. **15**(6), 1882–1899 (2013)
5. Shrestha, A.M.S., Frith, M.C.: An approximate Bayesian approach for mapping paired-end DNA reads to a reference genome. Bioinformatics **29**(8), 965–972 (2013)

Framework for Integration of Genome and Exome Data for More Accurate Identification of Somatic Variants

Vinaya Vijayan[1], Siu-Ming Yiu[2], and Liqing Zhang[3]

[1] Genetics Bioinformatics and Computational Biology, Virginia Tech,
Blacksburg, VA, USA
vinayav@vt.edu
[2] Department of Computer Science, The University of Hong Kong, Pokfulam,
Hong Kong
smyiu@cs.hku.hk
[3] Department of Computer Science, Virginia Tech, Blacksburg, VA, USA
lqzhang@cs.vt.edu

Somatic variants are novel mutations that occur within a cell population and are not inherited. Identification of somatic variants aids identification of significant genes and pathways that can then be used in predictive, prognostic, remission and metastatic analysis of cancer. High throughput sequencing technologies, efficient mappers and somatic variant callers have made identification of somatic variant callers relatively easy and cost-effective. In the past few years, a lot of methods such as SomaticSniper, VarScan2, MuTect, VCMM have been developed to identify somatic variants. These programs differ in the kind of statistics used and the parameters considered. Previous studies have shown that VCMM has the highest sensitivity while MuTect has the highest precision for detecting somatic variants.

The exome sequencing platform has been commonly used to identify somatic variants in cancer due to its low cost and high sequencing coverage as compared to the genome sequencing platform. However, a recent study has shown that compared to exome data, whole genome sequencing data enables identification of more germline variants, and is thus the better platform for identifying germline variations in exon regions [1]. Although the study is not for somatic variants, it nevertheless suggests that replying on exome data solely for somatic variant calling may miss many real somatic variants in exon regions. Therefore, integrating both platforms, i.e., whole genome and whole exome platforms, provides better strategy to identify a more complete set of somatic variants.

In this work, we develop a framework that integrates somatic variant calling results from both platforms. Integrating variant calling results from multiple variant callers has been used effectively in the past to identify more somatic variants for a single platform [2]. Here we integrated calling results from two callers, MuTect [3] and VCMM [4], the former shown to have the highest precision and the latter the highest sensitivity. We used the output of VCMM and the output derived from the call_stats option of MuTect to extract 108 features. The features that were selected include base quality, mapping quality, indel score, SNP quality, allele fraction, coverage of normal and tumor

© Springer International Publishing Switzerland 2016
A. Bourgeois et al. (Eds.): ISBRA 2016, LNBI 9683, pp. 330–331, 2016.
DOI: 10.1007/978-3-319-38782-6

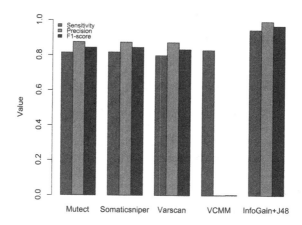

Fig. 1. The sensitivity, precision, and F1 score for somatic variant callers, Mutect, SomaticSniper, Varscan2, VCMM, and our ensemble model InfoGain+J48

samples, presence of the position in dbSNP or COSMIC database etc. These features were used as input to a machine learning algorithm J48 in WEKA [5], to correctly identify somatic variants from simulated data and real data.

Using InfoGain+J48 gives a sensitivity of 0.94, precision of 0.99 and an F1-score of 0.968 (Fig. 1). Using MuTect, SomaticSniper, VarScan2, and VCMM gives an F1-score of 0.84, 0.84, 0.83, and 0.002 respectively on simulated data. Our ensemble method, developed by integrating multiple tools, is better than individual callers in both sensitivity and precision (Fig. 1). Our ensemble method integrating whole genome and whole exome platforms also performs better than using variants from only exome or only genome platforms (results not shown). This was also verified using real data from TCGA (The Cancer Genome Analysis, results not shown).

References

1. Belkadi, A., Bolze, A., Itan, Y., Cobat, A., Vincent, Q.B., Antipenko, A., Shang, L., Boisson, B., Casanova, J.-L., Abel, L.: Whole-genome sequencing is more powerful than whole-exome sequencing for detecting exome variants. Proc. Natl. Acad. Sci. **112**, 5473–5478 (2015)
2. Fang, L.T., Afshar, P.T., Chhibber, A., Mohiyuddin, M., Fan, Y., Mu, J.C., Gibeling, G., Barr, S., Asadi, N.B., Gerstein, M.B., Koboldt, D.C., Wang, W., Wong, W.H., Lam, H.Y.K.: An ensemble approach to accurately detect somatic mutations using SomaticSeq. Genome Biol. **16** (2015)
3. Sensitive detection of somatic point mutations in impure and heterogeneous cancer samples. Nat. Biotechnol. **31**, 213–219 (2013)
4. Shigemizu, D., Fujimoto, A., Akiyama, S., Abe, T., Nakano, K., Boroevich, K.A., Yamamoto, Y., Furuta, M., Kubo, M., Nakagawa, H., Tsunodaa, T.: A practical method to detect SNVs and indels from whole genome and exome sequencing data. Sci. Rep. **3** (2013)
5. Hall, M., Frank, E., Holmes, G., Pfahringer, B., Reutemann, P., Witten, I.H.: The WEKA data mining software: an update. SIGKDD Explor. **11** (2009)

Semantic Biclustering: A New Way to Analyze and Interpret Gene Expression Data

Jiří Kléma, František Malinka, and Filip Železný

Department of Computer Science, Faculty of Electrical Engineering,
Czech Technical University in Prague, Karlovo náměstí 13,
12135 Prague 2, Czech Republic
{klema,malinfr1,zelezny}@fel.cvut.cz

Abstract. We motivate and define the task of semantic biclustering. In an input gene expression matrix, the task is to discover homogeneous biclusters allowing joint characterization of the contained elements in terms of knowledge pertaining to both the rows (e.g. genes) and the columns (e.g. situations). We propose two approaches to solve the task, based on adaptations of current biclustering, enrichment, and rule and tree learning methods. We compare the approaches in experiments with Drosophila ovary gene expression data. Our findings indicate that both the proposed methods induce compact bicluster sets whose description is applicable to unseen data. The bicluster enrichment method achieves the best performance in terms of the area under the ROC curve, at the price of employing a large number of ontology terms to describe the biclusters.

The objective of *biclustering* [8] is to find submatrices of a data matrix such that these submatrices exhibit an interesting pattern in their contained values; for example their values are all equal whereas the values in the containing matrix are non-constant. Biclustering has found significant applications in bioinformatics [5] and specifically in the context of gene expression data analysis [2, 7]. In the latter domain, biclustering can reveal special expression patterns of gene subsets in sample subsets.

By *semantic clustering* we refer to conventional clustering with a subsequent step in which the resulting clusters are described in terms of prior domain knowledge. A typical case of semantic clustering in gene expression analysis is clustering of genes with respect to their expression, followed by *enrichment* analysis where the clusters are characterized by Gene ontology terms overrepresented in them. In [3] the authors blend these two phases in that they directly cluster genes according to their functional similarities. The term semantic clustering was introduced on analogical principles in the software engineering domain [4]. It can be also viewed as an unsupervised counterpart of the *subgroup discovery* method [9]. The semantic descriptions provide an obvious value for interpretation of analysis results, as opposed to plain enumeration of cluster elements.

Here we explore a novel analytic technique termed *semantic biclustering*, combining the two concepts above. In particular, we aim at discovering biclusters

© Springer International Publishing Switzerland 2016
A. Bourgeois et al. (Eds.): ISBRA 2016, LNBI 9683, pp. 332–333, 2016.
DOI: 10.1007/978-3-319-38782-6

satisfying the usual biclustering desiderata, and also allowing joint characterization of the contained elements in terms of knowledge pertaining to both the rows (e.g. genes) and the columns (e.g. situations). This task is motivated by the frequent availability of formal ontologies relevant to both of the dimensions, as is the case of the publicly available *Dresden ovary table* dataset [1]. Informally, we want to be able to discover biclusters described automatically e.g. as *"sugar metabolism genes in early developmental stages"* whenever such genes exhibit uniform expression in the said stages (situations). In [6], the authors present a closely related approach, where the knowledge pertaining to both the matrix dimensions is directly applied to define constraints to filter biclusters (the authors use the more general term patterns). The user is provided only with the interpretable biclusters whose description is compact.

Besides the novel problem formulation stated above, our contributions described below include the proposal of two adaptations of existing computational methods towards the objective of semantic biclustering and their comparative evaluation on the mentioned publicly available dataset [1]. As usual in unsupervised data analysis, the way to validate the methods statistically is not fully obvious. Thus our proposed validation protocol represents a contribution on its own right.

Acknowledgments. This work was supported by Czech Science Foundation project 14-21421S.

References

1. Jambor, H., Surendranath, V., Kalinka, A.T., Mejstrik, P., Saalfeld, S., Tomancak, P.: Systematic imaging reveals features and changing localization of mRNAs in Drosophila development. eLife **4**, e05003 (2015)
2. Kluger, Y., Basri, R., Chang, J.T., Gerstein, M.: Spectral biclustering of microarray data: coclustering genes and conditions. Genome Res. **13**(4), 703–716 (2003)
3. Krejnik, M., Klema, J.: Empirical evidence of the applicability of functional clustering through gene expression classification. IEEE/ACM Trans. Comput. Biol. Bioinf. **9**(3), 788–798 (2012)
4. Kuhna, A., Ducasseb, S., Girbaa, T.: Semantic clustering: identifying topics in source code. Inf. Softw. Technol. **49**(3), 230–243 (2007)
5. Madeira, S.C., Oliveira, A.L.: Biclustering algorithms for biological data analysis: a survey. IEEE/ACM Trans. Comput. Biol. Bioinf. **1**(1), 24–45 (2004)
6. Soulet, A., Kléma, J., Crémilleux, B.: Efficient mining under rich constraints derived from various datasets. In: Džeroski, S., Struyf, J. (eds.) KDID 2006. LNCS, vol. 4747, pp. 223–239. Springer, Heidelberg (2007)
7. Tanay, A., Sharan, R., Shamir, R.: Discovering statistically significant biclusters in gene expression data. Bioinformatics **18**(suppl 1), S136–S144 (2002)
8. van Mechelen, I., Bock, H.H., De Boeck, P.: Two-mode clustering methods: a structured overview. Stat. Methods Med. Res. **13**(5), 363–394 (2004)
9. Zelezny, F., Lavrac, N.: Propositionalization-based relational subgroup discovery with RSD. Mach. Learn. **62**(1–2), 33–63 (2006)

Analyzing microRNA Epistasis in Colon Cancer Using Empirical Bayesian Elastic Nets

Jia Wen, Andrew Quitadamo, Benika Hall, and Xinghua Shi[(✉)]

University of North Carolina at Charlotte, Charlotte, NC 28223, USA
{jwen6,aquitada,bjohn157,x.shi}@uncc.edu

Abstract. Recent studies have shown that colon cancer progression involves epigenetic changes of small non-coding microRNAs (miRNAs). It is unclear how multiple miRNAs induce phenotypic changes via epistasis, even though epistasis is an important component in cancer research. We used an Empirical Bayesian Elastic Net to analyze both the main, and epistatic effects of miRNA on the pathological stage of colon cancer. We found that most of the epistatic miRNAs shared common target genes, and found evidence for colon cancer associations. Our pipeline offers a new opportunity to explore epistatic interactions among genetic and epigenetic factors associated with pathological stages of diseases.

Keywords: Colon cancer · Empirical Bayesian elastic net · Epistasis · MicroRNAs

1 Introduction

Colon cancer is the second leading cause of cancer-related death in the United States [1]. Some studies have shown that aberrant miRNA expression is involved in colon cancer development, however most studies have analyzed differential expression between tumor and non-tumor or between tumor stages. Only a few of these studies have investigated epistatic effects of miRNA on staging. Therefore, we focused on analyzing the impact of individual miRNAs, and epistasis between two miRNAs, on the pathological stage of colon cancer in this study. We used a scalable method, empirical Bayesian elastic net (EBEN) to identify the main and epistatic effects of miRNAs different pathological stages of colon cancer. We analyzed the miRNA data from TCGA [2] to evaluate our method, and the results provide potential for understanding the impact of miRNAs on colon cancer.

2 Methods

We downloaded miRNA expression profiles for colon cancer from The Cancer Genome Atlas (TCGA), and created an expression matrix, which was normalized using inverse quantile normalization. We transformed the pathological stage

© Springer International Publishing Switzerland 2016
A. Bourgeois et al. (Eds.): ISBRA 2016, LNBI 9683, pp. 334–336, 2016.
DOI: 10.1007/978-3-319-38782-6

using the natural log, and applied EBEN to detect both main and epistatic effects of the miRNA. EBEN selects one of the features that is most highly correlated with the dependent variable, and then applies a coordinate ascent method and two hierarchical prior distributions to estimate unknown parameters. Finally, EBEN tests the significance of non-zero coefficient using a t -test. Specially, we don't need to do multi-test correction, because the EBEN added drop features including individual miRNA and pair-wise miRNAs from each iteration using Bayesian method and was different with multi-correlation methods. Below is the linear regression model in this study:

$$y = \mu + X_h\beta_m + X_iX_j\beta_e + e \tag{1}$$

where y is the dependent variable (i.e., the natural log value of pathological stage in this study); X_i and X_j are two different miRNAs expression vectors; β_m and β_e are the main effect of individual miRNA and the epistasis between miRNAs. We used four step pipeline based on EBEN method for epistasis and main effect analysis: a) We ran a model including solely main effect $X\beta_m$ to select significant main effect, X'. The threshold value was set at 0.05; b) we eliminated these main effect from the original phenotype (y) as the formula $y' = y - X' \beta'$; c) The corrected y' was used as the new dependent variable to detect epistasis using EBEN. The threshold was still set at 0.05; d) All the significant main effect from the step a) and epistatic features identified in c) were included in Eq. 1 and estimated by EBEN again.

3   Results

We identified a total of 120 main effect miRNA, and 27 pairs of epistatic miRNAs that had an effect on the pathological stage. 26 of the 120 main effect miRNAs were found to be associated with colon cancer in the literature. For the 27 epistatic miRNAs, 26 pairs have common target genes in three miRNA target databases(miR2Disease [3], TargetScan [4] and miRDB [5]) When we queried Online Mendelian Inheritance in Man (OMIM) Disease for the target genes, we found 12 that were associated with colon cancer [6].

4   Conclusion

We used an empirical Bayesian elastic net to study main effects, and pair-wise epistatic effects on the pathological stage of colon cancer. It effectively selected significant features in the TCGA data. Changes in both the expression of mRNA and miRNA likely affect the pathological stages of tumor, and by incorporating mRNA expression levels into our model we could develop a better understanding of the molecular mechanisms of colon cancer.

References

1. Jemal, A., et al.: Cancer statistics, 2008. CA Cancer J. Clin. **58**(2), 71–96 (2008)
2. The Cancer Genome Atlas. Sample counts for TCGA data (2014). https://tcga-data.nci.nih.gov/datareports/sampleSummaryReport.htm
3. Jiang, Q.H., et al.: mir2disease: a manually curated database for microrna deregulation in human disease. Nucleic Acids Res. **37**(suppl 1), D98–D104 (2009)
4. Lewis, B.P., et al.: Conserved seed pairing, often anked by adenosines, indicates that thousands of human genes are microrna targets. Cell **120**(1), 15–20 (2005)
5. Wong, N., et al.: miRDB: an online resource for microRNA target prediction and functional annotations. Nucleic Acids Res. (2014). doi:10.1093/nar/gku1104
6. Online Mendelian Inheritance in Man (OMIM)

Tractable Kinetics of RNA–Ligand Interaction

Felix Kühnl[1], Peter F. Stadler[1,2,3,4,5,6,7], and Sebastian Will[1(✉)]

[1] Department of Computer Science, and Interdisciplinary Center for Bioinformatics, University of Leipzig, Härtelstr. 16-18, Leipzig, Germany
will@bioinf.uni-leipzig.de
[2] MPI Mathematics in the Sciences, Inselstr. 22, Leipzig, Germany
[3] FHI Cell Therapy and Immunology, Perlickstr. 1, Leipzig, Germany
[4] Department of Theoretical Chemistry, University of Vienna, Währingerstr. 17, Wien, Austria
[5] Bioinformatics and Computational Biology Research Group, University of Vienna, A-1090 Währingerstraße 17, Vienna, Austria
[6] RTH, University of Copenhagen, Grønnegårdsvej 3, Frederiksberg, Denmark
[7] Santa Fe Institute, 1399 Hyde Park Rd., Santa Fe, USA

The interaction of RNAs and their ligands strongly depends on folding kinetics and thus requires explanations that go beyond thermodynamic effects. Whereas the computational prediction of minimum energy secondary structures, and even RNA–RNA and RNA–ligand interactions, are well established, the analysis of their kinetics is still in its infancy. Due to enormous conformation spaces, the exact analysis of the combined processes of ligand binding and structure formation requires either the explicit modeling of an intractably large conformation space or—often debatable—simplifications. Moreover, concentration effects play a crucial role. This increases the complexity of modeling the interaction kinetics fundamentally over single molecule kinetics.

We present a novel tractable method for computing RNA–ligand interaction kinetics under the widely-applicable assumption of ligand excess, which allows the pseudo-first order approximation of the process. In the full paper, we rigorously outline the general macroprocess of RNA ligand interaction based on gradient basin macrostates (cf. [2]) and discuss how to derive the model parameters, including corresponding rate constants, from empirical measurements. Figure 1 illustrates important aspects of this coarse graining, based on gradient basins of the monomer and dimer states. This original description of the specific macrostate system is a fundamental prerequisite for RNA ligand interaction kinetics. Subsequently, we discuss this system under the assumption of excessive ligand concentrations, which is valid for a wide spectrum of biological systems. On this basis, we devise the first analytical approach for RNA ligand interaction kinetics enabling the computation of time-dependent macrostate probabilities based on solving the master equation of the interaction process. Finally, we discuss the interaction kinetics of the artificially designed theophylline riboswitch RS3 [1] at different concentrations and study the effect of co-transcriptional interaction in our model.

Our current model assumes only a single binding motif, while multiple binding motifs with different binding energies and multiple binding sites are plausible, in general. A corresponding generalization of the model naturally leads to

© Springer International Publishing Switzerland 2016
A. Bourgeois et al. (Eds.): ISBRA 2016, LNBI 9683, pp. 337–338, 2016.
DOI: 10.1007/978-3-319-38782-6

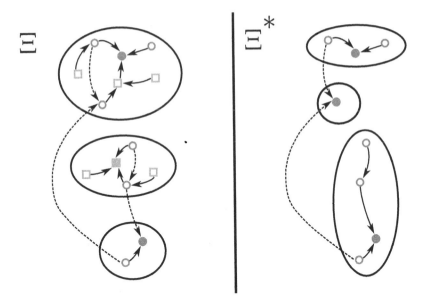

Fig. 1. Correspondence between the energy landscapes of the monomers (*left*) and the dimers (*right*). The dimer landscape is obtained from the monomer landscape by only retaining the structures possessing the binding pocket (*blue circles*) while removing the other ones (*green squares*). As the removed structures might lie on a gradient walk (*solid arrow*), rendering that path invalid in the dimer landscape, formerly suboptimal moves (*dashed arrows*) become gradient walks and new local minima (*filled squares and circles*) may arise. These effects might alter the mapping of the structures to their gradient basin.

multiple "ligand worlds" for the different binding modes. Moreover, riboswitches that control at the transcriptional level will strongly depend on the kinetics of transcription, i.e. the growth of the RNA chain itself. After all, growing RNA molecules are known to favor different local minima and thus to refold globally as the chain becomes longer. While the corresponding extensions of our framework are straightforward, experimental measurements are required to gauge additional thermodynamic parameters, detailed kinetic prefactors, and transcriptional speed – and these are very scarce at present.

Acknowledgments. SW is supported within de.NBI by the German Federal Ministry of Education and Research (BMBF; support code 031A538B).

References

1. Wachsmuth, M., Findeiss, S., Weissheimer, N., Stadler, P.F., Mörl, M.: De novo design of a synthetic riboswitch that regulates transcription termination. Nucleic Acids Res. **41**(4), 2541–2551 (2013)
2. Wolfinger, M.T., Svrcek-Seiler, W.A., Flamm, C., Hofacker, I.L., Stadler, P.F.: Efficient computation of RNA folding dynamics. J. Phys. A Math. Gen. **37**(17), 4731–4741 (2004). http://stacks.iop.org/0305-4470/37/4731

MitoDel: A Method to Detect and Quantify Mitochondrial DNA Deletions from Next-Generation Sequence Data

Colleen M. Bosworth, Sneha Grandhi, Meetha P. Gould,
and Thomas LaFramboise

Department of Genetics and Genome Sciences, Case Western Reserve University
School of Medicine, Cleveland, OH, USA
{colleen.bosworth, sneha.grandhi, meetha.gould,
thomas.laframboise}@case.edu

1 Introduction

Human genetic variation can be present in many forms, including single nucleotide variants, small insertions/deletions, larger chromosomal gains and losses, and inter-chromosomal translocations. There is a need for robust and accurate algorithms to detect all forms of human genetic variation from large genomic data sets. The vast majority of such algorithms focus exclusively on the 24 chromosomes (22 autosomes, X, and Y) comprising the nuclear genome. Usually ignored is the mitochondrial genome, despite the crucial role of the mitochondrion in cellular bioenergetics and the known roles of mitochondrial mutations in a number of human diseases [1], including cancer. The mitochondrial chromosome (mtDNA) may be present at up to tens of thousands of copies in a cell [2]. Therefore, variants may be present in a very small proportion of the cell's mtDNA copies, a condition known as heteroplasmy. Established computational tools used to identify biologically important nuclear DNA variants are often not adaptable to the mitochondrial genome. These tools have been developed to detect heterozygotic variants rather than heteroplasmic, so it is vitally important to develop new approaches to assess and quantify mtDNA genomic variation.

In this study, we focus on detecting deletions within the mitochondrial chromosome. We describe MitoDel, the computational procedure we have developed to extract predicted mtDNA deletions and their abundances from NGS data. We assess the theoretical sensitivity of our approach by using simulated data and also applying MitoDel to previously published data from a sequencing experiment involving aging human brain tissue. We also use our method to discover novel deletions in large publicly-available data from the DNA of healthy individuals. The accuracy of MitoDel compares favorably with an existing tool. Software implementing MitoDel is available at the LaFramboise laboratory website (http://mendel.gene.cwru.edu/laframboiselab/).

© Springer International Publishing Switzerland 2016
A. Bourgeois et al. (Eds.): ISBRA 2016, LNBI 9683, pp. 339–341, 2016.
DOI: 10.1007/978-3-319-38782-6

2 Methods

Unaligned.fastq files were aligned to a modified human genome build hg19 using BWA [3]. Hg19 was modified by removing the original chrM and replacing it with the revised Cambridge Reference Sequence (rCRS; NC_012920.1) [4]. The mitochondrial reference genome is described a circular chromosome 16,569 bases in length. The base positions are numbered in a clock-like manner, from 3' to 5' on the "light" strand, from base position 1 to base position 16569. Suppose that the region from mitochondrial base position $s + 1$ to base position $e - 1$ is deleted in proportion p of mtDNA copies such that positions s and e are ligated together in the mitochondrial chromosome, and suppose that the NGS experiment generates reads of length l bases. In our study, we restrict attention to deletions larger than 10 bp, *i.e.* $(e - s + 1) > 10$, since standard aligners like BWA are able to align reads harboring the smaller deletions wholly encompassed by the read. Suppose further that n reads harbor the deletion fusion point. For the ith of these reads, let x_i $(i = 1,...,n)$ denote the position in the read (oriented from lower mtDNA base position to higher) harboring base position s in the mitochondrial genome $(1 \leq x_i \leq l)$. Most of these reads will not align to anywhere in the reference genome, and will be therefore be marked as "unaligned" in the resulting .bam file output by BWA. We extract these unaligned reads and align them using BLAT [5].

BLAT's output for split reads includes the start and end read positions of each aligned segment of the read. In the above notation, this would correspond to two segments with (start, end) positions $(1, x_i)$ and $(x_i + 1, l)$ for read i harboring the deletion fusion point. BLAT's output also includes the beginning genomic coordinates (mtDNA base position) to which each segment aligns. In the above notation, this would correspond to mtDNA positions $(s - x_i + 1)$ and e for the two read segments. It follows that we may mine the BLAT output for a set of n split reads that each split into two segments and pass various quality filters. If the number of such reads harboring precisely the same breakpoint is sufficiently large, enough evidence is deemed to have been produced to report the breakpoint as biologically real. Finally, we can estimate the abundance of a discovered deletion (i.e. the proportion p of mtDNA copies harboring the deletion) by using the number of reads harboring the 16569/1 "artificial" fusion point as a proxy for 100 % abundance.

3 Results

Our simulation results shows that MitoDel is very sensitive at read depths of the magnitude produced by typical NGS runs, and has an extremely low false-positive rate. We are able to detect deletions present in as few as 0.1 % of mtDNA copies of the cell. Comparisons with an existing method, MitoSeek [6], show improved detection of deletions, particularly at low levels of heteroplasmy. We verified the results of Williams et al. using MitoDel. We also applied MitoDel to data from the 1000 Genomes Project [7] and discovered a number of novel deletions in several individuals. Regarding runtime, a .fastq file with 3.4 million reads took approximately 69 min to run.

References

1. Wallace, D.C.: Mitochondrial DNA mutations in disease and aging. Environ. Mol. Mutagen. **51**(5), 440–450 (2010)
2. Robin, E.D., Wong, R.: Mitochondrial DNA molecules and virtual number of mitochondria per cell in mammalian cells. J. Cell. Physiol. **136**(3), 507–513 (1988)
3. Li, H., Durbin, R.: Fast and accurate short read alignment with Burrows-Wheeler transform. Bioinformatics **25**(14), 1754–1760 (2009)
4. Andrews, R.M., et al.: Reanalysis and revision of the Cambridge reference sequence for human mitochondrial DNA. Nat. Genet. **23**(2), 147 (1999)
5. Kent, W.J.: BLAT–the BLAST-like alignment tool. Genome Res. **12**(4), 656–664 (2002)
6. Guo, Y., et al.: MitoSeek: extracting mitochondria information and performing high-throughput mitochondria sequencing analysis. Bioinformatics **29**(9), 1210–1211 (2013)
7. Genomes Project: C., et al., A map of human genome variation from population-scale sequencing. Nature **467**(7319), 1061–1073 (2010)

TRANScendence: Transposable Elements Database and *De-novo* Mining Tool Allows Inferring TEs Activity Chronology

Michał Startek[1], Jakub Nogły[1], Dariusz Grzebelus[2], and Anna Gambin[1(✉)]

[1] Institute of Informatics, University of Warsaw, Banacha 2, 02-097 Warsaw, Poland
aniag@mimuw.edu.pl
[2] Institute of Plant Biology and Biotechnology, University of Agriculture in Kraków,
29 Listopada 54, 31-425 Kraków, Poland

It has recently come to general attention that TEs may be a major (if not the main) driving force behind speciation and evolution of species. Thus understanding of TE behavior seems crucial to deepening our knowledge on evolution. However, the lack of general, easily-usable and free available tools for TEs detection and annotation, hinders scientific progress in this area. This is especially apparent with the advent of next-generation sequencing techniques, and the resultant abundance of genomic data, most of which has not been scanned for transposable elements yet.

The task of searching for TEs is not an easy one. It has been split into many smaller sub-tasks, such as searching for repeated genomic sequences, annotation of structural elements of TEs, clustering TEs into families, annotating TEs in genomes, de-novo searching for repeatable elements, classifying repeatable elements, and so on. A multitude of tools that can perform each of these steps have been written: the exhaustive list of tools and resources for TE analysis compiled by Bergman Lab (http://bergmanlab.smith.man.ac.uk/) contains about 120 items.

Therefore a need has arisen to merge the appropriate tools into one generic pipeline, which would be capable of performing a complete, *de-novo* annotation of TEs in a whole genome, from the ground up, starting only from the sequence of an organism's genome. One such pipeline has been created, called REPET [1]. It combines several different programs for the clustering of interspersed repeats and the annotation phase requires the use of multiple mechanisms, mainly based on comparisons to TEs stored in the Repbase [3]. However, in order to use the pipeline, each one of its components has to be separately installed and configured.

Here we have decided to integrate REPET pipeline with a flexible, relational TE repository, along with a web interface to benefit from a wide range of efficient services, but at the same time to eliminate the inconvenience of the lack of user-friendly interface. Our tool in contrast with previous ones, which either require manual assistance (and thus, are unsuitable for high-throughput analyses), or require deep programmer's experience to set-up and use, is fully automatic (though it is possible to manually curate the results if desired). To complement

This work was partially supported by Polish National Science Center grant 2012/06/M/ST6/00438.

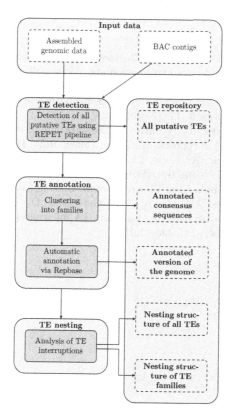

Fig. 1. Overview of the TRANScendence tool.

the functionality of specialized tools designed for specific TE families (see [2] and discussion therein) TRANScendence is developed to study the genome-wide landscape of TEs families.

In addition to tagging of TEs in genomes, our tool is capable of performing different qualitative and quantitative analyses. It classifies TEs into families, superfamilies and orders, allowing us to estimate relative abundances of TEs in selected genomes, and to perform comparative genomic studies. It also performs searches for TE clusters, i.e. regions containing high concentration of TEs nested in one another. Based on detected TE nesting structure the methods to reconstruct evolutionary history of TE families have been developed. Proposed algorithms yield the chronology of TE families activity.

The main objective of the proposed solution is the support of TE evolutionary studies. We put special emphasis on the design of the web-interface to assure simplicity and flexibility of data manipulation process. The tool is freely available for use by the general public at: http://bioputer.mimuw.edu.pl/transcendence. It is worth to mention that it has already been used in several scientific projects [4, 5].

References

1. Flutre, T., Duprat, E., Feuillet, C., Quesneville, H.: Considering transposable element diversification in de novo annotation approaches. PLoS ONE **6**(1):15 (2011)
2. Gambin, T., Startek, M., Walczak, K., Paszek, J., Grzebelus, D., Gambin, A.: Tirfinder: a web tool for mining class II transposons carrying terminal inverted repeats. Evol. Bioinform. **9**, 17–27 (2013)
3. Kapitonov, V.V., Jurka, J.: A universal classification of eukaryotic transposable elements implemented in repbase. Nat. Rev. Genet. **9**(5), 411–412 (2008)
4. Startek, M., Le Rouzic, A., Capy, P., Grzebelus, D., Gambin, A.: Genomic parasites or symbionts? Modeling the effects of environmental pressure on transposition activity in asexual populations. Theoret. Popul. Biol. **90**, 145–151 (2013)
5. Stawujak, K., Startek, M., Gambin, A., Grzebelus, D.: MuTAnT: a family of Mutator-like transposable elements targeting TA microsatellites in Medicago truncatula. Genetica **143**(4), 433–440 (2015)

Phylogeny Reconstruction from Whole-Genome Data Using Variable Length Binary Encoding

Lingxi Zhou[1,4], Yu Lin[2], Bing Feng[1], Jieyi Zhao[3], and Jijun Tang[1,4]([✉])

[1] Tianjin Key Laboratory of Cognitive Computing and Application,
Tianjin University, Tianjin 300072, People's Republic of China
jtang@cse.sc.edu
[2] Department of Computer Science and Engineering,
University of California, San Diego, USA
[3] University of Texas School of Biomedical Informatics, Houston, USA
[4] Department of Computer Science and Engineering,
University of South Carolina, Columbia, USA

Abstract. In this study, we propose three simple yet powerful maximum-likelihood (ML) based methods which take into account both gene adjacency and gene content information for phylogenetic reconstruction. We conducted extensive experiments on simulated data sets and our new method derives the most accurate phylogenies compared to existing approaches. We also tested our method on real whole-genome data from eleven mammals and six plants.

1 Variable Length Binary Encoding

A genome can then be represented as a multiset of adjacencies and genes.

Variable Length Binary Encoding 1 ($VLBE_1$): Given a data set D of n genomes, screen over it, collect all unique adjacencies and get a list A of m adjacencies; for each adjacency $a \in A$, count the maximum state number t of it among all the genomes. We encode each adjacency a in that column as follows: if genome D_i has n copies of adjacency a, we append $t - n$ 0's and n 1's to the sequence. We encode every adjacency a in list A.

Variable Length Binary Encoding 2 ($VLBE_2$): for each adjacency $a \in A$, count the maximum state number t of it among all the genomes. We encode each adjacency a in that column as follows: if genome D_i has n copies adjacency a, we append $t - n$ 0's and n 1's to the sequence. We encode every adjacency a in list A. We also append content encoding in this way, for each unique gene, if it present this genome D_i, append 1, otherwise append 0 to the sequence.

Variable Length Binary Encoding 3 ($VLBE_3$): $VLBE_3$ is designed to encode the multiplicity of both adjacencies and gene content. for each adjacency $a \in A$, count the maximum state number t of it among all the genomes. We encode each adjacency a in that column: if genome D_i has n copies adjacency a, we append $t - n$ 0's and n 1's to the sequence. We encode every adjacency a in list A. We also append content encoding in the same way as for adjacency.

© Springer International Publishing Switzerland 2016
A. Bourgeois et al. (Eds.): ISBRA 2016, LNBI 9683, pp. 345–346, 2016.
DOI: 10.1007/978-3-319-38782-6

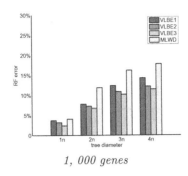

1, 000 genes

Fig. 1. RF error rates for different approaches for trees with 100 species, with genomes of 1,000 genes and tree diameters from 1 to 4 time the number of genes, under the evolutionary events with both segmental and whole genome duplications.

Since fliping a state (from 1 to 0 or from 0 to 1) depends on the transition model within the encoding scheme, we use the same transition model as described in MLWD [1], in order to perform a fair comparison with MLWD. Once we have encoded input genomes into binary sequences and have computed the transition parameters, we use RAxML [2] to build a tree from them.

2   Results

Our simulation studies follow the standard practice in [1]. We generate model trees under various parameter settings, then use each model tree to evolve an artificial root genome from the root down to the leaves, by performing randomly chosen evolutionary events on the current genome, finally obtaining data sets of leaf genomes for which we know the complete evolutionary history. We then reconstruct trees for each data set by applying different reconstruction methods and compare the results against the model tree. Figure 1 shows the comparison of the accuracy of three new approaches, $VLBE_1$, $VLBE_2$, $VLBE_3$ and MLWD [1].

3   Conclusion

Our new encoding schemes successfully make use of the multiplicity information of gene adjacencies and gene content in different genomes, and apply maximum-likelihood method designed for sequence data to reconstruct phylogenies for whole-genome data. As shown in the expetimental results, this new approach is particularly useful in handling haploid or polyploid species.

References

1. Lin, Y., Fei, H., Tang, J., Moret, B.M.E.: Maximum likelihood phylogenetic reconstruction from high-resolution whole-genome data and a tree of 68 eukaryotes. In: Pacific Symposium on Biocomputing, pp. 285–296 (2013)
2. Stamatakis, A.: RAxML-VI-HPC: maximum likelihood-based phylogenetic analyses with thousands of taxa and mixed models. Bioinformatics **22**(21), 2688–2690 (2006)

Author Index

Aganezov, Sergey 237
Al-Aqel, Haifa 323
Alekseyev, Max A. 237
Aluru, Srinivas 3
Anishchenko, Ivan 95

Bachega, José Fernando Ruggiero 163
Badal, Varsha 95
Badr, Ghada 323
Bonizzoni, Paola 27
Bosworth, Colleen M. 339
Bu, Hongda 312

Cai, Zhipeng 316
Chan, Ting-Fung 67
Chauve, Cedric 200
Chen, Xiaopei 106
Chiba, Kenichi 40
Chockalingam, Sriram P. 3

Das, Madhurima 95
Dauzhenka, Taras 95
Davidovskii, Alexander 304
Davydov, Iakov I. 253
Della Vedova, Gianluca 27
Dong, Ming 15
Duval, Béatrice 117

Egorova, V.P. 52
Eshleman, Ryan 136
Eskin, Eleazar 80
Eulenstein, Oliver 211

Feng, Bing 345
Flynn, Emily 151

Gabrielian, Andrei 258
Gambin, Anna 342
Gan, Yanglan 312
Gao, Ke 297
Gelfand, Mikhail S. 253
Geng, Yu 314
Górecki, Paweł 189
Gould, Meetha P. 339

Grandhi, Sneha 339
Grushevskaya, H.V. 52
Grzebelus, Dariusz 342
Guan, Jihong 301, 312, 321

Hall, Benika 334
Hormozdiari, Farhad 80
Hou, Aiju 316
Huang, Yukun 309

Imoto, Seiya 40

Jiang, Qijia 224

Kavaliou, Ivan 258
Kirkpatrick, B. 269
Kléma, Jiří 332
Krot, V.I. 52
Krylova, N.G. 52
Kühnl, Felix 337
Kundrotas, Petras J. 95
Kwok, Pui-Yan 67

LaFramboise, Thomas 339
Lam, Ernest T. 67
Lam, Tak-Wah 309
Lan, Wei 127
Legeay, Marc 117
Lei, Jikai 326
Lei, Peng 297
Leung, Henry Chi-Ming 309
Leyi, Wei 299
Li, Dinghua 309
Li, Jia-Xin 297
Li, Menglu 67
Li, Min 106, 307
Li, Yan 15
Lin, Hongfei 319
Lin, Yu 345
Lipinski-Paes, Thiago 163
Lipnevich, I.V. 52
Liu, Xiaowen 175
Luhmann, Nina 200
Luo, Ruibang 309

Mak, Angel C.Y.   67
Malinka, František   332
Mangul, Serghei   80
Marczyk, Michal   284
Markin, Alexey   211
Miyano, Satoru   40
Moret, Bernard M.E.   224
Moriyama, Takuya   40
Mykowiecka, Agnieszka   189

Nogły, Jakub   342
Ni, Peng   106
Norberto de Souza, Osmar   163

Orekhovskaja, T.I.   52
Ouangraoua, Aïda   200

Pan, Yi   106, 127, 307
Paszek, Jarosław   189
Peng, Wei   127
Pirola, Yuri   27
Previtali, Marco   27

Qingge, Letu   175
Quan, Zou   299
Quitadamo, Andrew   334

Racz, Gabriela C.   224
Renou, Jean-Pierre   117
Rizzi, Raffaella   27
Rosenthal, Alex   258

Sergeev, Roman   258
Shi, Jian-Yu   297
Shi, Xinghua   334
Shiraishi, Yuichi   40
Shulitski, B.G.   52
Singh, Rahul   136
Stadler, Peter F.   337
Startek, Michał   342
Streinu, Ileana   151
Sun, Yanni   326

Tang, Jijun   345
Tanus, Michele dos Santos da Silva   163
Techa-Angkoon, Prapaporn   326
Thankachan, Sharma V.   3
Thévenin, Annelyse   200
Tian, Kai   321

Ting, Hing-Fung   309
Tseng, Elizabeth   80
Tuzikov, Alexander V.   95, 258

Vakser, Ilya A.   95
Veresov, Valery   304
Vijayan, Vinaya   330

Wang, Jian   319
Wang, Jianxin   106, 127, 307
Wang, Jiayin   314
Wang, Lu   15
Wang, Wenke   314
Wang, Yang   301, 312, 321
Wen, Jia   334
Will, Sebastian   337
Wittler, Roland   200
Wu, Fang-Xiang   307

Xiao, Ming   67
Xiao, Xiao   314
Xu, Bin   301
Xu, Dechang   316

Yamaguchi, Rui   40
Yan, Xiaodong   307
Yang, Harry (Taegyun)   80
Yang, Zhihao   319
Ye, Min   224
Yip, Kevin Y.   67
Yiu, Siu-Ming   67, 297, 330
Yu, Zeng   127

Zaika, Andrey V.   253
Železný, Filip   332
Zelikovsky, Alex   80
Zeng, Xiangmiao   316
Zhang, Liqing   330
Zhang, Xiuwei   224
Zhang, Xuanping   314
Zhang, Yijia   319
Zhao, Jieyi   345
Zhao, Lingchen   307
Zhao, Zhongmeng   314
Zheng, Xiaoxiang   321
Zhong, Farong   175
Zhou, Lingxi   345
Zhou, Shuigeng   301, 312, 321
Zhu, Binhai   175
Zhu, Dongxiao   15